INTRODUCTION TO
QUANTUM MECHANICS

INTRODUCTION TO QUANTUM MECHANICS

Henrik Smith
*University of Copenhagen
Denmark*

World Scientific
Singapore • New Jersey • London • Hong Kong

Published by

World Scientific Publishing Co. Pte. Ltd.
P O Box 128, Farrer Road, Singapore 9128
USA office: 687 Hartwell Street, Teaneck, NJ 07666
UK office: 73 Lynton Mead, Totteridge, London N20 8DH

Library of Congress Cataloging-in-Publication data is available.

INTRODUCTION TO QUANTUM MECHANICS

Copyright © 1991 by World Scientific Publishing Co. Pte. Ltd.

All rights reserved. This book, or parts thereof, may not be reproduced in any form or by any means, electronic or mechanical, including photocopying, recording or any information storage and retrieval system now known or to be invented, without written permission from the Publisher.

ISBN 981-02-0475-2
      981-02-0476-0 (pbk)

Printed in Singapore by JBW Printers and Binders Pte. Ltd.

PREFACE

This book is an introduction to quantum mechanics at the undergraduate level. It is based on notes which have been used since 1988 in the second year of the undergraduate program at the University of Copenhagen. In writing the book a major concern was to keep its length below 300 pages. By its very nature an introductory text must be brief. It should include the essentials but must necessarily leave out many topics of importance for the applications.

The book is written by a solid state physicist and therefore has a natural tendency to emphasize applications of quantum theory to solid state physics, in addition to atomic and molecular physics. The first chapter is a description of the framework in which quantum mechanics was developed. Though it provides a background and a historical perspective, it is not essential for the understanding of the following chapters. The reader may start at Chapter 2.

Any elementary text on quantum mechanics must deal with the few exactly solvable problems, including the harmonic oscillator and the hydrogen atom. In presenting these important topics I have put some emphasis on making the reader familiar with different methods of solution. For instance, in the case of the hydrogen atom the reader is encouraged to think about it in an approximate manner as a motion in a harmonic oscillator potential, while the general solution involving Laguerre polynomials is treated in a problem.

The matrix formulation of quantum mechanics is emphasized throughout the text, and the Dirac notation is introduced from the start. The advantage of doing this is that the reader develops a better feel for how quantum mechanics is employed in practice, for instance with regard to the use of approximation methods such as perturbation theory. The quantization described in Chapter 2 is based on the matrix formulation and its relation to classical analytical mechanics. The reader may, however, consult the first chapter for an exposition of Schrödinger's wave mechanics. In Chapters 3 and 4 the two approaches are merged with the intention that the reader acquire a working knowledge of both, as well as an understanding of their equivalence.

To illustrate some applications of quantum mechanics, ten major examples are included in the text. They represent a selection which could have been made in many other ways. These examples may be supplemented with computer exercises involving for instance barrier penetration (Example 5), variational determination of ground state energies (Example 8) or calculation of band structures (Example 10).

While Chapters 2-9 contain the basic tools to be mastered in an introduction to quantum mechanics, Chapter 10 is devoted to a discussion of its interpretation. The aim of this chapter is to present some of the high points in the debate over the meaning of quantum mechanics.

In the course of writing this book I have received much valuable advice from my students and colleagues. In particular, I thank Vinay Ambegaokar, C. J. Ballhausen, Per Hedegård, Antti-Pekka Jauho, H. Højgaard Jensen, Allan Mackintosh, Morten Bo Madsen, Anders Smith and Georg Ole Sørensen for their help and encouragement.

University of Copenhagen,
March 1991.
Henrik Smith

Contents

1 **Introduction**   1
   1.1 Fundamental constants 2
       1.1.1 The velocity of light 3
       1.1.2 The mass and charge of the electron 5
       1.1.3 The Planck constant 7
   1.2 The birth of quantum mechanics 9
       1.2.1 The Bohr model 12
       1.2.2 Matrix mechanics 15
   1.3 Dimensions and fundamental constants 17
       1.3.1 The hydrogen atom 18
       1.3.2 Angular momentum 21
       1.3.3 Harmonic oscillator 22
       1.3.4 Molecular spectra 24
       1.3.5 Gravitation 26
   1.4 Particles and waves 27
       1.4.1 A free particle 30
       1.4.2 Particle in a box 31
       1.4.3 Harmonic oscillator 33
       1.4.4 Wave packets 37
       1.4.5 A relativistic wave equation 38
   1.5 Problems 42

2 **Quantization**   46
   2.1 Analytical mechanics 46
       2.1.1 The Lagrange formalism 46
       2.1.2 Hamilton's equations 52
       2.1.3 Poisson brackets 54
   2.2 From classical mechanics to quantum mechanics 57
       2.2.1 Matrices and eigenvalues 59
       2.2.2 Energy quanta 60
   2.3 Problems 67

3 **Basic principles**   72
   3.1 Linear operators 72
       3.1.1 A simple example: The harmonic oscillator 77
   3.2 The postulates of quantum mechanics 83
   3.3 Probability amplitudes 85
       3.3.1 Expectation value 87
       3.3.2 The uncertainty relations 90
       3.3.3 Probability amplitudes and the Schrödinger equation . . 97
   3.4 Problems 99

4 The Schrödinger equation — 106
- 4.1 Ehrenfest's theorem 108
- 4.2 The superposition principle 109
- 4.3 Boundary conditions 113
- 4.4 Summary 120
- 4.5 Problems 121

5 Tunnelling — 124
- 5.1 Bound states 124
- 5.2 Probability current 126
- 5.3 Barrier transmission 128
- 5.4 The golden rule 133
  - 5.4.1 Reflection from a barrier 134
  - 5.4.2 Time-dependent perturbation theory 136
- 5.5 Problems 139

6 Electron in a magnetic field — 142
- 6.1 The classical Hamiltonian 142
- 6.2 Quantization 144
  - 6.2.1 The Hamiltonian and its eigenvalues 145
  - 6.2.2 Degeneracy 146
- 6.3 Problems 151

7 Angular momentum — 152
- 7.1 Quantization of angular momentum 152
  - 7.1.1 Spherical harmonics 154
  - 7.1.2 Eigenstates of $\hat{\mathbf{L}}^2$ and $\hat{L}_z$ 157
- 7.2 Spin 162
- 7.3 Addition of angular momentum 163
- 7.4 Motion in a central field 167
  - 7.4.1 The hydrogen atom 168
- 7.5 The spin of the electron 175
  - 7.5.1 The magnetic moment of the electron 176
- 7.6 Problems 178

8 Symmetries — 184
- 8.1 Parity 184
  - 8.1.1 Time-independent perturbation theory 187
- 8.2 Permutation 190
- 8.3 Translation 205
- 8.4 Gauge transformation 207
- 8.5 Rotation 208
- 8.6 Problems 209

9 Fermions and bosons — 212
- 9.1 Free electron gas — 213
  - 9.1.1 Sound in metals — 216
- 9.2 Periodic potential — 218
- 9.3 Lattice vibrations — 223
  - 9.3.1 Normal modes — 223
  - 9.3.2 Phonons — 226
  - 9.3.3 Lattice specific heat — 227
- 9.4 Spin waves — 234
- 9.5 Blackbody radiation — 238
  - 9.5.1 The background radiation of the universe — 239
- 9.6 Quantum liquids and superconductors — 239
  - 9.6.1 Superfluidity — 240
  - 9.6.2 Superconductivity — 245
- 9.7 Problems — 247

10 Physical reality — 249
- 10.1 The Bohr-Einstein discussion — 249
- 10.2 Bell's inequalities — 253
- 10.3 Elements of reality — 255

11 More Problems — 258
- 11.1 Problems with solutions — 258
- 11.2 Problems — 268

Appendices — 275
- A. Table of fundamental constants — 275
- B. Polar and cylindrical coordinates — 276
- C. Fourier transformation and the delta function — 278

Index — 282

Examples
1. A moveable pendulum — 50
2. Poisson brackets and the two-dimensional harmonic oscillator — 56
3. Quantization of the two-dimensional harmonic oscillator — 91
4. The classical limit for a harmonic oscillator — 109
5. Quantum wells — 131
6. The symmetric gauge — 148
7. Zeeman effect and spin-orbit coupling — 173
8. The He atom and the H^- ion — 193
9. The ammonia molecule, the maser and the chemical bond — 197
10. A one-dimensional lattice — 221

1 INTRODUCTION

The natural philosophers of Greek antiquity imagined the world to be built from indivisible parts, the atoms. The Greek word *atomos* means indivisible. Philosophical considerations, rather than experiments, led Democritus and Leucippus (about 400 BC) to assume the existence of such indivisible parts. The physics and chemistry of the nineteenth century put the atomic theory on solid ground. The British chemist and school teacher John Dalton (1766-1844) showed that the atomic theory yields a natural explanation of the constitution of chemical compounds.

During the last decade of the nineteenth century and the beginning of the twentieth, it became clear that the atom was not indivisible but had a structure of its own. The British physicist Ernest Rutherford (1871-1937), who was born in New Zealand, and his collaborators showed by their scattering experiments that the atom resembles a solar system, with practically the entire mass of the atom residing in a positively charged nucleus surrounded by negatively charged electrons. As we shall see in the following chapters, the similarity between the solar system and the atom is only superficial. While the orbits of the planets may be calculated from the laws of classical mechanics, the description of the atom requires a different framework, that of quantum mechanics. The laws of quantum mechanics have more general validity than those of classical mechanics, which is contained in quantum mechanics as a limiting case. Though classical mechanics accounts extremely well for the planetary orbits, it is unable to explain the stability and structure of the atoms.

The first indication of the divisibility of the atom was the discovery of natural radioactivity in the latter part of the nineteenth century. The element radium was found to emit three kinds of radiation, which were named α-, β- and γ-rays. As a result of the α-radiation, which consists of positively charged particles, the element radium is transformed into a different element called radon. It turned out that the positively charged α-particles combined with two electrons to form another element, helium. Elements could thus be transformed into other elements. By bombarding a metal foil with fast α-particles and investigating their deflection Rutherford was able to conclude that the overwhelming majority of the atomic mass had to be concentrated in a nucleus of very small size compared to the size of the atom.

A radioactive process such as α-decay consists in the nucleus splitting in two by giving off four of its elementary particles, two protons and two neutrons. The charge of the proton has the same magnitude as that of the electron but opposite sign, while the neutron - as the name suggests - is electrically neutral. The four particles leave the nucleus as one particle, the α-particle, because they are tightly bound to each other.

Protons and neutrons have almost the same mass, about 2000 times the mass of the electron. A neutron moving outside the nucleus is unstable, with a mean lifetime of 15 minutes. Such a freely-moving neutron decays into an electron and a proton. Despite their different charge, the neutron and the proton show many similarities. Their mass is nearly the same, the mass of the unstable neutron slightly exceeding that of the proton. The proton appears to be stable, but in recent years a number of experiments have been carried out to test whether it is in fact unstable. It is now known that the mean lifetime of the proton is at least 10^{31} years. This is much longer than the age of the universe, which is estimated to be 10^{10} years.

The proton and the neutron have a structure of their own. Each one is made up of three smaller constituents, the quarks. But these constituents differ from those of the atom - the electrons, protons and neutrons - in that they have never been observed as freely-moving particles. There are strong reasons to believe that the nature of the forces which bind the quarks together, prevent their observation as freely-moving particles. While the neutron and the proton both have inner structure, the electron is structureless. Today one considers the proton and the neutron to be members of a large family of particles, the hadrons, comprising several hundred so-called elementary particles. The electron belongs to a much smaller family of particles named leptons.

In summary, the atom consists of smaller entities, the hadrons and the leptons. The hadrons in turn are built from quarks. The quarks always appear to be bound together, either in pairs or three at a time as in the neutron and the proton. The development from Rutherford's simple model of the atom based on electrons surrounding a positively charged nucleus to the present-day families of hadrons and leptons represents an enormous expansion of our knowledge. This expansion has been achieved by a combination of experiment and theory. The advances made during this century have brought about radical changes in our understanding of the structure of matter, but the conceptual framework, the quantum theory, has not changed. All experimental observations which have been carried out since the advent of quantum theory have supported its foundations. Quantum mechanics not only forms the starting point for the description of atoms and molecules. It is also the basis for explaining the characteristic properties of solids, whether they are metals, insulators or semiconductors. Quantum mechanics has thus played a crucial role in much of the technological development of this century.

1.1 Fundamental constants

Values of physical constants may be looked up in tables. The mass density of aluminum is for instance seen to be 2.7 gram per cubic centimeter, while the sound velocity in water is 1500 metres per second. Neither of these quantities are fundamental constants. They depend on temperature and pressure, and both the mass density and the velocity of sound are different for different elements and chemical compounds.

By contrast the mass of an atomic electron is the same, regardless of which particular atom it belongs to. The same holds for its electric charge. Quantities such as the charge or the mass of an electron are named fundamental constants.

Introduction

The velocity of light in vacuum is a fundamental constant. Its value was first measured by the Danish astronomer Ole Rømer in 1675. Major contributions to the determination of the mass and the charge of the electron were made by J. J. Thomson and R. A. Millikan at the turn of this century. The fourth fundamental constant to be discussed in the following was introduced by Max Planck in 1900 in order to explain the radiation of black bodies. The Planck constant is the fundamental constant of the quantum theory. Together with the mass and charge of the electron it determines the size of the atom and the characteristic wavelength of the light emitted by atoms.

In the following we describe the discovery of these four fundamental constants. Subsequently we shall see how the constants determine the characteristic size of the atoms as well as the binding energy of an atomic electron.

1.1.1 The velocity of light

In 1671 the French astronomer Picard arrived in Denmark. He was sent on this mission by the French Academy with the purpose of measuring the precise location of Uranienborg, which was the observatory used by Tycho Brahe and located on Hven, an island in the strait between Denmark and Sweden. The academy intended to publish the original observations by Tycho Brahe in France. It was therefore important to know the exact location of his observatory. The Danish astronomer Ole Rømer (1644-1710) became Picard's assistant in this work. When Picard returned to France the following year, he was accompanied by Ole Rømer, who stayed in Paris for ten years.

During Ole Rømer's stay in Paris he worked together with Picard doing observations of planetary orbits. In addition he observed the eclipses of Io, one of the moons of the planet Jupiter. The eclipse of Io is a result of the moon entering the shadow cast by Jupiter once during each revolution. The entrance into the shadow can only be observed when the earth approaches Jupiter in its own orbit around the sun. This is illustrated in Fig. 1.1, which is taken from Ole Rømer's original publication in *Journal des Scavans*. On the figure A marks the sun and B Jupiter. The lower circle represents the orbit of the earth. The direction of revolution is given by LKEFGH. The upper circle denotes Io's orbit, which is followed in the same direction - from C to D along the shortest arc CD. Correspondingly, the exit of Io from the shadow can only be observed when the earth moves away from Jupiter.

In 1675 Ole Rømer presented his epoch-making discovery to the academy: The observed difference in time between two consecutive eclipses differed, depending on whether the moon was entering or leaving the shadow cast by Jupiter. He explained the phenomenon as a consequence of the finite velocity of light. Because of the motion of the earth, the difference in time between two consecutive entrances of Io into the shadow, as observed from the earth, would be smaller than the corresponding difference in time between two consecutive exits. Since the velocity of the earth is about 30 km/s, it moves 5 million km during one revolution of Io, which lasts nearly two days. On the basis of his

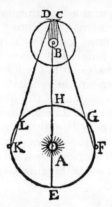

Figure 1.1: Ole Rømer's illustration of the eclipse of Io.

observations Ole Rømer concluded that light takes 22 minutes to go through a distance corresponding to the diameter in the orbit of the earth around the sun. The corresponding value of the velocity of light is 230 000 km/s, since the diameter of the orbit of the earth is 300 million km.

It has not been possible to reconstruct, how Ole Rømer obtained the result of 22 minutes, since his publication contained no details on this. Later measurements showed that the velocity of light is close to 300 000 km/s. In 1978 the velocity of light was determined to be 299 792 458 m/s with an uncertainty of ±1 in the last digit. Since 1983 the velocity of light is by definition 299 792 458 m/s, and this value is then used to define the unit of length, one metre, from the unit of time, one second.

When Ole Rømer returned to Denmark in 1681 he was appointed professor of astronomy. He invented an instrument for astronomical observations as well as several instruments for use in physics. He was also a practical man, as evidenced by his extraordinary range of activities. He was entrusted with the task of publishing a new land register, which was kept in use until 1844. He prepared the King's new declaration on weights and measures. In addition he was a member of the Supreme Court and became chief of police as well as mayor of Copenhagen. He also served as rector of the University of Copenhagen. Many of his astronomical observations were lost during the fire in Copenhagen in 1728. During his life as an astronomer he was preoccupied with the possibility of measuring the parallax of the fixed stars due to the earth's motion around the sun. Today we know that this effect is too small to be measured with the instruments that were available to Ole Rømer. He was aware of the numerous sources of error involved in such measurements and never published his results, although they were much more accurate than those of his contemporaries.

Introduction

1.1.2 The mass and charge of the electron

Joseph John Thomson (1856-1940) was born in the surroundings of Manchester as the son of the owner of a bookstore. Originally he wanted to become an engineer, but a long waiting list prevented him from getting a job as an apprentice. At the age of 14 he started his studies at Owens College, which later became the University of Manchester. At the age of 20 he was admitted to Trinity College, Cambridge, where he remained for the rest of his life. Thomson was the head of the Cavendish Laboratory from 1884 to 1919. During that period the Cavendish was the dominant centre for research on the properties of atoms.

Throughout most of the nineteenth century there had been a continued debate over the nature of electric charge. Was it possible to divide electric charge into lumps of arbitrarily small size, or did a minimum charge exist? Michael Faraday's experiments on electrolysis from 1833 suggested the existence of a minimum amount of electric charge. This atomistic point of view was not easily reconciled with the established laws of electrodynamics due, in particular, to James Clerk Maxwell. Maxwell's theory, which appeared in final form in 1873, could account successfully for the variation of electric and magnetic fields in space and time. The electric charge enters the Maxwell equations as a quantity which varies continuously and may in principle be divided into lumps of arbitrary size.

J. J. Thomson and his co-workers were the first to prove directly the existence of a minimum electric charge, an electron. Thomson investigated the influence of electric and magnetic fields on a beam of charged particles. Such a beam of particles will be deflected by an electric field perpendicular to the direction of the beam. The measured deflection depends on the magnitude E of the electric field, the particle velocity v and the ratio m/e between the mass m of the particles and their charge e. If in addition a magnetic field **B** is applied perpendicular to both the direction of the electric field and the velocity of the beam, the effect of the electric and magnetic fields will cancel each other if $E = vB$. In this manner it was possible to measure the velocity v and hence the ratio m/e determining the deflection in the electric field. Thomson's measured value of m/e for electrons was published in 1897. His results showed that the ratio was about 1000 times smaller than the corresponding ratio for the hydrogen ion, the lightest of the charged particles that had previously been known. If one assumes that the hydrogen ion carries the same minimum charge as an electron, only with opposite sign, one had to conclude that the electron must be about a thousand times lighter than the hydrogen ion.

Thomson belonged to the period of transition between classical nineteenth century physics and the quantum physics of the twentieth century. His proof of the existence of the electron and the determination of m/e represented a departure from classical electrodynamics. His later attempts to explain the structure of atoms were, however, based on the classical concepts of the nineteenth century. Thomson assumed that the negative charge of the electrons in an atom was compensated by a positive charge distributed uniformly over the whole atom. He used the laws of classical electrodynamics to analyze this model in an attempt to explain the periodic system.

Throughout this period Cambridge was a center for research on physics. It was therefore natural for the young Danish physicist Niels Bohr (1885-1962) to choose Cambridge as the first goal of his travels abroad in 1911. In his doctoral thesis from 1911 on the electron theory of metals, Bohr had pointed out some errors in Thomson's publications. He had therefore reason to expect Thomson to read his work carefully, but it turned out - perhaps not surprisingly - that Thomson was not interested in discussing his earlier work. The visit was in some sense a failure, and Niels Bohr decided in the beginning of 1912 to go to Ernest Rutherford in Manchester. The previous year Rutherford had obtained decisive evidence that the positive charge of the atom is concentrated in a small nucleus. This was the starting point for Bohr. In 1913 Bohr proposed his famous model of the atom based on a set of assumptions that involved a radical departure from the concepts of classical physics. Thomson attempted in the beginning of the twenties to develop his classical model of the atom by making special assumptions about the nature of the forces between the electric charges, but the general formulation of the quantum theory in 1925-26 by Heisenberg, Born, Schrödinger and others put an end to further attempts of describing the atoms by means of classical physics.

Some of Thomson's theoretical ideas proved to be incorrect, but his experiments had far-reaching influence. Although he never fully identified himself with quantum physics, his direct proof of the existence of the electron and his measurement of m/e was a significant factor in the development of the new physics.

Thomson's original experiment determined only m/e. In the following years he and his associates attempted to measure the value of the elementary charge itself, without getting very accurate results. A dramatic improvement of the measurement of e was achieved by the American physicist Robert Andrews Millikan (1868-1953).

Millikan grew up in Illinois and Iowa, where his father was a preacher. He was admitted to Oberlin College in Ohio. In the second year of his studies he was asked by his teacher in Greek whether he would do some teaching of physics. At the time Millikan did not have much knowledge of the subject, but the teacher argued that whoever does well in Greek should also be able to teach physics. It was not a coincidence that Millikan's interest in physics was stimulated by teaching. Throughout his career Millikan remained strongly committed to education. Millikan's graduate studies were carried out at Columbia University in New York City, where he remained the only physics student for a period of two years. His teachers encouraged him to go to Europe, and he therefore spent one year in Berlin and Göttingen. After graduation he obtained his first job at the University of Chicago, where he contributed significantly to the teaching. Millikan did not like the form of the traditional lecture. His teaching utilized experimental laboratory work much more than had been done before. He wrote several textbooks, which obtained wide usage. In 1921 he moved to the California Institute of Technology, where he played a major role in building up the research and the teaching of physics.

During his stay in Chicago, Millikan had been interested in Thomson's attempts - subsequent to the measurement of m/e - to determine the magnitude of the electron charge itself. Thomson and his co-workers had produced clouds of charged droplets and watched the change with time of the cloud surface in order to determine the number of droplets and their charge. The method was not very accurate - the uncertainty could be as large as 100%. Millikan got the idea of using oil droplets instead of water droplets. Due to friction the oil droplets became charged when they were produced. Millikan could vary the charge of the droplets by ionizing the surrounding air, using X-rays. Some of the droplets fell through a small hole into a region between two condenser plates. The potential difference between the condenser plates could be varied, and the motion of the droplets was followed in a microscope. In the absence of a potential difference between the plates, a drop would fall with uniform speed, the effect of gravity being cancelled by the frictional force, which is proportional to the velocity of the droplet. In the presence of a potential difference the force on a droplet depends on its charge. In his experiment Millikan could follow a single droplet and measure its charge directly. By changing the droplet charge through X-ray ionization of the surrounding air he determined e, the smallest unit of electric charge. His method was so accurate that the measured value only differed by 1/10 % from the value accepted today, $e = 1.602\,177 \cdot 10^{-19}$ C. The present value of the mass of the electron is $m = 9.109\,39 \cdot 10^{-31}$ kg (see also Appendix A).

1.1.3 The Planck constant

A black body is defined as a body which absorbs all radiation incident on it. This implies that the emission from a black body only depends on its temperature. The blackbody emission (which may be realized in practice as the radiation escaping a cavity through a small hole) was a subject of considerable interest during the latter part of the nineteenth century. Experimentally the dependence of the intensity on temperature and wavelength was investigated. The German physicist Wilhelm Wien had succeeded in finding an empirical law which described the observations accurately at short wavelengths, while the English physicist Lord Rayleigh had given a theory for long wavelengths which also described the experiments well in this limit. There was, however, no understanding of how these two separate regions should be joined together, nor was it clear why Rayleigh's theory only worked at long wavelengths.

The German physicist Max Planck (1858-1947) arrived in Berlin in 1889 as a successor of Gustav Robert Kirchhoff, who had formerly been his teacher. Max Planck was born in Kiel, but spent part of his school days and university studies in München. He obtained his doctorate at the age of 21. In Berlin, Planck's colleagues measured the blackbody emission in order to understand why and how Rayleigh's theory lost its validity.

One Sunday afternoon on October 7, 1900 the Planck family had guests for tea. Planck's colleague Rubens from the Friedrich Wilhelm University was paying a visit together with his wife, and Rubens took the opportunity to tell Planck about his latest measurements of the blackbody radiation. After hearing about these new results Planck attempted later in the day to construct an empirical formula, which would combine the regions of long and short wavelengths corresponding respectively to Rayleigh's and Wien's expressions. The result, Planck's radiation law, was made public only twelve days later at a talk given by Planck in Deutsche Physikalische Gesellschaft.

On empirical grounds Planck had found a simple formula that worked in the sense that it agreed with the measurements. It contained a new fundamental constant, which today is called the Planck constant. Two months later Planck published a derivation of the radiation law. His derivation contained assumptions that were quite foreign to the physics of the nineteenth century[1].

Planck asssumed that the radiation escaping the cavity originated in oscillations in the wall of the cavity. According to classical physics an oscillator may have any value of its total energy, depending on the way in which the oscillation is started. An example is furnished by a pendulum executing small oscillations. The frequency of the oscillation depends on the length of the pendulum and the gravitational acceleration, while its energy is determined by the departure of the pendulum from vertical position when started from rest. In classical physics there is no basis for arguing that it is only possible for the pendulum to oscillate if it has certain discrete values of its total energy. This, however, was the assumption which Planck made concerning the oscillators in the wall. By the use of statistical arguments he arrived at a derivation of the law of radiation. His model associated the frequency ν of the oscillators (and hence the frequency ν of the radiation) with a definite energy $h\nu$, where h is the Planck constant. By comparing with experiment, Planck determined the constant h to be $6.55 \cdot 10^{-34}$ m^{2}kg/s. The most accurate modern determination of the Planck constant gives $h = 6.626\,076 \cdot 10^{-34}$ m^{2}kg/s. Today one usually employs the constant $\hbar$, which is the Planck constant h divided by 2π, so that $\hbar = 1.054\,572 \cdot 10^{-34}$ m^{2}kg/s. This constant is also named the Planck constant.

Planck's radiation formula describes how the radiation energy in a cavity is distributed in frequency. Let us denote by $u_\nu(T)\Delta\nu$ the energy density of radiation with frequency between ν and $\nu + \Delta\nu$. The function $u_\nu(T)$ describes how the energy density depends on frequency ν and temperature T. This distribution is observed when a small part of the radiation energy in the cavity is emitted to the surroundings through the hole. The frequency distribution of the emitted radiation is a direct measure of u_ν. Planck showed in 1900 that the function

$$u_\nu(T) = \frac{8\pi h \nu^3}{c^3(e^{h\nu/kT} - 1)} \qquad (1.1)$$

yields an accurate description of the experimental observations. The value of the dimensionless ratio $x = h\nu/kT$, where k is the Boltzmann constant, clearly plays an important role. The function $x^3/(e^x - 1)$ is for small x approximately equal to x^2. When x is large, the function may be approximated by $x^3 e^{-x}$. These two limits

---

[1] A description of Planck's argument may be found in an article by A. Pais, Reviews of Modern Physics **51**, 861, 1982.

correspond to long wavelengths ($h\nu$ much less than kT) and short wavelengths ($h\nu$ much greater than kT), respectively. In his first communication Planck showed that the expression (1.1) yields an accurate description of the observations. In his second communication he derived the radiation law (1.1) by statistical arguments which represented a radical departure from the classical thinking of the nineteenth century.

It is readily seen from (1.1) that the radiation law in the classical limit (kT much greater than $h\nu$) is given by $u_\nu(T) = 8\pi\nu^2 kT/c^3$. This is the result of Rayleigh's theory mentioned above. It is easy to understand why Rayleigh's result cannot be valid for all frequencies ν. If the energy density per unit frequency, u_ν, increases in proportion to ν^2 for all ν, then the total energy density must be infinitely large, since the total energy density is obtained by integrating u_ν over all frequencies. Since the total energy density cannot be infinitely large, the classical result must be modified at high frequencies. This in itself points to the existence of a fundamental constant such as the Planck constant, which according to (1.1) ensures that u_ν decreases like $\nu^3 e^{-h\nu/kT}$ instead of growing like ν^2 for large ν.

Planck's discovery of the radiation law marked the beginning of the era of quantum physics. In order to explain the radiation law, he postulated the existence of quanta of oscillation energy in units of $h\nu$, where ν is the oscillator frequency. The observed disagreement between the predictions of classical physics and the measured blackbody radiation was a decisive factor in this development. It is difficult to imagine anyone in 1900 departing so radically from established concepts without the support given by experiment.

Planck was the founder of quantum physics, but he did not take much part in its further development. He grew up in the classical tradition and never fully accepted the revolutionary ideas which were formulated in 1925-6 by Werner Heisenberg and others, who laid the conceptual basis for quantum physics as we know it today.

1.2 The birth of quantum mechanics

The experimental observation of discrete lines in the light emitted by incandescent gases dates back to the eighteenth century. The development and use of diffraction gratings in the nineteenth century made it possible to obtain detailed information on the line spectra of different elements. In 1885 the Swiss physicist J. J. Balmer discovered that the frequencies ν of a series of lines in the visible part of the spectrum of atomic hydrogen could be represented by the empirical formula

$$\nu = cR(\frac{1}{n_1^2} - \frac{1}{n_2^2}). \tag{1.2}$$

Here n_1 and n_2 are positive integers ($n_2 > n_1$), while R is a positive constant, known as the Rydberg constant, and c is the velocity of light. The series with $n_1 = 2$ and $n_2 = 3, 4, 5 \cdots$ is called the Balmer series. The first Balmer line with $n_2 = 3$ is a red line with a wavelength of 6563 Å, while the next two lines corresponding to $n_2 = 4$ and $n_2 = 5$ lie in the blue and violet part of the spectrum. It was later discovered that the empirical formula (1.2) is an accurate representation of the entire discrete spectrum of atomic hydrogen. The series with $n_1 = 1$ and $n_2 = 2, 3, 4 \cdots$ is named the Lyman series, the three following series with $n_1 = 3, 4$ and 5 are named after Paschen, Brackett and Pfund, respectively.

When Niels Bohr went from Cambridge to Manchester University in the middle of March 1912 to visit Rutherford's laboratory, he did not know[2] of the empirical regularities given by (1.2). The starting point for his atomic model was the experimental discovery by Rutherford and his associates of the existence of a positively charged nucleus, and the apparent contradiction with the observed stability of atoms that arose from applying the laws of classical electrodynamics to such a model of the atom.

Planck's theory of radiation was an essential element of Bohr's model. Planck had assumed in his derivation of the radiation law that an oscillator emits or absorbs energy in units of $h\nu$. In 1905 Albert Einstein (1879-1955) made the revolutionary proposal that electromagnetic waves in free space also exist in discrete quanta. His arguments were based on considerations of the entropy of the radiation in connection with the use of the Planck law (1.1). As an application of his theory of light quanta Einstein proposed an explanation of the photoelectric effect, which had been discovered by Heinrich Hertz in 1886-87. The photoelectric effect consists in the ejection of electrons from a metal surface irradiated with high frequency electromagnetic waves. According to Einstein's theory the maximum energy $mv^2/2$ of the ejected electrons is independent of the radiation intensity and given by $mv^2/2 = h\nu - W$, where W - the so-called work function - is a constant[3] characteristic of the metal under consideration. This relation was in good agreement with experiment and later confirmed by a series of accurate measurements carried out by Millikan in 1914-16. Since the frequency ν of an electromagnetic wave is inversely proportional to its wavelength λ,

$$\nu = \frac{c}{\lambda}, \tag{1.3}$$

where c is the velocity of light, the energy $E = h\nu$ of a light quantum (a photon) may be expressed in terms of the wavelength as follows

$$E = h\nu = h\frac{c}{\lambda}. \tag{1.4}$$

The energy of a photon is thus inversely proportional to the wavelength of the electromagnetic radiation.

---

[2] Bohr learned about the Balmer formula in the beginning of 1913, a few weeks before submitting the first of his three papers to Philosophical Magazine.

[3] The work function W is the minimum energy needed to allow an electron to escape from the surface.

Introduction

The concept of electromagnetic radiation existing in discrete quanta and Rutherford's nuclear atom together formed the basis for Niels Bohr's epoch-making model of the hydrogen atom. We shall present the Bohr model in a form which follows closely the first of his trilogy of papers from 1913. Before doing so we quote the following passage from the introduction of the paper:

'Whatever the alteration in the laws of motion of the electrons may be, it seems necessary to introduce in the laws in question a quantity foreign to the classical electrodynamics, i. e. Planck's constant, or as it often is called the elementary quantum of action. By the introduction of this quantity the question of the stable configuration of the electrons in the atoms is essentially changed, as this constant is of such dimensions and magnitude that it, together with the mass and charge of the particles, can determine a length of the order of magnitude required.' [Niels Bohr, Philosophical Magazine **26**, no. 151, 6, 1913.]

To illustrate this point, let us consider three of the fundamental constants discussed in the previous section, the elementary charge e, the electron mass m and the Planck constant $\hbar$. The magnitude F of the force between the nucleus and the electron is given by the Coulomb law

$$F = \frac{e_0^2}{r^2}, \tag{1.5}$$

where r is the distance between the electron and the proton. The quantity e_0^2 is defined by

$$e_0^2 = \frac{e^2}{4\pi\epsilon_0}, \tag{1.6}$$

where ϵ_0 is the vacuum permittivity. Its value is $\epsilon_0 = 8.854 \cdot 10^{-12}$ farad/m. Since a force has the same dimension as energy divided by length, the constant e_0^2 is measured in units of m^{3}kg/s^2.

We now want to investigate whether it is possible to combine the quantities $e_0, \hbar$ and m in such a way that they result in a length. As an example we may form the quantity

$$e_0^2 m \hbar.$$

It is readily seen that this cannot be a length, since the product of e_0^2, m and $\hbar$ is proportional to a mass to the third power. The combination

$$\frac{\hbar^2}{m e_0^2}$$

is a more promising candidate. The numerator in this expression has the unit m^{4}kg^2/s^2, while the denominator has the unit m^{3}kg^2/s^2, and the result is therefore seen to be a length. The quantity $\hbar^2/me_0^2$ is called the Bohr radius and denoted by the symbol a_0.

When the values of the constants m, e, ϵ_0 and $\hbar$ are inserted, the length a_0 becomes

$$a_0 = \frac{\hbar^2}{me_0^2} = 0.529 \cdot 10^{-10}\,\text{m}. \tag{1.7}$$

It will be shown in Section 1.3 below that the Bohr radius is the only length that can be made from the constants m, e_0 and $\hbar$. The fundamental constants m, e_0 and $\hbar$ determine a length, which is of the same order of magnitude as the size of the atoms, as well as an energy, which is of the same order of magnitude as the ionization energy of the atoms. This indicates that the Planck constant forms the key to the understanding of the properties of the atom.

After this interlude we now return to the discussion of the Bohr model of the hydrogen atom.

1.2.1 The Bohr model

Bohr's atomic model contained two basic assumptions (later called postulates), which represent a radical departure from classical physics. Textbooks on quantum mechanics often present the Bohr model in a form which is historically inaccurate as regards the content of the postulates which Bohr used. In the first of the trilogy of papers Bohr wrote:

'The principal assumptions used are: (1) That the dynamical equilibrium of the systems in the stationary states can be discussed by help of the ordinary mechanics, while the passing of the systems between different stationary states cannot be treated on that basis. (2) That the latter process is followed by emission of a *homogeneous* radiation, for which the relation between the frequency and the amount of energy emitted is the one given by Planck's theory.'

The quantization of orbital angular momentum, $L = n\hbar$, is not a separate postulate, but a consequence of Bohr's theory when applied to electrons moving in circular orbits.

The purpose of this section is to present Bohr's arguments based on his two assumptions stated above and a condition that the classical and quantum mechanical description are in agreement for large quantum numbers. To facilitate the discussion of the general case we shall, however, first treat the special case of circular orbits combined with an additional assumption of the quantization of angular momentum.

a) Special case: circular orbits

The potential energy of the electron due to the attractive Coulomb force

Introduction

from the hydrogen nucleus is

$$V(r) = -\frac{e_0^2}{r}. \tag{1.8}$$

In classical mechanics, for a particle of mass m and speed v following a circular orbit with radius r, the centripetal force mv^2/r must equal the attractive Coulomb force e_0^2/r^2, or

$$\frac{mv^2}{r} = \frac{e_0^2}{r^2}. \tag{1.9}$$

Let us now combine this classical equation with the additional assumption that the orbital angular momentum $L = mvr$ is quantized in units of $\hbar$,

$$L = mvr = n\hbar, \quad \text{where} \quad n = 1, 2, 3 \cdots. \tag{1.10}$$

From this it follows that the radius r of the n'th stationary orbit is

$$r = n^2 a_0 \tag{1.11}$$

while the velocity v becomes

$$v = \frac{e_0^2}{n\hbar}. \tag{1.12}$$

The total energy equals the sum of the kinetic and potential energy

$$E = \frac{mv^2}{2} - \frac{e_0^2}{r}. \tag{1.13}$$

By combining (1.11), (1.12) and (1.13) we obtain the energy of the n'th stationary state as

$$E_n = -\frac{me_0^4}{2\hbar^2 n^2}, \tag{1.14}$$

from which the frequencies ν of the various families of spectral lines are derived by use of the condition

$$h\nu = E_{n_1} - E_{n_2}. \tag{1.15}$$

Bohr's value of the Rydberg constant R obtained by comparing (1.15) with (1.2) is

$$R = \frac{me_0^4}{4\pi\hbar^3 c}. \tag{1.16}$$

Bohr compared his calculated value of the Rydberg constant with the one determined experimentally and obtained perfect agreement within the uncertainties[4] with which the values of e, m and h were known.

---

[4] When corrections are made for the finite mass M of the proton, the electron mass m is replaced by the reduced mass $\mu = mM/(M+m)$, see Chapter 7. The relative correction is thus 0.0005, which is considerably less than the uncertainties with which the fundamental constants were known at the time.

b) General case: elliptic orbits

We now present Bohr's argument for the general case of elliptic orbits. The classical description of an electron moving under the influence of the electrostatic attraction from the nucleus is given by the theory of Kepler motion, the reason being that the gravitational attraction has the same $1/r^2$-dependence as the Coulomb force. When the total energy E is negative, the electron moves classically along closed elliptic orbits with a semi-major axis a, which is related to the total energy E by

$$E = -\frac{e_0^2}{2a}. \tag{1.17}$$

The time of revolution T is given by Kepler's law

$$T^2 = \frac{4\pi^2 m a^3}{e_0^2}. \tag{1.18}$$

According to (1.17) and (1.18) the frequency of revolution $\Omega = 2\pi/T$ is then

$$\Omega = 2\sqrt{\frac{2}{m}}\frac{|E|^{3/2}}{e_0^2}. \tag{1.19}$$

On the basis of the postulate concerning the existence of stationary states, Bohr related the energy of the n'th stationary state to the frequency of revolution Ω by

$$|E| = f(n)\hbar\Omega, \tag{1.20}$$

where $f(n)$ is a function to be determined. It follows from (1.19) and (1.20) by elimination of Ω that the energy $E = E_n$ of the n'th stationary state may be written as

$$E_n = -\frac{A}{f^2(n)}, \quad A = \frac{me_0^4}{8\hbar^2}. \tag{1.21}$$

According to the second postulate, radiation is emitted by a transition from one stationary state numbered n_1 to a stationary state numbered n_2, with the frequency ω given by

$$\hbar\omega = E_{n_1} - E_{n_2}, \tag{1.22}$$

where it is assumed that $E_{n_1} > E_{n_2}$.

By forming the energy difference $E_n - E_{n-1}$, it is seen that the radiation frequency ω for large n may be written as

$$\hbar\omega = E_n - E_{n-1} = 2A\frac{f'(n)}{f^3(n)}, \tag{1.23}$$

where f' denotes the derivative of the function $f(x)$ with respect to the variable x. Note that (1.20) and (1.21) imply that the classical period of revolution Ω can be written as

$$\hbar\Omega_n = \frac{A}{f^3(n)}. \tag{1.24}$$

The stationary states are numbered in such a way that E_n and therefore $f(n)$ are increasing functions of n. In the limit of large n the ratio Ω_n/Ω_{n-1} approaches 1. In order to obtain agreement with the measured Balmer formula, Bohr took f to be linear in n, $f = cn$, where c is a numerical constant. He determined the constant c by imposing the condition that the radiation frequency ω given by (1.23) for large n should be equal to the frequency of revolution $\Omega_n \simeq \Omega_{n-1}$ in (1.24). This condition expresses the requirement that the classical and quantum mechanical description should agree for large quantum numbers. Such conditions were used extensively by Bohr in the following years and are generally referred to as the correspondence principle.

On the basis of (1.23) and (1.24) it is seen that the two frequencies ω and Ω agree for large quantum numbers, if f' is a constant, equal to 1/2, corresponding to $f = n/2$. By inserting $f = n/2$ into (1.21) one obtains the energies of the stationary states as given by (1.14).

The success of the Bohr model in accounting for the spectrum of hydrogen and relating the Rydberg constant to the fundamental constants marked the beginning of the period of the 'old quantum mechanics'. During this period many attempts were made to generalize Bohr's quantization rules to other elements. The German physicist Arnold Sommerfeld (1868-1951) played a key role in this development. Accurate observations of the hydrogen spectrum had revealed fine structure not accounted for by the Bohr model. Sommerfeld was able in 1916 to generalize the Bohr model and explain quantitatively the observed fine structure in the spectrum of hydrogen. Measurements of Paschen showed that Sommerfeld's formula represented the observed spectra with an accuracy of seven digits. It turned out later that Sommerfeld had obtained the correct formula by an accident, since his treatment did not take into account the existence of the electron spin (which was found several years later). The period of the 'old quantum mechanics' ended in 1925, when Heisenberg, Schrödinger and others founded the 'new quantum mechanics', which is quantum mechanics as we know it today.

1.2.2 Matrix mechanics

The German physicist Werner Heisenberg (1901-1976) was one of the founders of quantum mechanics. His 1925 article in Zeitschrift für Physik ranks among the most important papers in the history of physics.

Heisenberg's ideas on quantum mechanics were developed during a stay on the island of Helgoland in June 1925. Because of an attack of hay fever he was forced during this period to leave Göttingen for Helgoland. His own description many years later[5] gives a rare insight into the creative process: 'Im Helgoland war ein Augenblick, in dem es mir wie eine Erleuchtung kam, als ich sah, dass die Energie zeitlich konstant war. Es war ziemlich spät in der Nacht. Ich rechnete es mühsam aus, und es stimmte.

---
[5] "Sources of Quantum Mechanics", ed. by B. L. van der Waerden, Dover, 1967, p. 25.

Da bin ich auf einen Felsen gestiegen und habe den Sonnenaufgang gesehen und war glücklich.'[6]. On July 29 of the same year Zeitschrift für Physik received the paper by Heisenberg. It gave rise to papers by M. Born and P. Jordan (received on September 27, 1925), by P.A.M. Dirac (received on November 7, 1925) and by Born, Heisenberg and Jordan (received on November 16, 1925) in a development which is unparalleled in the history of science. These authors were the founders of quantum theory in its modern form. They built on the foundation which was given by Niels Bohr with his atomic model from 1913 and the quantization rules developed by Niels Bohr and Arnold Sommerfeld in the following decade.

Heisenberg's paper 'Über quantentheoretischen Umdeutung kinematischer und mechanischer Beziehungen' from July 1925 (Zeitschrift für Physik **33**, 879, 1925) laid the foundations of quantum mechanics. The abstract of the paper reads: 'In der Arbeit soll versucht werden, Grundlagen zu gewinnen für eine quantentheoretische Mechanik, die ausschliesslich auf Beziehungen zwischen prinzipiell beobachtbaren Grössen basiert ist'[7]. This point of view is elaborated in the first paragraph of the paper:

'Bekanntlich lässt sich gegen die formalen Regeln, die allgemein in der Quantentheorie zur Berechnung beobachtbarer Grössen (z. B. der Energie in der Wasserstoffatom) benutzt werden, der schwerwiegende Einwand erheben, dass jene Rechenregeln als wesentlichen Bestandteil Beziehungen enthalten zwischen Grössen, die scheinbar prinzipiell nicht beobachtet werden können (wie z. B. Ort, Umlaufzeit des Elektrons), dass also jenen Regeln offenbar jedes anschauliche physikalische Fundament mangelt, wenn mann nicht immer noch an der Hoffnung festhalten will, dass jene bis jetzt unbeobachtbaren Grössen später vielleicht experimentell zugänglich gemacht werden könnten. Diese Hoffnung könnte als berechtigt angesehen werden, wenn die genannten Regeln in sich konsequent und auf einen bestimmt umgrenzten Bereich quantentheoretischer Probleme anwendbar wären. Die Erfahrung zeigt aber, dass sich nur das Wasserstoffatom und der Starkeffekt dieses Atoms jenen formalen Regeln der Quantentheorie fügen, dass aber schon beim Problem der 'gekreuzten Felder' (Wasserstoffatom in elektrischem und magnetischem Feld verschiedener Richtung) fundamentale Schwierigkeiten auftreten, dass die Reaktion der Atome auf periodisch wechselnde Felder sicherlich nicht durch die genannten Regeln beschrieben werden kann, und dass schliesslich eine Ausdehnung der Quantenregeln auf die Behandlung der Atome mit mehreren Elektronen sich als unmöglich erwiesen hat' [8].

---

[6] *Translation*: 'At Helgoland there was a inspired moment when I saw that the energy was constant in time. It was rather late in the night. I worked it out laboriously, and it was right. Then I went to the top of a hill and watched the sun rise and was happy.'

[7] *Translation*: 'In this paper an attempt will be made to obtain the foundation of a quantum-theoretical mechanics, which is based solely on relations between quantities that in principle are observable.'

[8] *Translation*: 'As is well known, a serious objection may be raised against the formal rules, which are generally used in the quantum theory to calculate observable quantities (e. g. the energy in the hydrogen atom), namely that these rules contain as an essential element relations between quantities which apparently are not observable in principle (such as the position and the period of revolution of the electron), so that these rules evidently lack a clear physical foundation, unless one is prepared to maintain the hope that these so far unobservable quantities will be available experimentally at a later time. This hope might be justified, if the above-mentioned rules by themselves could be used consistently and were

In the introduction to the paper Heisenberg noted that the Einstein-Bohr frequency condition represents a complete departure from the kinematics of classical mechanics. This condition associates a frequency $\nu(n, n-\alpha)$ with the energy difference $W(n) - W(n-\alpha)$ divided by the Planck constant, and thus depends on *two* variables n and α. In a similar way Heisenberg attempted to represent physical variables such as $x(t)$, which enters the classical equation of motion for the harmonic oscillator,

$$\ddot{x} + \omega^2 x(t) = 0,$$

by quantities depending on two sets of variables. He noted in the paper: 'Während klassisch $x(t).y(t)$ stets gleich $y(t).x(t)$ wird, brauch dies in der Quantentheorie im allgemeinen nicht der Fall zu sein. - In speziellen Fällen, z. B. bei der Bildung von $x(t).x(t)^2$ tritt diese Schwierigkeit nicht auf'[9]. At this time Heisenberg did not know of matrices and the rules of matrix multiplication. It was left to Max Born and his student Pascual Jordan to point out that matrices form the natural mathematical framework for Heisenberg's ideas. As Born and Jordan acknowledged in their paper, they had the opportunity to discuss with Heisenberg, before he had submitted his paper, and they were therefore able already in September 1925 to complete their paper 'Zur Quantenmechanik'. This work was followed by 'Zur Quantenmechanik II' by Born, Heisenberg and Jordan in November of the same year, completing the general formulation of quantum mechanics.

Many fundamental contributions followed the breakthrough in 1925. It rapidly became clear that the new formulation of the quantum theory removed the difficulties which had plagued the attempts to generalize the Bohr model of the hydrogen atom. A new mechanics was created, which turned out to be the key not only to the understanding of the atoms and the elementary particles, but also to macroscopic systems such as metals or semiconductors.

1.3 Dimensions and fundamental constants

In the present subsection we shall temporarily abandon the historical setting used elsewhere in this chapter to discuss the role of dimensional analysis. In Section 1.2 we simply guessed the form of the characteristic length for a hydrogen atom. Now we shall use dimensional analysis in a systematic fashion to determine the order of magnitude of various characteristic lengths, times and energies. As we shall see, the result of a dimensional analysis is not necessarily

---

applicable to a well-defined range of quantum-theoretical problems. Experience shows, however, that only the hydrogen atom and the Stark effect in this atom obey these formal rules of the quantum theory, and that already in the problem of 'crossed fields' (hydrogen atom in an electric and a magnetic field pointing in different directions) fundamental difficulties appear, that the reaction of the atoms to periodically varying fields can certainly not be described according to the rules mentioned, and finally that an extension of the quantum rules to atoms with more electrons has proved impossible'.

[9] *Translation*: 'While from a classical point of view $x(t).y(t)$ is always equal to $y(t).x(t)$, this is in general not the case in the quantum theory. - In special cases, e. g. with the formation of $x(t).x(t)^2$ this difficulty does not arise.'

unique, depending on the number of fundamental constants considered. In spite of this, dimensional analysis remains a useful tool for estimating orders of magnitude and for checking the results of a specific calculation. In the following we shall use dimensional analysis to estimate the size of the hydrogen atom and the characteristic energy differences that are involved in the emission and absorption of electromagnetic radiation. The analysis will be based on the Planck constant $\hbar$, the velocity of light c, the mass of the electron m and the elementary charge e.

1.3.1 The hydrogen atom

A hydrogen atom consists of a positively charged nucleus, a proton, surrounded by a negatively charged electron. A measure of the size of the atom may be taken to be the distance between the two protons in the hydrogen molecule, which is approximately 10^{-10} m. The positively charged nucleus in the hydrogen atom gives rise to an attractive force on the negatively charged electron. This force is obtained from the potential energy $V(r)$ given by

$$V(r) = -\frac{e^2}{4\pi\epsilon_0 r}, \tag{1.25}$$

where ϵ_0 is the permittivity of vacuum, while r is the distance from the proton to the electron. In order to avoid that $4\pi\epsilon_0$ appears in the formulas derived below we define as before the quantity e_0 by

$$e_0^2 = \frac{e^2}{4\pi\epsilon_0}. \tag{1.26}$$

Since $V(r)$ has the dimension of energy, the dimension of e_0^2 is that of energy times length.

The dimensional analysis below is based on the fundamental constants e_0, $\hbar$ and m. Later on we shall include the velocity of light. In order to determine how e_0, $\hbar$ and m should be combined to a length we form the expression

$$e_0^x m^y \hbar^z \tag{1.27}$$

where the unknown exponents x, y and z are determined by the requirement that (1.27) is a length. If A is an arbitrary physical quantity, we shall use the symbol $\dim A$ to denote its dimension. The dimensions of mass, length and time are denoted by M, L and T, respectively. The dimension of A can then be written as

$$\dim A = M^\alpha L^\beta T^\gamma, \tag{1.28}$$

where α, β and γ are definite exponents. For the fundamental constants $\hbar$, e_0 and m we get

$$\dim \hbar = ML^2T^{-1} \tag{1.29}$$

Introduction

and
$$\dim e_0 = M^{1/2}L^{3/2}T^{-1} \tag{1.30}$$

together with
$$\dim m = M. \tag{1.31}$$

Since (1.27) must be a length, the unknown exponents are determined from

$$\dim e_0^x m^y \hbar^z = M^{x/2}L^{3x/2}T^{-x}M^y M^z L^{2z}T^{-z} = L, \tag{1.32}$$

which results in the three equations

$$x + z = 0 \tag{1.33}$$

and
$$\frac{x}{2} + y + z = 0 \tag{1.34}$$

together with
$$\frac{3x}{2} + 2z = 1. \tag{1.35}$$

It is readily seen that the solution to (1.33), (1.34) and (1.35) is

$$x = -2, \quad y = -1, \quad z = 2. \tag{1.36}$$

It follows that the characteristic length (1.27), which we denote by a_0, is given by the expression

$$a_0 = \frac{\hbar^2}{me_0^2}, \tag{1.37}$$

which is the Bohr radius.

The determination of a characteristic energy proceeds in a similar way. The only difference is that the right hand side of (1.32) is replaced by the dimension of energy, which is ML^2T^{-2}. This results in the solution $x = 4, y = 1, z = -2$ and the corresponding characteristic energy E_0 given by

$$E_0 = \frac{me_0^4}{\hbar^2}. \tag{1.38}$$

Since the hydrogen atom is stable, the electron must have a positive binding energy in the sense that energy must be supplied to remove the electron from the nucleus. This binding energy may then be expected to be given by (1.38) except for a numerical constant.

It is worth noting that the dimensional expressions (1.37) and (1.38) are unique. From the constants e_0, $\hbar$ and m we can form one and only one length or energy. The corresponding time is found by dividing $\hbar$ with E_0. It should be emphasized that the method does not allow one to determine the magnitude of a numerical constant in front of (1.37) and (1.38), but such a constant will usually be of order of magnitude 1.

When the values of the fundamental constants are inserted, the expressions given above become

$$a_0 = 0.529 \cdot 10^{-10} \,\text{m} \tag{1.39}$$

and

$$E_0 = 27.2 \,\text{eV}. \tag{1.40}$$

As we shall see, the binding energy for the hydrogen atom is only one half of E_0, but it is encouraging that the simple dimensional analysis gives results that are close to the experimental ones. The result (1.38) gives no detailed information about the hydrogen spectrum, that is the wavelengths of the electromagnetic radiation emitted in the various transitions, but it is consistent with the known fact that electromagnetic radiation is emitted with an angular frequency ω of order of magnitude $E_0/\hbar$ corresponding to a wavelength of order of magnitude 10^{-7} m.

If we include the velocity of light c in the fundamental constants under consideration, the result of the dimensional analysis is no longer unique. The quantity mc^2 is an energy, the rest energy of the electron. The ratio between E_0 as given by (1.38) and mc^2 is

$$\frac{E_0}{mc^2} = \alpha^2, \tag{1.41}$$

where we have introduced the dimensionless constant α by the definition

$$\alpha = \frac{e_0^2}{\hbar c} = \frac{1}{137.036}. \tag{1.42}$$

The constant α is called the fine-structure constant. It is so named because it enters the small corrections to the spectral lines due to relativistic effects, since α according to (1.42) is the ratio between the velocity $\sqrt{E_0/m}$ and the velocity of light.

That α is a measure of relativistic effects is apparent from the relativistic expression for the energy of a freely-moving particle,

$$E = \sqrt{m^2c^4 + p^2c^2}, \tag{1.43}$$

where p is the momentum of the particle. For small velocities it is possible to expand (1.43) in the ratio p^2/m^2c^2,

$$E = mc^2 + \frac{p^2}{2m} - \frac{p^4}{8m^3c^2} + \cdots \tag{1.44}$$

The third term in this expansion contributes to the fine structure. This can be seen by using results from classical mechanics: Let us assume that a particle

Introduction

of mass m moves in a circular orbit with radius r under the influence of the Coulomb potential $-e_0^2/r$. By equating the Coulomb force with the centripetal force it is seen that

$$\frac{mv^2}{r} = \frac{e_0^2}{r^2}, \qquad (1.45)$$

where v is the constant speed of the particle. It follows from (1.45) that the kinetic energy $mv^2/2$, except for the sign, equals one half the potential energy, which is $-e_0^2/r$. The binding energy $e_0^2/2r$ is therefore equal to the kinetic energy. The second term in (1.44) is the kinetic energy of a non-relativistic particle with velocity $v = p/m$, while the contribution of the third term is seen to be α^2 times that of the second term, which is of order e_0^2/a_0.

This kinematic correction is not the only relativistic effect. Another consequence of relativity is the occurrence of the electron spin. According to the relativistic theory of the electron it behaves as if it possessed an intrinsic angular momentum, called spin. In the following we shall briefly discuss the orbital angular momentum of a particle and the influence of external fields on its motion.

1.3.2 Angular momentum

According to classical physics a particle moving in a circular orbit of radius r has an angular momentum L of magnitude

$$L = mrv, \qquad (1.46)$$

where m is the mass of the particle and v its velocity. The dimension of the Planck constant is seen to be the same as the dimension of angular momentum. In Chapter 7 we shall see how quantum theory restricts a given component of the angular momentum vector to assume values that are an integral or a half-integral number times $\hbar$. Any angular momentum may of course be expressed in units of $\hbar$. The angular momentum of the earth in its orbit around the sun is about 10^{74} times $\hbar$. Such a result indicates that quantum theory is of no consequence for the calculation of planetary orbits. The quantization of angular momentum is however essential for the understanding of the properties of atoms and molecules.

A charged particle moving in a circular orbit constitutes a current loop, which gives rise to a magnetic field. The current is equal to the charge per unit time moving through a plane perpendicular to the orbit. If ν is the frequency of revolution, equal to $\nu = v/2\pi r$, one sees that the current I through an (imaginary) tube surrounding the moving particle is given by

$$I = e\nu. \qquad (1.47)$$

The area A of the current loop, $A = \pi r^2$, times the current I is called the magnetic moment M.

Let us now form the ratio between the magnetic moment and the angular momentum of the particle in its orbit. The ratio is seen to be independent of the radius of the orbit as well as the magnitude of the angular momentum,

$$\frac{M}{L} = \frac{e(v/2\pi r)\pi r^2}{mrv} = \frac{e}{2m}. \tag{1.48}$$

This ratio is called the gyromagnetic ratio. Its value given in (1.48) has been found on the basis of classical physics. When the angular momentum originates in the intrinsic motion, which we have called spin, the gyromagnetic ratio differs from value $e/2m$ given in (1.48). Since the ratio must have dimension of charge divided by mass, one expects that it is given by (1.48) multiplied by a numerical constant for an electron of charge $-e$ and mass m. The numerical constant is called the g-factor and denoted by g. For an electron $g = 2$ (except for a correction of order α), provided the angular momentum only originates in the spin of the electron.

When a charged particle moves in a constant magnetic field with flux density B, it executes a circular orbit in the plane perpendicular to the direction of the magnetic field. The reason for this is that the force caused by the magnetic field is perpendicular to both the field and the direction of the velocity of the particle. The frequency of revolution only depends on e, m and B. The product of e/m with B has the same dimension as frequency. By means of dimensional analysis it is readily seen that this is the only frequency that can be formed from these three quantities (apart from a numerical constant). This frequency is named the cyclotron frequency ω_c. In Chapter 6 we show that the classical frequency of revolution equals $\omega_c = eB/m$. If this frequency is multiplied by the Planck constant $\hbar$, one obtains an energy, which is the characteristic energy for motion in a magnetic field. Note that $\hbar/e$ has the dimension of magnetic flux, while $\hbar/eB$ is an area.

On the basis of the fundamental constants $\hbar$, e, ϵ_0, m and B, it is possible to form not only the energy $\hbar\omega_c$, but also the energy e_0^2/a_0. As long as we neglect the repulsion between the electrons (or the attraction from positively charged nuclei) there is, however, no physical significance associated with the energy e_0^2/a_0. This illustrates the necessity of using physical arguments together with dimensional analysis. As another example of this we shall now discuss molecular spectra, and show how the ratio between the mass of the electron and that of the nucleus enters in a way which may be determined from a combination of dimensional analysis and physical arguments.

1.3.3 Harmonic oscillator

Let us consider the motion of a particle in a harmonic oscillator potential given by the potential energy

$$V(x) = \frac{1}{2}Kx^2, \tag{1.49}$$

Introduction

where K is the force constant entering Hooke's law. We denote the mass of the particle by M. It is readily seen that it is not possible to form a characteristic length or energy on the basis of K and M alone. This is in accordance with what we know about the classical harmonic oscillator; the maximum departure from its equilibrium position can take any value, depending on the initial conditions. There is no characteristic length, nor is there any characteristic energy. If we now consider the Planck constant $\hbar$ along with K and M it becomes possible to form both a characteristic length and a characteristic energy. Since the classical angular frequency ω equals

$$\omega = \sqrt{\frac{K}{M}}, \tag{1.50}$$

the characteristic energy must be

$$\hbar\sqrt{\frac{K}{M}}. \tag{1.51}$$

By repeating the arguments leading to (1.37) one obtains the characteristic length

$$\sqrt{\frac{\hbar}{M\omega}}. \tag{1.52}$$

In Section 1.4.3 and in the following chapter we show that the possible energy values of a harmonic oscillator are given by (1.51) times $(n + \frac{1}{2})$, where n is an integer greater than or equal to zero. This causes the separation between adjoining energy values to be given by (1.51). Such a result cannot be obtained by dimensional analysis; the latter can only lead to the conclusion that the possible energy values are a number multiplied by (1.51).

Let us suppose instead that the particle is moving in a potential given by

$$V(x) = ax^4, \tag{1.53}$$

where a is a positive constant. Then the characteristic energy becomes

$$\sqrt[3]{\frac{a\hbar^4}{M^2}}, \tag{1.54}$$

since this is the only energy that can be formed from the constants a, $\hbar$ and M. In this case the possible energy values are not equally separated, and (1.54) is therefore only an estimate of the separation of the lowest levels.

When a particle of mass M moves freely without being influenced by external forces, it is not possible to form a characteristic length unless one includes the velocity of light c besides the Planck constant $\hbar$. Since the Planck constant has dimension of momentum times length, it follows that

$$\frac{\hbar}{Mc} \qquad (1.55)$$

is a length. The characteristic length (1.55) may be used to estimate the range of the strong attractive force that holds the nuclei together by counteracting the effect of the repulsive Coulomb forces between the protons. The attractive force between the nucleons comes about by the interchange of pions, these being about 300 times as heavy as an electron. When such a value of M is inserted into (1.55), the length scale comes out to be about 10^{-15} m, consistent with the known size of an atomic nucleus. The electromagnetic forces, which give rise to the Coulomb potential of (1.25), can be interpreted as arising from the exchange of a light quantum, a photon, between the charged particles. This is in close analogy with the nuclear forces arising from the exchange of pions between the neutrons and protons constituting the atomic nucleus. A light quantum, however, has no mass, which implies that the range of electromagnetic forces is infinite, in agreement with the fact that the Coulomb potential of (1.25) does not contain any characteristic length.

In addition to the gravitational, the electromagnetic and the strong interaction there exists in nature a fourth force, the weak interaction. Weak interactions are responsible for the decay of the neutron. As mentioned earlier the mean lifetime of a freely-moving neutron is about 15 minutes. The existence of vector bosons which mediate the weak interaction was shown experimentally in 1983 at the European center for particle physics, CERN. Three different vector bosons, W_+, W_- and Z_0 were found, with masses between 80 and 90 GeV/c^2 (the rest mass of a proton is 0.94 GeV/c^2). The vector bosons are thus several hundred times heavier than a pion, and the range $\hbar/Mc$ of the weak interaction is therefore much smaller than that of the strong nuclear force.

1.3.4 Molecular spectra

We have argued above, on the basis of dimensional analysis, that the characteristic energies of atoms involve the energy E_0 given in (1.38), as long as we neglect fine structure, which is characterized by energies given roughly by α^2 times E_0, and the effect of external magnetic fields giving rise to energy differences of order $\hbar eB/m$. Since molecules contain protons and electrons they must also be characterized by energies such as E_0. However, the nuclei in a molecule may also vibrate relative to each other. In addition molecules may rotate as (almost) rigid bodies. The two types of motion, vibration and rotation, are characterized by different energies, which we shall now seek to determine. It is reasonable to assume that these energies, besides E_0, involve the ratio m/M, where m is the mass of an electron, while M is the mass of a

Introduction

nucleus. For simplicity we shall consider the hydrogen molecule, for which the two nuclei are protons, thus making the ratio m/M equal to $1/1836$. For more complicated molecules the ratio is even smaller.

It is obvious that we can form any energy from E_0 and m/M, for instance by raising the latter to an arbitrary power involving an exponent, which is not necessarily an integer. It is therefore necessary to use dimensional analysis together with physical arguments. Let us first consider vibration. For small amplitudes of vibration the motion is similar to that of a harmonic oscillator. For a harmonic oscillator the characteristic energy is given by (1.51). The mass M occurring in this expression should be identified with the mass of a proton, while the force constant K can be determined by the following argument: If the change in the separation of the two protons during their motion becomes as large as a_0, the corresponding increase in the potential energy must be of order E_0. It follows that the force constant K is of order E_0/a_0^2, and the characteristic energy is therefore

$$\sqrt{\frac{m}{M}} E_0 = \frac{m^{3/2} e_0^4}{M^{1/2} \hbar^2}. \qquad (1.56)$$

The motion of the two protons relative to each other therefore gives rise to spectral lines with wavelengths that are about $50 - 100$ times as long as those corresponding to electronic transitions characterized by energy differences of order E_0.

The rotation of the hydrogen molecule is characterized by even smaller energies. The moment of inertia of the molecule is approximately given by Ma_0^2, since the distance between the protons is of the same order of magnitude as the Bohr radius. The total angular momentum of the molecule may be assumed to be a multiple of $\hbar$. The kinetic energy associated with the rotation is the square of the angular momentum divided by (twice) the moment of inertia corresponding to a characteristic energy which is of the order of magnitude

$$\frac{\hbar^2}{Ma_0^2} = \frac{m}{M} E_0 = \frac{m^2 e_0^4}{M\hbar^2}. \qquad (1.57)$$

It is seen from this expression that the characteristic energies involved in the rotation of the molecule are much less than the energies of vibration corresponding to the factor $\sqrt{m/M}$, which also determines the ratio between the vibrational energies and the energies associated with electronic transitions, cf. (1.56). This estimate of the various energies is in qualitative agreement with both experimental observations and the results of detailed calculations based on quantum theory. A closer comparison of the predictions from quantum theory with the observed spectra requires a different starting point from that given here. The purpose of these considerations is only to illustrate the usefulness as well as the limitations of dimensional analysis.

1.3.5 Gravitation

So far the discussion of the different characteristic quantities has primarily been based on Coulomb's law (1.25). In concluding this section we construct characteristic quantities from the Planck constant $\hbar$, the gravitational constant G and the velocity of light c.

The method is the same as the one described above. We form the quantity

$$\hbar^x G^y c^z \tag{1.58}$$

and seek to determine those values of the exponents x, y and z that make (1.58) into a length, an energy or a time. The only length that can be formed is given by

$$\sqrt{\frac{\hbar G}{c^3}} \tag{1.59}$$

while the corresponding energy is

$$\sqrt{\frac{\hbar c^5}{G}}. \tag{1.60}$$

Finally the characteristic time is obtained by dividing $\hbar$ with the energy (1.60), yielding the result

$$\sqrt{\frac{\hbar G}{c^5}}. \tag{1.61}$$

The order of magnitude of these results may be determined by inserting the values of the fundamental constants. From (1.59) we obtain that the characteristic length is

$$10^{-35} \, \text{m}. \tag{1.62}$$

The characteristic energy is

$$10^{28} \, \text{eV}, \tag{1.63}$$

while the characteristic time is

$$10^{-43} \, \text{s}. \tag{1.64}$$

These quantities are named the Planck length, the Planck energy and the Planck time. Note that the Planck energy is about 10^{15} times as large as the highest energies available in existing and planned particle accelerators.

Contemporary physics has identified four different forces of nature: Electromagnetic forces, weak interactions (such as those giving rise to β-decay), strong interactions (such as nuclear forces) and gravitational forces. Within the framework of relativity and quantum theory it has been possible to achieve a certain unity in the description of the first three of these forces. In the last few years new theories have been proposed with the aim of unifying all four forces of nature, but the size of the Planck energy relative to those available in the laboratory makes it very difficult to test these theories experimentally.

Though the Planck energy so far represents an inaccessible range for experiments that can test the validity of such theories, the Planck energy or the corresponding Planck mass $\sqrt{\hbar c/G}$ may of course occur in different contexts. For the type of stars known as white dwarfs it can be shown (Chapter 9) that their greatest possible mass is of the order of the Planck mass raised to the third power divided by the nucleon mass to the second power. This corresponds to a mass which is one and a half times the solar mass or roughly 10^{30} kg.

1.4 Particles and waves

The interplay of the concepts of particles and waves has left its mark on the development of quantum theory since the turn of the century. Einstein's theory of the photoelectric effect from 1905 associated particles - photons - of energy $h\nu$ and momentum h/λ with electromagnetic radiation of frequency ν and wavelength λ. This enabled Einstein to explain the existence of a lower limit ν_m on the frequency of the electromagnetic radiation needed to remove electrons from the surface of a metal. In 1923 Louis de Broglie suggested the converse; a particle of momentum p should be associated with a wave of wavelength $\lambda = h/p$.

In the early part of the nineteenth century, during the years 1828-1837, William Hamilton had noted a formal similarity between the description of a light ray in optics and the motion of a particle in classical mechanics.

> William Rowan Hamilton (1805-65) was an Irish astronomer and mathematician born in Dublin. He was a child prodigy, who was able to translate from Latin and Greek at the age of five, and mastered 13 languages at the age of 13. He studied at Trinity College, Dublin and became a professor of astronomy at the age of 22, while he was still a student. Besides mechanics and optics he worked on algebra and discovered quaternions, which generalize the complex numbers to a non-commutative algebra. He wrote poems and counted Wordsworth among his friends, though Wordsworth advised him to write mathematics instead of poetry.
>
> In spite of these achievements it may seem strange that Hamilton's name has been immortalized through its association with the energy operator in quantum mechanics, in view of the fact that Hamilton died 35 years before Planck announced his radiation law. The explanation of this will be given in the next chapter, where we introduce the basic concepts of analytical mechanics and demonstrate their relation to those of mechanics.

In 1657 Fermat discovered a minimum principle describing the path of a light ray. Consider a light ray moving between two points A and B in a medium, where the index of refraction $n = n(x, y, z)$ is a function of the spatial coordinates. The path of the ray between the two points is then determined by the condition that the time of traversal is a minimum. If the phase velocity of the ray is[10] $u = c/n$, the time of traversal for the distance ds equals ds/u. Since $u = \nu\lambda$, where λ is the wavelength, the time of traversal is given by the line integral

$$I = \frac{1}{\nu} \int_A^B \frac{ds}{\lambda}. \qquad (1.65)$$

The minimum condition $\delta I = 0$ becomes

$$\delta \int_A^B ds \frac{1}{\lambda(x, y, z)} = 0, \qquad (1.66)$$

with δI denoting the change in the value of the integral due to an infinitesimal change of the path of integration from A to B.

A similar minimum principle for the motion of a particle with mass m moving under the influence of conservative forces was put forward by Maupertuis in 1744. The principle of Maupertuis says that the path of the particle between two points A and B is determined by the condition that the time integral of the kinetic energy T has its least possible value,

$$\delta \int_A^B dt\ T = 0. \qquad (1.67)$$

With the use of $2T dt = mv ds = p ds$ the principle of Maupertuis may also be expressed by the condition that the line integral of p has its least possible value,

$$\delta \int_A^B ds\ p = 0, \qquad (1.68)$$

in analogy with (1.66). The formal similarity, noted by Hamilton, between the path of a light ray and the motion of a particle was the starting point for the ideas of de Broglie nearly a hundred years later. In 1923 de Broglie proposed to associate with a particle of momentum p and energy E a wave of wavelength λ and frequency ν given by

$$\lambda = \frac{h}{p}, \quad \nu = \frac{E}{h}. \qquad (1.69)$$

This would imply that a particle with a well-defined momentum p would be able to undergo diffraction, as if it were a wave of wavelength $\lambda = h/p$.

In support of the identification (1.69), de Broglie compared the four-vector for the energy and momentum of a particle with mass m and velocity **v** with the

---

[10] The concepts of phase velocity and group velocity are discussed in Section 1.4.4 below.

form of a plane wave. The four-vector for energy and momentum is according to the special theory of relativity given by

$$\left(\frac{E}{c}, \mathbf{p}\right) = \left(\frac{mc}{\sqrt{1 - v^2/c^2}}, \frac{m\mathbf{v}}{\sqrt{1 - v^2/c^2}}\right), \quad (1.70)$$

which implies that the energy E is related to the momentum $\mathbf{p}$ by $E^2 = m^2c^4 + p^2c^2$. De Broglie compared the energy-momentum four-vector with the form of a plane wave,

$$\exp i(\mathbf{k} \cdot \mathbf{r} - \omega t), \quad (1.71)$$

which has its phase $(\mathbf{k}\cdot\mathbf{r} - \omega t)$ given by the scalar product[11] of the four-vectors $(\omega/c, \mathbf{k})$ and $(ct, \mathbf{r})$. Since $\lambda = 2\pi/k$ and $\nu = \omega/2\pi$, it is evident that (1.69) identifies the four-vector $(\hbar\omega/c, \hbar\mathbf{k})$ with $(E/c, \mathbf{p})$.

A direct experimental verification of de Broglie's ideas did not come until 1927 from experiments carried out by C. J. Davisson and L. H. Germer at Bell Laboratories in the U.S.A and, independently, by G. P. Thomson (son of J. J. Thomson) in England. These experiments showed that the reflection of an electron beam from the surface of a nickel crystal could be explained by associating a wave of wavelength $\lambda = h/p$ to an electron of momentum p, in agreeement with (1.69). Before that, however, de Broglie's hypothesis had become the starting point for Schrödinger's wave mechanics which was formulated in four pioneering papers during the first half of 1926.

Erwin Schrödinger (1887-1961) was trained in Vienna and taught for some years theoretical physics at the University of Zürich. At this time Peter Debye was a professor at the Eidgenössische Technische Hochschule in Zürich. The two institutions held a joint physics seminar led by Debye, who suggested to Schrödinger that he should present a report on the thesis of Louis de Broglie from 1924. As described above de Broglie had proposed to associate with a particle of momentum p and energy E a wave of wavelength λ and frequency ν. During Schrödinger's talk Debye made the remark[12] that the proper description of waves required a wave equation. Schrödinger immediately set out to find the wave equation that would describe the de Broglie waves. The result was a new seminar only a few weeks later - this time *with* a wave equation - and a series of four papers in Annalen der Physik with the title 'Quantizierung als Eigenwertproblem', submitted in January, February, May and June 1926. In these papers Schrödinger formulated both the time-independent and the time-dependent equation, and applied his formulation to the harmonic oscillator and other examples. He also developed perturbation theory, which he used to explain the influence of electric fields on the hydrogen spectrum. In the last of the four papers he found a relativistic wave equation valid for a spinless particle.

---

[11] The scalar product of the four-vectors $(k_0, \mathbf{k})$ and $(q_0, \mathbf{q})$ is defined by $(-k_0 q_0 + \mathbf{k}\cdot\mathbf{q})$.
[12] F. Bloch, Physics Today, December 1976.

Schrödinger arrived at the wave equation by using the analogy mentioned above between classical geometrical optics and classical particle dynamics. We shall not present his arguments here but only discuss the solution of the wave equation in some simple cases.

1.4.1 A free particle

For a freely-moving particle with mass m, the Schrödinger equation is

$$-\frac{\hbar^2}{2m}\left(\frac{\partial^2 \psi}{\partial x^2} + \frac{\partial^2 \psi}{\partial y^2} + \frac{\partial^2 \psi}{\partial z^2}\right) = i\hbar \frac{\partial \psi}{\partial t}, \quad (1.72)$$

where $\psi = \psi(\mathbf{r}, t)$ is the wave function.

The meaning of the wave function ψ was not clear to Schrödinger and his contemporaries, as is evident from the following verse, written by a colleague on the occasion of a boat trip held in connection with a conference in Zürich during the summer of 1926:

> Gar Manches rechnet Erwin schon
> Mit seiner Wellenfunktion.
> Nur wissen möcht' man gerne wohl
> Was man sich dabei vorstell'n soll.

The interpretation, due to Max Born, of the absolute square of the wave function, $\psi^*\psi$, as a probability density came later in 1926 in the form of two papers in Zeitschrift für Physik, **37**, 863 and **38**, 803. In the main text of the first paper Born identified the wave function itself with the probability, but he changed this to the square of the wave function in a footnote introduced during the proof-reading. Schrödinger had suggested that $|\psi|^2$ represents fluctuations in the density of electric charge. Despite the successes of quantum mechanics he remained sceptical about the probabilistic interpretation for the rest of his life. He was not alone: Neither Planck, Einstein nor de Broglie were willing to accept Born's interpretation as the final one, though every experiment carried out to date to test the predictions of quantum mechanics has given support to Born's interpretation.

As may be seen by insertion, a plane wave of the form (1.71) is a solution of (1.72) provided that

$$\hbar\omega = \frac{\hbar^2 k^2}{2m}. \quad (1.73)$$

When $\hbar\omega$ is identified with the energy E and $\hbar k$ with the magnitude of the momentum, p, it follows from (1.73) that

$$E = \frac{p^2}{2m}, \quad (1.74)$$

Introduction

which is the familiar relation between the energy and momentum of a freely-moving particle. The solution to the Schrödinger equation for a free particle is thus in agreement with de Broglie's hypothesis. This is hardly surprising. The Schrödinger equation was 'invented' as the wave equation describing de Broglie waves. The significance of the Schrödinger equation becomes evident by examination of its solutions for cases in which the particle is not moving freely.

Schrödinger used the similarity between geometrical optics and classical particle dynamics to propose a wave equation more general than (1.72), which is only valid for a free particle. On this basis he suggested that (1.72) in general should be replaced by

$$-\frac{\hbar^2}{2m}\left(\frac{\partial^2 \psi}{\partial x^2} + \frac{\partial^2 \psi}{\partial y^2} + \frac{\partial^2 \psi}{\partial z^2}\right) + V(\mathbf{r})\psi = i\hbar\frac{\partial \psi}{\partial t}, \quad (1.75)$$

which is the equation for a single particle moving under the influence of a conservative force described by the potential energy $V(\mathbf{r})$. Schrödinger did not *derive* (1.75). His arguments were based on a formal analogy taken from classical physics. Only experiments would be able to confirm or deny the validity of (1.75).

Schrödinger applied (1.75) to a number of physical problems. We shall briefly mention two of these in the subsequent sections of this chapter and return to a more thorough discussion of the Schrödinger equation in Chapters 3 and 4.

1.4.2 Particle in a box

Let us consider a particle moving in a potential V, which is independent of time. Under these circumstances it is possible to separate the solution to (1.75) into a product of two functions, one depending only on $\mathbf{r}$ and one depending only on t,

$$\psi(\mathbf{r}, t) = u(\mathbf{r})f(t). \quad (1.76)$$

By inserting ψ into (1.75) one finds that the equation is satisfied, provided $u(\mathbf{r})$ and $f(t)$ each satisfy the differential equations

$$-\frac{\hbar^2}{2m}\left(\frac{\partial^2 u}{\partial x^2} + \frac{\partial^2 u}{\partial y^2} + \frac{\partial^2 u}{\partial z^2}\right) + V(\mathbf{r})u = Eu \quad (1.77)$$

and

$$i\hbar\frac{df}{dt} = Ef. \quad (1.78)$$

Here E denotes the separation constant, which in the present case has the dimension of energy[13]. Since (1.78) is a differential equation of first order in

---

[13] To prove that (1.75) is satisfied by $\psi = uf$, when the two equations (1.77) and (1.78) both are satisfied, one simply multiplies (1.77) with f, and (1.78) with u, and subtracts the resulting equations.

time, its general solution is

$$f(t) = Ce^{-iEt/\hbar}, \tag{1.79}$$

where C is an arbitrary constant. The solution to (1.77) depends on the specific form of the potential. Mathematically (1.77) has the form of an eigenvalue equation with a differential operator H acting on u, while the right hand side is the eigenvalue E times u. For the reasons explained in the following chapter, H is called the Hamiltonian. Its eigenvalues E represent the possible results of a measurement of the energy of the system.

In the remainder of this section we shall only deal with situations where the particle moves in one dimension, in which case (1.77) becomes

$$-\frac{\hbar^2}{2m}\frac{d^2 u}{dx^2} + V(x)u(x) = Eu(x). \tag{1.80}$$

Let us assume that the particle moves in a potential $V(x)$ given by

$$V(x) = \infty \quad \text{for} \quad x > a \quad \text{and} \quad x < 0$$
$$V(x) = 0 \quad \text{for} \quad 0 < x < a. \tag{1.81}$$

A solution to (1.80) is then given by

$$u_n(x) = A_n \sin\left(\frac{n\pi x}{a}\right) \quad \text{for} \quad 0 < x < a, \quad u_n(x) = 0 \quad \text{otherwise}, \tag{1.82}$$

where n is a positive integer. We have assumed that ψ is everywhere continuous, and zero where the potential is infinite (this boundary condition is discussed in detail in Chapter 4).

The solution (1.82) is a normalized wave function for a *state* of the system, if the constant A_n is chosen such that

$$\int_{-\infty}^{\infty} dx\, u_n^* u_n = 1. \tag{1.83}$$

By inserting the solution (1.82) in (1.83) and performing the integral over x, one sees that (1.83) is satisfied if

$$A_n = \sqrt{\frac{2}{a}}. \tag{1.84}$$

Clearly (1.83) is also satisfied if A_n is multiplied by $\exp(i\phi)$ with an arbitrary phase angle ϕ.

The concept of a quantum mechanical state is discussed at length in the next two chapters. Here we merely determine the eigenvalues of the Hamiltonian and thus establish the possible results of a measurement of the energy of the particle in the box.

Introduction

The value of the constant E, which belongs to the solution (1.82), is seen by insertion to be

$$E_n = \frac{\hbar^2 \pi^2 n^2}{2ma^2}. \tag{1.85}$$

Physically, E_n is the particle energy in the state u_n. The result (1.85) agrees with what we would expect from dimensional considerations, since the only energy that may be formed from $\hbar$, m and a is the quantity $\hbar^2/ma^2$ (except for a numerical constant).

According to Fourier's theorem, which is discussed in Appendix C, the solutions (1.82) form a complete orthonormal basis for the general solution ψ to the Schrödinger equation (1.80) with the box potential (1.81), given that ψ vanishes at $x = a$ and $x = 0$, where the potential is infinite. Thus we have the general solution

$$\psi(x,t) = \sum_n a_n u_n(x) e^{-iE_n t/\hbar}, \tag{1.86}$$

where the constants c_n are determined by the value, $\psi(x,0)$, of the wave function at $t = 0$ according to

$$a_n = \int_0^a dx\, \psi(x,0) u_n^*(x), \tag{1.87}$$

as may be seen by multiplying (1.86) at $t = 0$ with $u_n^*(x)$ and using the condition that the functions u_n are normalized and orthogonal,

$$\int_{-\infty}^{\infty} dx\, u_n^* u_m = \delta_{nm}. \tag{1.88}$$

The symbol δ_{nm} is called a Kronecker delta. It is defined by

$$\begin{aligned}\delta_{nm} &= 1 \text{ if } n = m, \\ \delta_{nm} &= 0 \text{ if } n \neq m.\end{aligned} \tag{1.89}$$

Note that the normalization condition (1.83) is a special case of (1.88) corresponding to $n = m$.

1.4.3 Harmonic oscillator

As our next example of the use of the Schrödinger equation we shall discuss the motion of a particle with mass M in one dimension under the influence of the potential energy

$$V(x) = \frac{1}{2} K x^2. \tag{1.90}$$

From our discussion in Section 1.3.3 we know that the characteristic energy is $\hbar \omega$ and the characteristic length $(\hbar/M\omega)^{1/2}$, where $\omega = (K/M)^{1/2}$. We shall therefore simplify the Schrödinger equation (1.80) by writing it in terms of the dimensionless length variable ξ defined by

$$\xi = x\sqrt{\frac{M\omega}{\hbar}} \tag{1.91}$$

and the dimensionless energy variable λ defined by

$$\lambda = \frac{E}{\hbar\omega} \tag{1.92}$$

resulting in

$$-\frac{1}{2}\frac{d^2 u}{d\xi^2} + \frac{1}{2}\xi^2 u = \lambda u. \tag{1.93}$$

Our aim is to determine solutions to (1.93), which are normalizable, that is solutions for which

$$\int_{-\infty}^{\infty} dx\, u_n^*(x) u_n(x) = 1. \tag{1.94}$$

It is convenient to write the differential operator on the left hand side of (1.93) in terms of the annihilation and creation operators a and $a^\dagger$, which are differential operators defined by

$$a = \frac{1}{\sqrt{2}}(\xi + \frac{d}{d\xi}) \tag{1.95}$$

and

$$a^\dagger = \frac{1}{\sqrt{2}}(\xi - \frac{d}{d\xi}). \tag{1.96}$$

The reason why a is called an annihilation operator and $a^\dagger$ a creation operator will be elucidated in Chapter 2, where we shall see that a removes an energy quantum, while $a^\dagger$ adds one[14].

If we first let a and then $a^\dagger$ act on a function $f(\xi)$, the results become

$$af(\xi) = \frac{1}{\sqrt{2}}(\xi f + f') \tag{1.97}$$

and

$$a^\dagger f(\xi) = \frac{1}{\sqrt{2}}(\xi f - f'), \tag{1.98}$$

---

[14] In quantum physics particles may be created or made to disappear; an electron in a metal may absorb or emit a phonon (Chapter 9) or a quantum of light may turn into an electron and a positron (Section 1.4.5).

where f' denotes the derivative of f with respect to ξ. Next we let $a^\dagger$ act on $af(\xi)$ as given by (1.97). This yields

$$a^\dagger a f(\xi) = \frac{1}{2}(-f'' + \xi^2 f - f). \qquad (1.99)$$

Similarly, by letting a act on $a^\dagger f(\xi)$ as given by (1.98), we obtain

$$aa^\dagger f(x) = \frac{1}{2}(-f'' + \xi^2 f + f). \qquad (1.100)$$

A number of important conclusions may be drawn from (1.97)-(1.100):

- It matters in which order we put a and $a^\dagger$; according to (1.99) and (1.100) we have

$$aa^\dagger f(\xi) - a^\dagger a f(\xi) = f(\xi) \qquad (1.101)$$

or in operator form

$$aa^\dagger - a^\dagger a = 1. \qquad (1.102)$$

- The Schrödinger equation (1.93) may be expressed as

$$a^\dagger a\, u = (\lambda - \frac{1}{2})u \qquad (1.103)$$

since

$$-\frac{1}{2}\frac{d^2}{d\xi^2} + \frac{1}{2}\xi^2 = a^\dagger a + \frac{1}{2} \qquad (1.104)$$

according to (1.99).

- The solution to the differential equation

$$a\,u = 0, \qquad (1.105)$$

is denoted u_0 and given by

$$u_0 = Ce^{-\xi^2/2}, \qquad (1.106)$$

which may be shown by inserting u_0 in the differential equation (1.105). Here C is a constant of integration which is determined by the normalization condition

$$\int_{-\infty}^{\infty} dx u_0^*(x) u_0(x) = 1. \qquad (1.107)$$

It is seen by the use of (1.105), that u_0 is also a solution of the Schrödinger equation (1.103) belonging to the eigenvalue $\lambda = 1/2$.

- The function $a^\dagger u_0$ is a solution to the Schrödinger equation (1.103) belonging to the eigenvalue $\lambda = 3/2$, since (1.105) implies, that

$$a^\dagger a a^\dagger u_0 = a^\dagger(a^\dagger a + 1)u_0 = a^\dagger u_0. \qquad (1.108)$$

- If u_n is a solution to the Schrödinger equation (1.103) with eigenvalue $\lambda_n = (n + 1/2)$, then it follows that the function $a^\dagger u_n$ is a solution with eigenvalue $\lambda_{n+1} = \lambda_n + 1$, since

$$a^\dagger a a^\dagger u_n = a^\dagger(a^\dagger a + 1)u_n = (n+1)a^\dagger u_n. \qquad (1.109)$$

We may conclude from this that the energy eigenvalues for the harmonic oscillator are given by

$$E_n = \lambda_n \hbar \omega = (n + \frac{1}{2})\hbar \omega, \qquad (1.110)$$

where n is a non-negative integer. It is quite simple to prove that there are no other eigenvalues than those given by (1.110), but the proof itself will be deferred to Section 2.2.2 of the following chapter.

Let us return to the Schrödinger equation (1.77). The operator

$$-\frac{\hbar^2}{2m}\left(\frac{\partial^2}{\partial x^2} + \frac{\partial^2}{\partial y^2} + \frac{\partial^2}{\partial z^2}\right) \qquad (1.111)$$

represents the kinetic energy $p^2/2m$. This suggests that the momentum $\mathbf{p}$ should be represented by the differential operator $(\hbar/i)\nabla$. This identification is given further support by considering a plane wave $\exp(i\mathbf{k} \cdot \mathbf{r})$. When the momentum operator acts on this function the result becomes $\hbar\mathbf{k}$ times the function itself. The momentum eigenvalue $\hbar k$ is thus related to the wavelength $\lambda = 2\pi/k$ by the de Broglie relation $\lambda = 2\pi\hbar/p = h/p$.

For the particle moving in the one-dimensional harmonic oscillator potential, the momentum operator $p = (\hbar/i)d/dx$ may be expressed in terms of a and $a^\dagger$ with use of the definitions (1.95) and (1.96),

$$p = \frac{\sqrt{\hbar M \omega}}{i\sqrt{2}}(a - a^\dagger). \qquad (1.112)$$

In the following chapter we shall see how one represents physical quantities by matrices. These matrices are generally complex, but they have real eigenvalues which yield the possible results of a measurement for the corresponding physical quantity. In the case of the momentum of a particle moving in a harmonic oscillator potential the matrix is

$$p_{nm} = \int_{-\infty}^{\infty} dx\, u_n^*(x) p u_m(x). \qquad (1.113)$$

Introduction

Note that the diagonal elements of this infinite-dimensional matrix are zero. The only nonvanishing matrix elements are $p_{n\,n+1}$ and $p_{n\,n-1}$. This is a consequence of the fact that the functions resulting from a and $a^\dagger$ operating on u_n are proportional to u_{n-1} and u_{n+1}, respectively, provided that $n \geq 1$, together with the orthogonality of the functions u_n,

$$\int_{-\infty}^{\infty} dx\, u_n^*(x) u_m(x) = 0 \quad \text{for} \quad n \neq m. \tag{1.114}$$

1.4.4 Wave packets

Before we introduce the matrix formulation of quantum mechanics in the following chapter, we shall complete this brief introduction to wave mechanics by a discussion of the concepts of phase velocity and group velocity. For simplicity we only consider motion in one dimension. The results may easily be generalized to three dimensions.

A plane wave which propagates in one dimension has the form

$$e^{i(kx-\omega t)}. \tag{1.115}$$

It is seen from (1.115) that the phase $(kx - \omega t)$ is constant when x and t satisfy

$$x = \text{const.} + \frac{\omega}{k} t. \tag{1.116}$$

The ratio ω/k is called the phase velocity. It differs from the group velocity, to be defined below, if the relation between ω and k is non-linear. The group velocity is introduced as follows: Let us consider a superposition of wave functions of the form (1.115),

$$\psi(x,t) = \int_{-\infty}^{\infty} dk\, g(k) e^{i(kx-\omega t)}, \tag{1.117}$$

where $g(k)$ is a weight function. We shall assume that the weight function has a maximum at $k = k_0$. An example of such a weight function is the Gaussian function

$$g = e^{-(k-k_0)^2 a^2}, \tag{1.118}$$

where a and k_0 are constants. The superposition (1.117) consists in this case of plane waves with wavenumber[15] $k/2\pi$ primarily in an interval of width a^{-1} around $k_0/2\pi$. A superposition of this kind is called a wave packet, although the name does not imply that the weight function $g(k)$ necessarily has the precise form (1.118).

---

[15] We shall later on, in agreement with common usage, speak of k itself as the wave number, even though it should properly be called the cyclic wave number because of the factor of 2π.

Let us examine the development of the form of the wave packet (1.117) in time, on the basis of the assumption that the weight function g is peaked at $k = k_0$, corresponding to $g = \tilde{g}(k - k_0)$. We consider times that are not too long and expand $\omega(k)$ in the neighborhood of k_0,

$$\omega(k) \simeq \omega(k_0) + (k - k_0)\frac{d\omega}{dk}\Big|_{k=k_0}. \tag{1.119}$$

In carrying out the integration in (1.117) we introduce the variable $k - k_0$ and obtain

$$\psi(x,t) = e^{i(k_0 x - \omega(k_0)t)} f(x - v_g t), \tag{1.120}$$

where

$$v_g = \frac{d\omega}{dk}\Big|_{k=k_0} \tag{1.121}$$

denotes the group velocity and

$$f(x) = \int_{-\infty}^{\infty} dk\, e^{ikx} \tilde{g}(k). \tag{1.122}$$

In this approximation the wave packet keeps its form, the maximum of the function f being simply displaced in time according to

$$x = v_g t. \tag{1.123}$$

This is only valid for not too long times, since the approximation (1.119) does not retain its validity indefinitely.

In the three-dimensional case the group velocity is

$$\mathbf{v}_g = \frac{\partial \omega}{\partial \mathbf{k}}. \tag{1.124}$$

This shows that the group velocity for de Broglie waves with $\omega = \hbar k^2/2m$ becomes

$$\mathbf{v}_g = \frac{\hbar \mathbf{k}}{m}, \tag{1.125}$$

in agreement with the classical relation between velocity and momentum.

1.4.5 A relativistic wave equation

The Schrödinger equation (1.72) for a free particle is based on the nonrelativistic relation $E = p^2/2m$ between the energy and momentum of a free particle. According to the theory of relativity

$$E^2 = m^2 c^4 + p^2 c^2. \tag{1.126}$$

Introduction

Schrödinger suggested therefore that the wave equation (1.72) for a free particle should be replaced by

$$-\hbar^2 \frac{\partial^2 \psi}{\partial t^2} - m^2 c^4 \psi + \hbar^2 c^2 \nabla^2 \psi = 0 \qquad (1.127)$$

in a relativistic theory. This corresponds to the replacement of E by $i\hbar \partial/\partial t$, while $\mathbf{p}$ is replaced by $(\hbar/i)\nabla$.

Schrödinger's relativistic equation does not hold for an electron but only for a spinless particle. Goudsmit and Uhlenbeck had proposed in 1925 that the electron possesses a spin or 'internal' angular momentum. Their hypothesis was based on the observed influence of magnetic fields on the spectral lines of atoms, which was given by an empirical formula due to Landé from 1923. In 1928 Dirac published a relativistic equation which describes the electron spin. This equation also explained the fine structure of the hydrogen atom, which is the difference between the observed spectral lines and those predicted by the Bohr model, cf. (1.14). In this section we shall introduce the Dirac wave equation and later, in Chapter 7, we return to the discussion of the electron spin and the associated magnetic moment.

Paul Adrien Maurice Dirac (1902-84) was a graduate student in St. John's College at Cambridge University when Werner Heisenberg, who was only eight months older than Dirac, gave a lecture in Cambridge on July 28, 1925. Heisenberg spoke about the 'old' quantum mechanics, the complicated set of quantization rules invented by Niels Bohr and Arnold Sommerfeld to account for the atomic spectra. It is not clear whether Dirac himself was present at the lecture, but his advisor Ralph Fowler was able two weeks later to send him the proof sheets for Heisenberg's July 29 paper (see Section 1.2.2), and Dirac went to work immediately on Heisenberg's new formulation. In November 1925 Dirac submitted his paper 'The Fundamental Equations of Quantum Mechanics' to the Proceedings of the Royal Society, and the paper was followed by several others in the next two years. Many of Dirac's results from the years 1925-26 were obtained simultaneously and independently by others, but the relativistic theory of the electron from 1928 is entirely his own. Dirac's paper 'The Quantum Theory of the Electron' was received by the Proceedings of the Royal Society on January 2, 1928. Below we quote the introductory paragraph from the paper, which is another landmark in the history of physics:

'The new quantum mechanics, when applied to the problem of the structure of the atom with point-charge electrons, does not give results in agreement with experiment. The discrepancies consist of 'duplexity' phenomena, the observed number of stationary states for an electron in an atom being twice the number given by the theory. To meet the difficulty, Goudsmit and Uhlenbeck have introduced the idea of an electron with a spin angular momentum of half a quantum and a magnetic moment of one Bohr magneton. This model for the electron has been fitted into the new mechanics by Pauli,[*] and Darwin,[†] working with an equivalent theory, has shown that it gives results in agreement with experiment for hydrogen-like spectra to the first order of accuracy.

The question remains as to why Nature should have chosen this particular model for the electron instead of being satisfied with the point-charge. One would like to find some incompleteness in the previous methods of applying quantum mechanics to the point-charge electron such that, when removed, the whole of the duplexity phenomena follow without arbitrary assumptions. In the present paper it is shown that this is the case, the incompleteness of the previous theories lying in their disagreement with relativity, or, alternatively, with the general transformation theory of quantum mechanics. It appears that the simplest Hamiltonian for a point-charge electron satisfying the requirements of both relativity and the general transformation theory leads to an explanation of all duplexity phenomena without further assumption.'
[P. A. M. Dirac, Proc. Roy. Soc. **A** 117, 610 (1928)].

Dirac sought an equation in which the spatial and temporal derivatives enter linearly, since the temporal evolution of the wave function in this case is determined by the value of the wave function at some definite instant of time. For a free particle such an equation has the form

$$(E - c\boldsymbol{\alpha} \cdot \mathbf{p} - \beta mc^2)\psi = 0, \tag{1.128}$$

where

$$E \to i\hbar \frac{\partial}{\partial t}, \quad \mathbf{p} \to \frac{\hbar}{i}\boldsymbol{\nabla}. \tag{1.129}$$

By demanding that the solutions to (1.128) should also satisfy (1.127), Dirac was able to determine the quantities $\boldsymbol{\alpha}$ and β. The result of acting on (1.128) with the operator $(E + c\boldsymbol{\alpha} \cdot \mathbf{p} + \beta mc^2)$ is (1.127), *provided* that α_i and β satisfy the algebraic relations

$$\alpha_i \beta + \beta \alpha_i = 0 \tag{1.130}$$

for all i ($i = x, y, z$) and

$$\beta^2 = 1. \tag{1.131}$$

In addition we must require that

$$\alpha_i \alpha_j + \alpha_j \alpha_i = 0 \tag{1.132}$$

for $i \neq j$, while

$$\alpha_i^2 = 1 \tag{1.133}$$

for $i = x, y, z$, since these conditions make the cross terms vanish in

$$(E + c\boldsymbol{\alpha} \cdot \mathbf{p} + \beta mc^2)(E - c\boldsymbol{\alpha} \cdot \mathbf{p} - \beta mc^2)\psi = 0, \tag{1.134}$$

while the squares are in agreement with (1.127).

Clearly α_i and β cannot be ordinary numbers, if conditions like $\alpha_i \beta + \beta \alpha_i = 0$ are to be satisfied. If instead α_i and β are matrices, then it is possible to

Introduction

satisfy these conditions. An elementary example is furnished by the two 2×2 matrices

$$\begin{pmatrix} 1 & 0 \\ 0 & -1 \end{pmatrix} \quad \begin{pmatrix} 0 & 1 \\ 1 & 0 \end{pmatrix}.$$

The result of multiplying the first matrix with the second one is seen to be minus the result of multiplying the second one with the first one. Let us then try to represent α_i and β by

$$\alpha_i = \begin{pmatrix} 0 & \sigma_i \\ \sigma_i & 0 \end{pmatrix} \tag{1.135}$$

and

$$\beta = \begin{pmatrix} 1 & 0 \\ 0 & -1 \end{pmatrix}, \tag{1.136}$$

where σ_i should satisfy

$$\sigma_i \sigma_j + \sigma_j \sigma_i = 0 \tag{1.137}$$

for $i \neq j$, while

$$\sigma_i^2 = 1. \tag{1.138}$$

It follows that the quantities σ_i are also matrices, if (1.137) is to be satisfied. We choose them as the 2×2 matrices

$$\sigma_x = \begin{pmatrix} 0 & 1 \\ 1 & 0 \end{pmatrix}, \quad \sigma_y = \begin{pmatrix} 0 & -i \\ i & 0 \end{pmatrix}, \quad \sigma_z = \begin{pmatrix} 1 & 0 \\ 0 & -1 \end{pmatrix}. \tag{1.139}$$

These matrices are named Pauli matrices after the Austrian physicist Wolfgang Pauli, who had earlier (late 1924 and early 1925) pointed out the need to introduce an extra quantum number s with the values $\pm 1/2$ in order to understand the periodic system. Pauli also introduced the principle that no two electrons may have the same set of quantum numbers, which is known as the *Pauli principle*. We shall see in Chapter 8 how this principle is related to the requirement that the wave function for a system of several electrons is antisymmetric under interchange of any two electrons.

The quantities α_i ($i = x, y, z$) and β thus become 4×4 matrices, β being diagonal with elements 1, 1, -1 and -1 along the diagonal, cf. (1.136). To clarify this we write the wave equation in matrix form, the wave function ψ being represented by a column vector with four components $(\psi_1, \psi_2, \psi_3, \psi_4)$,

$$\begin{pmatrix} E - mc^2 & 0 & -cp_z & -c(p_x - ip_y) \\ 0 & E - mc^2 & -c(p_x + ip_y) & cp_z \\ -cp_z & -c(p_x - ip_y) & E + mc^2 & 0 \\ -c(p_x + ip_y) & cp_z & 0 & E + mc^2 \end{pmatrix} \begin{pmatrix} \psi_1 \\ \psi_2 \\ \psi_3 \\ \psi_4 \end{pmatrix}$$

$$= \begin{pmatrix} 0 \\ 0 \\ 0 \\ 0 \end{pmatrix} \tag{1.140}$$

As long as we discuss the motion of a free particle, the equation (1.140) only tells us that the connection between E and $\mathbf{p}$ is given by (1.126), if there are to exist solutions $(\psi_1, \psi_2, \psi_3, \psi_4)$ differing from zero. Each of the components is proportional to a plane wave $\exp i(\mathbf{k} \cdot \mathbf{r} - \omega t)$. However, as we shall see in Chapter 7, the inclusion of an external magnetic field in (1.140) will enable us to describe the occurrence of the electron spin and determine the associated magnetic moment.

In conclusion we mention another consequence of the Dirac equation, namely the existence of solutions corresponding to negative energies. The negative-energy states form a continuum starting at the energy $-mc^2$ and stretching to minus infinity. To make sense out of his theory Dirac assumed that these negative-energy states were all occupied in the ground state by an infinite sea of electrons. If one of these electrons were removed from the sea and put into a state with energy greater than mc^2, an 'antiparticle' with positive charge would thus be created along with the particle. In the case of electrons the antiparticle is called a positron. An electron-positron pair may thus be created by a light quantum with energy $h\nu$ greater than $2mc^2$.

1.5 Problems

Problem 1.1
Use the Planck law (1.1) to show that the total energy density u is given by $u = aT^4$ and express the constant a in terms of $\hbar$, c and the Boltzmann constant k.

Problem 1.2
Find an expression for the particular wavelength λ_m for which the distribution law (1.1) has its maximum, in terms of the absolute temperature and the other constants in the problem. How big is λ_m for a) $T = 3K$ and b) $T = 6000K$?

Problem 1.3
A muon is an elementary particle with the same charge as the electron, but with a mass which is 207 times that of the electron. Estimate the binding energy of a hydrogen atom in which the electron has been replaced by a muon. Determine the magnitude of the characteristic length and compare with the hydrogen atom.

Problem 1.4
Calculate the energy of a photon with wavelength 7000 Å. How large is the energy of a γ-ray photon with wavelength 10^{-13} m?

Introduction

PROBLEM 1.5
Determine the de Broglie wavelength of a particle with a mass of 10 g and a velocity of 1 m/s.

PROBLEM 1.6
Find the kinetic energy of a neutron with a de Broglie wavelength of 2 Å.

PROBLEM 1.7
Determine the de Broglie wavelength for a nitrogen molecule in atmospheric air at room temperature. The velocity of the molecule is assumed to be $\sqrt{<v^2>}$, where $<v^2>$ is the thermal average of the square of the velocity.

PROBLEM 1.8
Show that the differential operator

$$p = \frac{\hbar}{i}\frac{d}{dx}$$

has

$$e^{ikx}$$

as an eigenfunction, and determine the eigenvalue. Is $\sin kx$ an eigenfunction for p? For p^2?

PROBLEM 1.9
Estimate the order of magnitude of the rotational energies in a) the oxygen molecule and b) the nitrogen molecule. Discuss the significance of your result for the measured value of the ratio between the heat capacity at constant pressure and the heat capacity at constant volume in atmospheric air at room temperature.

PROBLEM 1.10
A particle of mass M moves in one dimension in a potential given by

$$V(z) = \infty \quad \text{for} \quad z < 0, \quad V(z) = az \quad \text{for} \quad z > 0, \tag{1.141}$$

where a is a positive constant. Use dimensional analysis to find an expression (within a numerical constant) for the energy of the particle in its ground state. Estimate the energy for $M = 1$ g, if az is the potential energy of the particle arising from the effect of gravity at the surface of the earth.

PROBLEM 1.11
An electron with speed v_F moves inside a metal along its surface. It is influenced by a magnetic field with magnetic flux density B. The magnetic field is

perpendicular to the surface normal, and the electron assumed to move perpendicular to the magnetic field. The magnitude of the force acting on the electron is approximately given by $ev_F B$, and the motion may therefore be characterized by the potential energy

$$V(z) = \infty \quad \text{for} \quad z < 0, \quad V(z) = ev_F Bz \quad \text{for} \quad z > 0, \tag{1.142}$$

where z denotes the distance to the metal surface ($z = 0$).

Use dimensional analysis to find an expression for the energy of the electron in its ground state (within a numerical constant) and determine the order of magnitude of the separation between the lowest energy levels for the following values of the parameters: $B = 1$ T and $v_F = 10^6$ m/s. Where in the electromagnetic spectrum is the corresponding frequency?

Problem 1.12
A particle of mass M moves in the potential given by

$$V(x) = \frac{1}{2} K x^2 + a x^4, \tag{1.143}$$

where K and a are positive constants.

Show that it is possible to form a characteristic energy proportional to a from the constants given above together with $\hbar$, and find an explicit expression for this energy except for a numerical constant.

Problem 1.13
The minimum possible value of the energy (1.85) is named the zero-point energy. Estimate the size of the zero-point energy for a ^{4}He-atom in solid ^{4}He, when a is assumed to be 10^{-8} cm, and convert the energy into a characteristic temperature.

Problem 1.14
Show for a one-dimensional harmonic oscillator that the diagonal elements of the matrix x_{nm} defined by

$$x_{nm} = \int_{-\infty}^{\infty} dx\, u_n^*(x) x u_m(x), \tag{1.144}$$

are zero. Determine x_{01} and x_{10}.

Problem 1.15
Octatetraen is an organic molecule which absorbs strongly at the wavelength $\lambda = 3020$ Å. In a simple model the molecule may be considered to be a one-dimensional chain, consisting of eight basic units. Each unit contributes one

electron that is free to move along the chain of the molecule. The size of each unit is 1.39 Å, and the total effective length a of the molecule is taken to be $a = 7 \cdot 1.39$ Å.

a) Find the spectrum of allowed energy values E_n for an electron moving in the one-dimensional box as a function of the quantum number n and determine the ground-state energy for a single electron.

b) Use the Pauli principle and the existence of the electron spin to determine how the states with energy E_n are occupied in the ground state of the molecule. Find the smallest energy involved in an absorption process and determine the corresponding wavelength. How much does the answer change if the effective length of the molecule is put equal to $a = 8 \cdot 1.39$ Å?

A similar model may be used to explain the yellow colour of carotenoides found in carrots and tomatoes.

2 QUANTIZATION

In the introductory chapter we discussed some simple applications of Schrödinger's wave mechanics, one example being the harmonic oscillator. We shall now make a fresh start and approach the world of quantum mechanics along a rather different route by taking classical analytical mechanics as our point of departure. We shall see how classical mechanics may be formulated in a way that makes the transition to quantum mechanics look like a simple generalization. This route is not the one taken by history, as we have seen in the introductory chapter. The revolutionary changes in the description of physical systems brought about by quantum mechanics were not due to formal generalizations of classical mechanics. The driving force was the increasing number of experimental observations, which could not be explained within the classical physics of the nineteenth century.

In the present chapter we shall use the harmonic oscillator to illustrate the quantization procedure. The reason for this is two-fold; the harmonic oscillator plays a very important role in quantum physics, and it is also one of the few physical systems which may be treated exactly. The harmonic oscillator forms the basis for understanding a wide range of different phenomena, such as absorption and emission of radiation, vibrations in molecules, specific heat of solids or collective oscillations in nuclei.

2.1 Analytical mechanics

The topic of analytical mechanics dates back to the latter part of the eighteenth century and the beginning of the nineteenth. The development of analytical mechanics is associated with the names of Leonhard Euler (1707-1783), Joseph Louis Lagrange (1736-1813), Siméon Denis Poisson (1781-1840), William Rowan Hamilton (1805-65) and others. It is essentially a reformulation of Newton's mechanics which allows many problems to be solved more simply. As we shall see in the present section, the equations of motion may be written in a form which makes the transition to quantum mechanics appear quite natural, despite the fundamental conceptual differences between classical and quantum mechanics.

2.1.1 The Lagrange formalism

The Lagrange formalism is a powerful tool for solving problems in mechanics. We shall introduce it by studying a simple example, the motion of a particle with mass M in a harmonic oscillator potential given by $V(x) = Kx^2/2$. According to Newton's second law the acceleration of the particle is determined by

$$M\frac{d^2x}{dt^2} = -Kx, \tag{2.1}$$

which has the well-known solution

$$x = x_0 \cos(\omega t + \phi), \tag{2.2}$$

where

$$\omega = \sqrt{\frac{K}{M}} \tag{2.3}$$

is the angular frequency. The constants x_0 and ϕ are determined by the initial conditions.

Let us consider two times, t_1 and t_2, which are defined as follows; at time t_1 the particle is at x_1, while at time t_2 the particle is at x_2. In the simple case we are considering, the values of x at times between t_1 and t_2 are given by the solution (2.2) to the equation of motion, when the constants x_0 and ϕ are expressed in terms of x_1 and x_2. Let us however pretend that we do not know this solution and attempt to determine the path $x(t)$ followed by the particle between times t_1 and t_2 by forming the quantity

$$S = \int_{t_1}^{t_2} dt\, L(x, \frac{dx}{dt}), \tag{2.4}$$

where

$$L = T - V \tag{2.5}$$

is called the Lagrangian. Here

$$T = \frac{1}{2}M(\frac{dx}{dt})^2 \tag{2.6}$$

is the kinetic energy, while

$$V = \frac{1}{2}Kx^2 \tag{2.7}$$

is the potential energy. The quantity S is named the *action*. Dimensionally the action is an energy times a time (or equivalently a momentum times a length). The action has thus the same dimension as that of Planck's constant.

According to (2.4) the action is a functional of x, that is a function of the function $x(t)$, which describes a path satisfying the two constraints that $x(t)$ at time $t = t_1$ assumes the value x_1, while $x(t)$ at time t_2 assumes the value x_2. Apart from these constraints the path may be arbitrary. In the following we shall impose the condition that the action S has an extremum and show that this condition results in the equation of motion (2.1). For brevity we shall often write the time derivative dx/dt as $\dot{x}$.

The action S given in (2.4) is thus a function of the different paths satisfying the two constraints. In order to formulate the extremum condition on S we consider a family of functions $x(t, \alpha)$ given by

$$x(t, \alpha) = x(t, 0) + \alpha \eta(t), \tag{2.8}$$

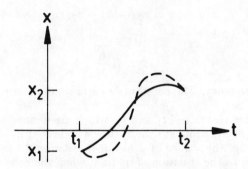

Figure 2.1: Two paths.

where the function $x(t,0)$ is the one corresponding to the extremum. The function $\eta(t)$ is arbitrary, except that it satisfies the constraints

$$\eta(t_1) = \eta(t_2) = 0. \tag{2.9}$$

In Fig. 2.1 we have illustrated two paths satisfying (2.9) corresponding to two different values of the parameter α. The action S is thus a function $S(\alpha)$ of the parameter α,

$$S(\alpha) = \int_{t_1}^{t_2} dt L(x(t,\alpha), \dot{x}(t,\alpha), t). \tag{2.10}$$

In writing (2.10) we have allowed for the possibility that the Lagrangian might depend explicitly on time, e. g. by the force constant K of the harmonic oscillator being a function of time, $K = K(t)$.

The extremum condition is

$$\frac{\partial S}{\partial \alpha}|_{\alpha=0} = 0. \tag{2.11}$$

By differentiating (2.10) we get

$$\frac{\partial S}{\partial \alpha} = \int_{t_1}^{t_2} dt \left(\frac{\partial L}{\partial x} \frac{\partial x}{\partial \alpha} + \frac{\partial L}{\partial \dot{x}} \frac{\partial \dot{x}}{\partial \alpha} \right) = \int_{t_1}^{t_2} dt \left(\frac{\partial L}{\partial x} \eta(t) + \frac{\partial L}{\partial \dot{x}} \dot{\eta} \right). \tag{2.12}$$

We now use integration by parts[1] in order to replace $\dot{\eta}$ by η. The result of the

---

[1] The method of integration by parts employs the identity $fg' = (fg)' - gf'$, where f and g both are functions of t, with $'$ denoting differentiation with respect to t. When both sides

integration by parts and the use of the boundary conditions (2.8) is thus

$$\frac{\partial S}{\partial \alpha} = \int_{t_1}^{t_2} dt (\frac{\partial L}{\partial x} - \frac{d}{dt}(\frac{\partial L}{\partial \dot{x}}))\eta(t),$$

since

$$\int_{t_1}^{t_2} dt \frac{\partial L}{\partial \dot{x}} \dot{\eta} = [\eta \frac{\partial L}{\partial \dot{x}}]_{t_1}^{t_2} - \int_{t_1}^{t_2} dt \eta(t) \frac{d}{dt}(\frac{\partial L}{\partial \dot{x}}) =$$
$$0 - \int_{t_1}^{t_2} dt \eta(t) \frac{d}{dt}(\frac{\partial L}{\partial \dot{x}}). \qquad (2.13)$$

The extremum condition (2.11) then becomes

$$\frac{d}{dt}(\frac{\partial L}{\partial \dot{x}}) - \frac{\partial L}{\partial x} = 0, \qquad (2.14)$$

since $\eta(t)$ is arbitrary except for the condition (2.9).

Equation (2.14) is named the Lagrange equation. It replaces Newton's equation (2.1) and has the same physical content. When the expression for L is inserted from (2.5-7), the resulting equation is Newton's second law (2.1). In this specific example, the one-dimensional harmonic oscillator, we have thus shown that Newton's second law is a consequence of the requirement that the action is stationary, but the derivation is easily generalized to an arbitrary potential $V(x)$ and to systems with more than one degree of freedom, as we shall see in the following.

It is readily seen that the Lagrange equation (2.14) is valid for motion in an arbitrary potential $V(x)$, when we identify L with $T - V = M\dot{x}^2/2 - V(x)$. When this Lagrangian is inserted into (2.14), it becomes $M\ddot{x} = -dV/dx$, which is Newton's second law.

It is also straightforward to generalize the Lagrange formalism to systems with more degrees of freedom. Let us consider a system (consisting, say, of several particles) described by the set of coordinates q_i, where i assumes the values $1, 2, \cdots, s$. As before we define the Lagrangian as the difference between the kinetic energy T and the potential energy V,

$$L = T - V. \qquad (2.15)$$

---

of this equation are integrated with respect to t over the interval from t_1 to t_2 one gets

$$\int_{t_1}^{t_2} dt f g' = [fg]_{t_1}^{t_2} - \int_{t_1}^{t_2} dt f' g$$

where $[fg]_{t_1}^{t_2} = f(t_2)g(t_2) - f(t_1)g(t_1)$ is zero, if $f(t_2)g(t_2) = f(t_1)g(t_1) = 0$.

The Lagrange equations are derived by requiring the action

$$S = \int_{t_1}^{t_2} dt L(q_1, q_2, \cdots, q_s; \dot{q}_1, \dot{q}_2, \cdots, \dot{q}_s, t)$$

to have an extremum. By generalizing (2.8) to

$$q_i(t, \alpha) = q_i(t, 0) + \alpha \eta_i(t), \quad i = 1, 2, \cdots, s,$$

we obtain the s equations

$$\frac{d}{dt}\left(\frac{\partial L}{\partial \dot{q}_i}\right) - \frac{\partial L}{\partial q_i} = 0, \tag{2.16}$$

since we are allowed to vary the functions η_i one at a time, while setting the others equal to zero. The coordinates q_i are named the generalized coordinates, since they are not necessarily identical to the cartesian coordinates for each of the particles in the system. If we wish to describe the motion of a particle in three dimensions, we may choose its cartesian coordinates x, y and z as the generalized coordinates, but we might find it more convenient to use the spherical coordinates r, θ and ϕ. When we consider a system consisting of N particles, the number s of generalized coordinates is thus $3N$. The choice of generalized coordinates may in practice be dictated by the dependence of the potential energy on these coordinates. The Lagrange formalism is especially convenient for solving problems with constraints, i. e. particular conditions between variables which apply at all times, such that the mechanical state of the system is specified by a relatively small number of generalized coordinates.

The Lagrange formalism is the starting point for the introduction of generalized momenta and the subsequent reformulation of the laws of classical mechanics in their Hamiltonian form. Prior to doing this we illustrate the usefulness of the Lagrange formalism by an example.

EXAMPLE 1. A MOVEABLE PENDULUM.

The Lagrange formalism will be used to determine the motion of two masses, m_1 and m_2, in a gravitational field (acceleration g) as illustrated in Fig. 2.2. The two masses are connected by a massless rod of length l. The mass m_1 moves without friction along a horizontal line, while the mass m_2 moves in the plane of the figure. The system of coordinates is indicated on the figure. The coordinates for the mass m_1 are $(x_1, y_1) = (x, 0)$, while those for the mass m_2 are $(x_2, y_2) = (x + l\sin\phi, l\cos\phi)$. We shall therefore use x and ϕ as our generalized coordinates. In terms of their time-derivatives $\dot{x}$ and $\dot{\phi}$ the kinetic energy T is

$$T = \frac{1}{2}m_1\dot{x}^2 + \frac{1}{2}m_2((\dot{x} + l\dot{\phi}\cos\phi)^2 + l^2\dot{\phi}^2\sin^2\phi) =$$
$$\frac{1}{2}(m_1 + m_2)\dot{x}^2 + \frac{1}{2}m_2 l^2 \dot{\phi}^2 + m_2 l \dot{x}\dot{\phi}\cos\phi. \tag{2.17}$$

Quantization

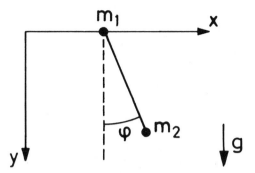

Figure 2.2: The moveable pendulum.

The potential energy is
$$V = m_2 g l (1 - \cos\phi), \tag{2.18}$$
since the potential energy is chosen to be equal to zero in equilibrium, when $\phi = 0$.

The Lagrange equations are
$$\frac{d}{dt}\frac{\partial L}{\partial \dot{x}} - \frac{\partial L}{\partial x} = 0 \tag{2.19}$$

and
$$\frac{d}{dt}\frac{\partial L}{\partial \dot{\phi}} - \frac{\partial L}{\partial \phi} = 0. \tag{2.20}$$

The equations of motion for ϕ and x are obtained by inserting $L = T - V$ in (2.19-20). In the following we only consider small oscillations about equilibrium and consequently approximate the Lagrangian by an expression which is quadratic in the small quantities $\dot{x}, \phi, \dot{\phi}$. This allows us to replace $\cos\phi$ by 1 in the kinetic energy T, while $\cos\phi$ in the potential energy is approximated by $1 - \phi^2/2$.

The Lagrange equations (2.19-20) then become
$$(m_1 + m_2)\ddot{x} + m_2 l \ddot{\phi} = 0, \tag{2.21}$$

and
$$m_2 l^2 \ddot{\phi} + m_2 l \ddot{x} = -m_2 g l \phi. \tag{2.22}$$

These two coupled linear differential equations have solutions which vary harmonically in time with angular frequency ω,
$$x = a_1 \sin(\omega t + \phi_1), \quad \phi = a_2 \sin(\omega t + \phi_2), \tag{2.23}$$
where $a_1, a_2, \phi_1,$ and ϕ_2 are constants. Such solutions are called normal modes. We shall show that the equations of motion are satisfied by solutions of this type with $\phi_1 = \phi_2$ and determine ω in terms of the constants l, g, m_1 and m_2.

For solutions of the form (2.23) the derivatives $\ddot{x}$ and $\ddot{\phi}$ in (2.21-22) may be replaced by $-\omega^2 x$ and $-\omega^2 \phi$, respectively. This simplifies the two coupled differential equations to two ordinary algebraic equations in the variables x and ϕ,

$$-(m_1 + m_2)\omega^2 x - m_2 l \omega^2 \phi = 0, \qquad (2.24)$$

and

$$-m_2 \omega^2 x - m_2 l (\omega^2 - \omega_0^2)\phi = 0. \qquad (2.25)$$

We have introduced ω_0 by the definition $\omega_0^2 = g/l$, and assumed that $\phi_1 = \phi_2$, which results in x and ϕ having the same time dependence.

Since (2.24-25) is a homogeneous set of equations, the corresponding determinant must vanish to ensure the existence of solutions different from zero. The resulting second-order equation in ω^2 determines the two possible frequencies ω_1 and ω_2 to be

$$\omega = \omega_1 = 0 \qquad (2.26)$$

and

$$\omega = \omega_2 = \sqrt{\frac{m_2 + m_1}{m_1}} \omega_0. \qquad (2.27)$$

Note that ω_2 is larger than the frequency ω_0 of the mathematical pendulum, which would result if the mass m_1 were held in a fixed position. It should also be pointed out that the system considered is special in the sense that the coordinate x does not enter the Lagrange function (such a coordinate is named cyclic). The corresponding generalized momentum, which is defined below in (2.31), is therefore a constant of the motion.

2.1.2 Hamilton's equations

In order to introduce the concept of a generalized momentum we now return to the consideration of a system described by only one generalized coordinate. We shall assume that the Lagrangian has no explicit dependence on time. The Lagrangian $L = M\dot{x}^2/2 - V(x)$ depends on the generalized coordinate

$$q = x \qquad (2.28)$$

and the corresponding generalized velocity

$$\dot{q} = \dot{x}. \qquad (2.29)$$

The differential dL is

$$dL = \frac{\partial L}{\partial q} dq + \frac{\partial L}{\partial \dot{q}} d\dot{q}. \qquad (2.30)$$

The generalized momentum p is now introduced by the definition

$$p = \frac{\partial L}{\partial \dot{q}}. \qquad (2.31)$$

Quantization

In the case under consideration the generalized momentum equals the mass times the velocity, $p = M\dot{x}$, but in other cases the generalized momentum does not have the same dimension as an ordinary momentum. When the generalized coordinate is an angle, the corresponding generalized momentum has the same dimension as an angular momentum.

The Lagrangian depends on the variables q and $\dot{q}$. We now introduce the Hamiltonian H by the transformation

$$H = p\dot{q} - L, \tag{2.32}$$

considering H to be a function of p and q. The differential dH is

$$dH = pd\dot{q} + \dot{q}dp - dL = -\dot{p}dq + \dot{q}dp \tag{2.33}$$

since

$$\dot{p} = \frac{\partial L}{\partial q} \tag{2.34}$$

according to (2.14).

It follows from (2.33) that the equation of motion for the system may be written in the following form

$$\dot{q} = \frac{\partial H}{\partial p} \tag{2.35}$$

and

$$\dot{p} = -\frac{\partial H}{\partial q}, \tag{2.36}$$

which constitute Hamilton's equations. The change of variables from $(q, \dot{q})$ to (q, p) brought about by the transformation (2.32) is used in other contexts. It is called a Legendre transformation[2].

Since L is assumed not to depend explicitly on time, it follows that $\partial H/\partial t = 0$. We therefore have $dH/dt = (\partial H/\partial q)\dot{q} + (\partial H/\partial p)\dot{p}$. The use of (2.35-36) allows one to conclude that $dH/dt = 0$. When $p = M\dot{q}$ and $L = M\dot{q}^2/2 - Kq^2/2$ are inserted into (2.32), one finds that H is the total energy, equal to the sum of the kinetic and potential energy. When expressed in terms of the generalized coordinates q and p, the so-called canonically-conjugate coordinates, the Hamiltonian is

$$H = \frac{p^2}{2M} + \frac{1}{2}Kq^2. \tag{2.37}$$

In the qp-plane the curves of constant energy are seen to be ellipses. The Hamilton equations for a system described by s generalized coordinates are

$$\dot{q}_i = \frac{\partial H}{\partial p_i} \tag{2.38}$$

---

[2] The transition in thermodynamics from energy to enthalpy is one example of the use of a Legendre transformation.

and
$$\dot{p}_i = -\frac{\partial H}{\partial q_i}, \qquad (2.39)$$

where $i = 1, \cdots, s$. The Hamiltonian H is given by

$$H = -L + \sum_{i=1}^{s} p_i \dot{q}_i. \qquad (2.40)$$

The equations (2.38-39) with the Hamiltonian (2.40) are equivalent to Newton's equations. In the following we shall reformulate these equations in order to establish a formal basis for the transition to quantum mechanics.

2.1.3 Poisson brackets

Hamilton's equations can be reformulated using the so-called Poisson brackets. For a system described by s generalized coordinates and s generalized momenta, the Poisson bracket $\{,\}$ is defined by

$$\{u, v\} = \sum_{i=1}^{s} \left(\frac{\partial u}{\partial q_i} \frac{\partial v}{\partial p_i} - \frac{\partial v}{\partial q_i} \frac{\partial u}{\partial p_i} \right) \qquad (2.41)$$

with u and v being functions of the variables $q_1, q_2, \ldots, q_s; p_1, p_2, \ldots, p_s$.

With this definition the Hamilton equations (2.38-39) may be written in the symmetric form

$$\dot{q}_i = \{q_i, H\} \qquad (2.42)$$

and

$$\dot{p}_i = \{p_i, H\}. \qquad (2.43)$$

We also note that

$$\{q_i, p_j\} = \delta_{ij}, \qquad (2.44)$$

where the Kronecker delta δ_{ij} equals 1, if $i = j$ and zero, if $i \neq j$. The Poisson brackets involving only q or p are seen to vanish,

$$\{q_i, q_j\} = \{p_i, p_j\} = 0. \qquad (2.45)$$

It follows from the definition (2.41) that the Poisson bracket of a function u with itself must vanish,

$$\{u, u\} = 0. \qquad (2.46)$$

Furthermore one sees that the Poisson bracket of a function u with a constant c equals zero,

$$\{u, c\} = 0, \qquad (2.47)$$

and

$$\{u, v\} = -\{v, u\}. \qquad (2.48)$$

Quantization

By using the ordinary rules of differentiation one finds that

$$\{u+v,w\} = \{u,w\} + \{v,w\} \tag{2.49}$$

and

$$\{u,vw\} = \{u,v\}w + v\{u,w\}. \tag{2.50}$$

The use of Poisson brackets allows one to form a compact expression for the time derivative of an arbitrary function u,

$$\frac{du}{dt} = \frac{\partial u}{\partial t} + \sum_{i=1}^{s}(\frac{\partial u}{\partial q_i}\dot{q}_i + \frac{\partial u}{\partial p_i}\dot{p}_i) = \frac{\partial u}{\partial t} + \{u,H\}, \tag{2.51}$$

where we have used the Hamilton equations (2.38-39) and the definition (2.41) of the Poisson bracket. For any quantity which does not depend explicitly on time, the relation

$$\{u,H\} = 0 \tag{2.52}$$

therefore implies that $du/dt = 0$. Such a quantity is called a constant of the motion. The energy of a harmonic oscillator is obviously a constant of the motion, since H has no explicit time dependence according to (2.37).

The equation of motion (2.1) for a harmonic oscillator may be expressed in terms of Poisson brackets. The Hamilton equations (2.38-39) become

$$\dot{q} = \{q,H\} \tag{2.53}$$

and

$$\dot{p} = \{p,H\}, \tag{2.54}$$

where H is given by (2.37). The Poisson brackets are evaluated by use of the rules (2.49-50) together with

$$\{q,p\} = 1. \tag{2.55}$$

This yields

$$\dot{q} = \frac{p}{M} \tag{2.56}$$

and

$$\dot{p} = -Kq. \tag{2.57}$$

The two equations (2.56-57) imply that

$$\frac{d^2q}{dt^2} + \omega^2 q = 0 \tag{2.58}$$

in agreement with (2.1), since ω is given by (2.3).

EXAMPLE 2. POISSON BRACKETS AND THE TWO-DIMENSIONAL HARMONIC OSCILLATOR.

As an example of the use of Poisson brackets we shall discuss the motion of a particle of mass m in a two-dimensional harmonic-oscillator potential characterized by the force constants K_1 and K_2.

The classical Hamiltonian is

$$H = \frac{1}{2m}(p_x^2 + p_y^2) + \frac{1}{2}(K_1 x^2 + K_2 y^2). \tag{2.59}$$

We introduce the quantity L by

$$L = xp_y - yp_x, \tag{2.60}$$

which is seen to be the z-component of the angular momentum of the particle.

Since $p_x(p_y)$ is the generalized momentum corresponding to $x(y)$, we have

$$\{x, p_x\} = \{y, p_y\} = 1, \tag{2.61}$$

and

$$\{x, p_y\} = \{y, p_x\} = \{x, y\} = \{p_x, p_y\} = 0. \tag{2.62}$$

Furthermore it is readily verified that

$$\{x, p_x^2\} = 2p_x, \quad \{x^2, p_x\} = 2x \tag{2.63}$$

together with the analogous relations for the y-coordinate. It follows that

$$\{H, L\} = (K_2 - K_1)xy. \tag{2.64}$$

If the two force constants are equal, $K_2 = K_1$, the right hand side of (2.64) vanishes, and the z-component of the angular momentum, L, is therefore a constant of the motion.

We conclude that L is a constant of the motion if the two force constants K_1 and K_2 are equal to each other, which means that the potential energy is invariant under rotations about the z-axis. In the remaining part of this example we shall only treat the case in which the two force constants are the same, corresponding to the motion of the particle in the xy-plane for a harmonic oscillator potential given by $V(x, y) = K(x^2 + y^2)/2$. The Hamiltonian is thus

$$H = \frac{1}{2m}(p_x^2 + p_y^2) + \frac{m\omega^2}{2}(x^2 + y^2), \tag{2.65}$$

where $\omega = \sqrt{K/m}$.

We have already seen that the quantity L given by (2.60) is a constant of the motion for an isotropic two-dimensional harmonic oscillator. Due to the simple quadratic form of the potential energy there exist, however, two additional constants of the motion M and N given by

$$M = \frac{m\omega}{2}(x^2 - y^2) + \frac{1}{2m\omega}(p_x^2 - p_y^2) \tag{2.66}$$

and
$$N = m\omega xy + \frac{1}{m\omega}p_x p_y. \qquad (2.67)$$

This may be verified by working out the Poisson brackets $\{H, M\}$ and $\{H, N\}$, both of which are seen to be equal to zero. The existence of the constants of the motion M and N is a special feature of the harmonic oscillator potential. If, for example, one adds to the classical Hamiltonian (2.65) a term of the form $a(x^2 + y^2)^2$, where a is a constant, the Poisson brackets $\{H, M\}$ and $\{H, N\}$ no longer vanish, meaning that neither M nor N are constants of the motion. By contrast, one finds that $\{L, H\} = 0$, provided that the potential energy as in the case considered only depends on the length $\sqrt{x^2 + y^2}$. The vanishing of the Poisson bracket $\{L, H\}$ reflects the symmetry of the potential with respect to rotations about an axis through $(x, y) = (0, 0)$, perpendicular to the xy-plane. As long as the potential energy only depends on the length $\sqrt{x^2 + y^2}$, the Hamiltonian is invariant under rotations about this axis, and the component of the angular momentum along the symmetry axis is a constant of the motion.

Let us finally work out the Poisson brackets $\{L, M\}$, $\{M, N\}$ and $\{N, L\}$. By using (2.61-63) we find that

$$\{L, M\} = 2N; \; \{M, N\} = 2L; \; \{N, L\} = 2M. \qquad (2.68)$$

We shall see later on (Chapter 7) that there exists a connection between this result and the description of the electron spin.

2.2 From classical mechanics to quantum mechanics

The path is now cleared for the great leap forward: The transition to quantum mechanics. By using Hamilton's equations as the starting point for the quantization we bypass decades of careful experimental investigations of the atomic spectra. We also ignore the historical importance of the difficulties inherent in the attempts to reconcile Rutherford's model of the atom with the laws of classical electrodynamics, difficulties which led Bohr to the postulates formulated in 1913 and the subsequent development of quantum mechanics.

In the following u and v represent two different physical quantities such as position and momentum. We introduce a commutator [,] by the definition

$$[u, v] = uv - vu. \qquad (2.69)$$

If u and v were simply functions, the commutator (2.69) would be equal to zero. In order that the commutator $[u, v]$ may be different from zero (just as the classical Poisson bracket $\{u, v\}$ may be different from zero for a given pair of functions u and v), we shall represent the physical quantities u and v by matrices. This allows us to introduce a commutator algebra which corresponds closely to the algebra of the Poisson brackets, since the commutator defined by (2.69) satisfies the same relations (2.49-50) as those of the Poisson brackets (cf. (2.72-73) below).

Since q times p has the dimension of an action, it is not possible to relate $\{,\}$ to $[,]$ without introducing a constant with the dimension of action. We shall choose this to be the Planck constant $\hbar$. The choice of the constant as $\hbar = 1.054 \cdot 10^{-34}$ J·s is dictated by the need to obtain agreement with experiment.

The correspondence between Poisson brackets and commutators is thus determined by the rule

$$\{u,v\} \to \frac{1}{i\hbar}[u,v]. \qquad (2.70)$$

Besides the Planck constant we have introduced the imaginary unit i. As will be evident later on, this is necessary in order that the quantum theory contains classical mechanics as a limiting case.

According to the rule (2.70) the quantization consists in the replacement of the classical Poisson bracket $\{q_i, p_j\}$ by the commutator $[q_i, p_j]$ divided by $i\hbar$. We may consider (2.70) to be a postulate consistent with dimensional analysis, but otherwise only subject to verification by comparison with experiment. In the following we show that this postulate leads us to the quantization of the energy of the harmonic oscillator.

When (2.55) is compared to (2.70), we see that q and p must satisfy the commutation relation

$$qp - pq = i\hbar. \qquad (2.71)$$

Evidently q and p do not commute. It is clear that q must commute with itself, just as p must. It is tempting to represent such quantities by matrices, since matrix multiplication is non-commutative. If u, v and w denote matrices, they satisfy the rules of matrix multiplication

$$[u+v, w] = [u,w] + [v,w] \qquad (2.72)$$

and

$$[u, vw] = [u,v]w + v[u,w], \qquad (2.73)$$

in analogy with (2.49-50).

Rather than using matrices it is possible, as we shall see in the following chapter, to describe the lack of commutativity by representing the physical quantities by operators. As examples of operators which do not commute we may take q and d/dq. Their commutator is

$$[q, \frac{d}{dq}] = -1. \qquad (2.74)$$

To establish the validity of (2.74) we let the operator $[q, d/dq]$ act on a function $u(q)$ of q, using $q(du/dq) - d(uq)/dq = -u$. It is evident that the commutation relation (2.71) is satisfied, if we represent the momentum by

$$p = \frac{\hbar}{i} \frac{d}{dq}. \qquad (2.75)$$

Quantization

A more detailed discussion of the operator concept will be deferred to the next chapter, where we also establish the close relation between the use of matrices and differential operators.

2.2.1 Matrices and eigenvalues

The lack of commutativity between physical quantities may be expressed mathematically by representing these by matrices. Matrix multiplication is not commutative; the product AB of the square matrices A and B is in general different from BA.

An elementary example of matrices which do not commute is furnished by the three 2×2 matrices, σ_x, σ_y and σ_z, which were introduced in Section 1.4.5 in connection with the discussion of the relativistic wave equation. These *Pauli matrices* are given by

$$\sigma_x = \begin{pmatrix} 0 & 1 \\ 1 & 0 \end{pmatrix}, \quad \sigma_y = \begin{pmatrix} 0 & -i \\ i & 0 \end{pmatrix}, \quad \sigma_z = \begin{pmatrix} 1 & 0 \\ 0 & -1 \end{pmatrix}. \tag{2.76}$$

It is immediately evident that the Pauli matrices do not commute. For instance we have

$$[\sigma_x, \sigma_y] = 2i\sigma_z, \tag{2.77}$$

while the two other commutators are obtained from (2.77) by cyclic permutation of x, y and z. We note that the eigenvalues of each of the three matrices are 1 and -1. Although the matrices (in the case of σ_y) contain complex numbers, the eigenvalues are real. This property is a consequence of the fact that the matrices satisfy the condition

$$A_{ij} = A^*_{ji}, \tag{2.78}$$

where $*$ denotes complex conjugation[3] of the elements in the matrix A. A matrix which satisfies (2.78) is said to be Hermitian. The proof that the eigenvalues of a Hermitian matrix are real, will be given in the following chapter. It is not always possible to represent physical quantities by matrices of a finite rank, i. e. with a finite number of rows and columns. As shown in the following, the position coordinate for the harmonic oscillator is - just as its momentum and energy - associated with a matrix of infinite rank.

The eigenvectors associated with σ_z may be chosen to be

$$\begin{pmatrix} 1 \\ 0 \end{pmatrix} \tag{2.79}$$

---

[3]Note that this symbol, which is generally used in physics, differs from that used in mathematics, where the complex conjugate of c is denoted by $\bar{c}$.

AUGUSTANA UNIVERSITY COLLEGE
LIBRARY

corresponding to the eigenvalue 1, and

$$\begin{pmatrix} 0 \\ 1 \end{pmatrix} \tag{2.80}$$

corresponding to the eigenvalue -1. *It is a basic postulate of quantum theory that physical quantities are represented by matrices (or operators), which do not in general commute, and that the eigenvalues of these are the possible results of a measurement of the corresponding physical quantity*[4]. In the case considered the possible results of a measurement are evidently 1 and -1. It should be added that we have used in this example a dimensionless quantity, the vector $\sigma = (\sigma_x, \sigma_y, \sigma_z)$. It will be shown in Chapter 7 that the electron spin **S**, which has the dimension of angular momentum, is given by

$$\mathbf{S} = \frac{1}{2}\hbar\boldsymbol{\sigma}. \tag{2.81}$$

It follows from (2.77) that the components of **S** satisfy the commutation relations

$$[S_x, S_y] = i\hbar S_z \tag{2.82}$$

together with those obtained by cyclic permutation. The possible results of a measurement of a component of the electron spin are thus $\hbar/2$ and $-\hbar/2$ according to our interpretation of the eigenvalues of the matrices.

After this brief interlude about Hermitian matrices and their properties we return to the harmonic oscillator.

2.2.2 Energy quanta

The commutation rule (2.71) for p and q will now be used to determine the possible values of the energy of the harmonic oscillator with the classical Hamiltonian given by (2.37). In order to make it explicit that p and q should be considered to be matrices, we write them as $\hat{p}$ and $\hat{q}$ and similarly for other matrices[5]. As already noted the possible values of the oscillator energy are given by the eigenvalues of the energy matrix. These may be determined by introducing the matrices $\hat{a}$ and $\hat{a}^\dagger$ according to the definition

$$\hat{a} = \sqrt{\frac{M\omega}{2\hbar}}(\hat{q} + i\frac{\hat{p}}{M\omega}) \tag{2.83}$$

and

$$\hat{a}^\dagger = \sqrt{\frac{M\omega}{2\hbar}}(\hat{q} - i\frac{\hat{p}}{M\omega}). \tag{2.84}$$

---

[4] The basic postulates of quantum theory are summarized in Section 3.2.

[5] This notation, which in general is used to represent operators, has been chosen to indicate that the following matrix equations also hold for the corresponding operators, cf. Chapter 3.

Quantization

The inverse relations are

$$\hat{q} = \sqrt{\frac{\hbar}{2M\omega}}(\hat{a} + \hat{a}^\dagger) \qquad (2.85)$$

and

$$\hat{p} = \frac{1}{i}\sqrt{\frac{\hbar M\omega}{2}}(\hat{a} - \hat{a}^\dagger). \qquad (2.86)$$

Note that these definitions exactly parallel those given in (1.95-96), where a and $a^\dagger$ were introduced as differential operators. Since both $\hat{q}$ and $\hat{p}$ must be Hermitian matrices in order that their eigenvalues are real, it follows that $\hat{a}^\dagger$ is the Hermitian conjugate of $\hat{a}$. The general definition of the Hermitian conjugate matrix $A^\dagger$ of a given matrix A is as follows:

$$A^\dagger_{ij} = A^*_{ji}. \qquad (2.87)$$

A matrix A is Hermitian when $A^\dagger = A$, which is equivalent to the relation (2.78).

Since $\hat{q}$ and $\hat{p}$ satisfy the commutation relation

$$[\hat{q}, \hat{p}] = i\hbar, \qquad (2.88)$$

while $[\hat{q}, \hat{q}] = [\hat{p}, \hat{p}] = 0$, we conclude that

$$[\hat{a}, \hat{a}^\dagger] = 1. \qquad (2.89)$$

The energy matrix is obtained from the classical Hamiltonian (2.37) by replacing p with $\hat{p}$ and q with $\hat{q}$,

$$\hat{H} = \frac{\hat{p}^2}{2M} + \frac{1}{2}K\hat{q}^2. \qquad (2.90)$$

By inserting (2.85-86), which express $\hat{q}$ and $\hat{p}$ in terms of $\hat{a}$ and $\hat{a}^\dagger$, and using (2.89), one finds that $\hat{H}$ assumes the form

$$\hat{H} = \hbar\omega(\hat{a}^\dagger \hat{a} + \frac{1}{2}). \qquad (2.91)$$

The eigenvalues of $\hat{H}$ are therefore obtained from the eigenvalues of the matrix $\hat{N}$ defined by

$$\hat{N} = \hat{a}^\dagger \hat{a}.$$

It is an immediate consequence of (2.89) and (2.73) that

$$[\hat{N}, \hat{a}] = -\hat{a} \qquad (2.92)$$

and
$$[\hat{N}, \hat{a}^\dagger] = \hat{a}^\dagger. \tag{2.93}$$

We consider $\hat{H}$ to be a matrix with eigenvalues E determined by

$$\hat{H}|E\rangle = E|E\rangle. \tag{2.94}$$

The notation $|E\rangle$ is due to Dirac. The eigenvector $|E\rangle$ is called a *state vector*. It is a column vector, which is analogous to (2.79-80). The state vector is labelled by E, which is the corresponding eigenvalue of the energy matrix. The name *ket* is used for the column vector $|E\rangle$, a name which originates in the word bracket, which may be separated as bra-(c)ket. A *bra* is symbolized by $\langle E|$. It is an eigenvector 'from the left', i. e. a row vector satisfying the equation

$$\langle E|\hat{H} = \langle E|E.$$

If $|\alpha\rangle$ denotes the column vector

$$\begin{pmatrix} a_1 \\ a_2 \\ a_3 \\ \vdots \end{pmatrix} \tag{2.95}$$

then $\langle \alpha |$ is the row vector

$$\begin{pmatrix} a_1^* & a_2^* & a_3^* & \cdots \end{pmatrix}. \tag{2.96}$$

The inner product $\langle \alpha | \alpha \rangle$ is then

$$\langle \alpha | \alpha \rangle = \sum_i a_i^* a_i = |a_1|^2 + |a_2|^2 + |a_3|^2 + \cdots. \tag{2.97}$$

Evidently $\langle \alpha |$ is the Hermitian conjugate of $|\alpha\rangle$. If the sum of the absolute squares in (2.97) is 1, the state is said to be normalized, $\langle \alpha | \alpha \rangle = 1$.

The Hermitian conjugate of the matrix product AB is obtained from the rule

$$(AB)^\dagger = B^\dagger A^\dagger. \tag{2.98}$$

The proof of (2.98) is obtained by using

$$(AB)^\dagger_{ik} = \sum_j A^*_{kj} B^*_{ji} = \sum_j B^\dagger_{ij} A^\dagger_{jk}. \tag{2.99}$$

It is seen from (2.98) that the commutator $[A, B]$ of two Hermitian matrices A and B is non-Hermitian, the Hermitian conjugate being $[B, A] = -[A, B]$, while $i[A, B]$ is Hermitian and therefore has real eigenvalues.

Quantization

The eigenvalues E will now be determined by finding the eigenvalues of $\hat{N}$. Let us write the eigenvalue equation for $\hat{N}$ in the form

$$\hat{N}|\lambda\rangle = \lambda|\lambda\rangle \tag{2.100}$$

and seek to determine λ together with the corresponding normalized state vectors $|\lambda\rangle$ satisfying $\langle\lambda|\lambda\rangle = 1$.

The argument runs as follows:

1. First we shall prove that the eigenvalues of $\hat{N}$ are greater than or equal to zero. This may be seen by writing λ as the inner product of $|\lambda\rangle$ and $\hat{N}|\lambda\rangle$, $\lambda = \langle\lambda|\hat{N}|\lambda\rangle$, since $|\lambda\rangle$ is assumed to be normalized, and transforming the expression according to

$$\lambda = \langle\lambda|\hat{N}|\lambda\rangle = \langle\lambda|\hat{a}^\dagger\hat{a}|\lambda\rangle = \langle\alpha|\alpha\rangle \geq 0.$$

Here $|\alpha\rangle = \hat{a}|\lambda\rangle$, and we have used the definition (2.97) of the inner product and the rule (2.98). The eigenvalue $\lambda = 0$ appears if and only if $\hat{a}|\lambda\rangle = 0$.

2. Next we prove that, if $|\lambda\rangle$ is an eigenstate, then so is $|\lambda - 1\rangle$, provided $\lambda \neq 0$. This follows from the commutation relation (2.92) in connection with the eigenvalue equation (2.100), since

$$\hat{N}\hat{a}|\lambda\rangle = (\hat{a}\hat{N} - \hat{a})|\lambda\rangle = (\lambda - 1)\hat{a}|\lambda\rangle, \tag{2.101}$$

which shows that the state $\hat{a}|\lambda\rangle$ is an eigenstate associated with the eigenvalue $\lambda - 1$ and therefore proportional to $|\lambda - 1\rangle$, unless $\lambda = 0$.

3. We are now able to conclude that the series of eigenstates $|\lambda\rangle$, $|\lambda - 1\rangle$, $|\lambda - 2\rangle$ must terminate, since the eigenvalues as shown in 1. should be greater than or equal to zero. The eigenvalue at which the series of eigenstates must terminate is zero, since $\hat{a}^\dagger\hat{a}$ acting on the corresponding eigenstate yields zero. Thus

$$\hat{a}|0\rangle = 0. \tag{2.102}$$

There is only one linearly-independent eigenvector $|0\rangle$, which is a solution of (2.102), namely the column vector given below in (2.105).

4. Finally we observe that if $|\lambda\rangle$ is an eigenstate, then so is $|\lambda + 1\rangle$. This follows from the commutation relation (2.93) in connection with the eigenvalue equation (2.100), since

$$\hat{N}\hat{a}^\dagger|\lambda\rangle = (\hat{a}^\dagger\hat{N} + \hat{a}^\dagger)|\lambda\rangle = (\lambda + 1)\hat{a}^\dagger|\lambda\rangle,$$

which shows that the state $\hat{a}^\dagger|\lambda\rangle$ is an eigenstate associated with the eigenvalue $\lambda + 1$. Therefore it must be proportional to $|\lambda + 1\rangle$. Note that $\hat{a}^\dagger|\lambda\rangle$ cannot be equal to zero, since $\hat{a}\hat{a}^\dagger|\lambda\rangle = (1+\hat{a}^\dagger\hat{a})|\lambda\rangle = (\lambda+1)|\lambda\rangle \neq 0$.

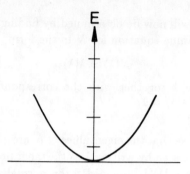

Figure 2.3: Energy eigenvalues for the harmonic oscillator.

We have thus shown that the eigenvalues of $\hat{N}$ are the non-negative integers. There are no other eigenstates than those associated with the non-negative integers, since if there were, we could generate states with negative eigenvalues by letting $\hat{a}$ operate on such a state a sufficient number of times, in disagreement with 1. above (this point is further elucidated in Problem 2.6). It is evident from the eigenvectors determined below (see (2.105) and (2.106)) together with the expression (2.104) for energy matrix, that there is only one linearly-independent state vector $|\lambda\rangle$ associated with each eigenvalue λ.

Since the eigenvalues of $\hat{N}$ are the non-negative integers, it follows that the energy eigenvalues are given by

$$E_n = (n + \frac{1}{2})\hbar\omega, \quad n = 0, 1, 2, \ldots \quad (2.103)$$

The smallest possible energy, which is called the ground-state energy of the oscillator, is thus $\hbar\omega/2$. The spectrum of eigenvalues (2.103) has been illustrated in Fig. 2.3, where we have also shown the potential energy $V(x)$. From a classical point of view it is of course strange that the smallest possible energy differs from zero. The result is consistent with the Heisenberg uncertainty relations discussed in the following chapter, since a localization of the particle to the minimum of the potential energy would require infinite kinetic energy.

Let us now return to the matrix representation of the physical quantities. Having obtained the eigenvalues of the energy we may immediately write down the energy matrix, which is diagonal. The matrix has infinite rank, since the

Quantization

spectrum (2.103) has no upper limit. It is given as

$$\hat{H}/\hbar\omega : \begin{pmatrix} \frac{1}{2} & 0 & 0 & 0 & \cdots \\ 0 & \frac{3}{2} & 0 & 0 & \cdots \\ 0 & 0 & \frac{5}{2} & 0 & \cdots \\ 0 & 0 & 0 & \frac{7}{2} & \cdots \\ \vdots & \vdots & \vdots & \vdots & \ddots \end{pmatrix} \qquad (2.104)$$

The eigenvectors associated with these eigenvalues have the form

$$|0\rangle : \begin{pmatrix} e^{i\phi_0} \\ 0 \\ 0 \\ \vdots \end{pmatrix} \qquad (2.105)$$

corresponding to the lowest eigenvalue,

$$|1\rangle : \begin{pmatrix} 0 \\ e^{i\phi_1} \\ 0 \\ \vdots \end{pmatrix} \qquad (2.106)$$

corresponding to the second lowest eigenvalue etc. The phases $\phi_0, \phi_1, \cdots$ are arbitrary real numbers, which we choose to be equal, $\phi_0 = \phi_1 = \phi_2 = \cdots$. The eigenvectors are seen to be normalized and orthogonal, since

$$\langle m|n\rangle = \delta_{nm}. \qquad (2.107)$$

While the energy matrix (2.104) is diagonal, the matrices corresponding to $\hat{a}$ and $\hat{a}^\dagger$ are not. When the matrix $\hat{a}^\dagger$ acts on an eigenstate $|n\rangle$ for the energy, the resulting vector is the state $|n+1\rangle$ times a constant c, that is $\hat{a}^\dagger|n\rangle = c|n+1\rangle$. Since $\langle n|\hat{a} = \langle n+1|c$, if the constant c is chosen to be real, it follows that

$$\langle n|\hat{a}\hat{a}^\dagger|n\rangle = c^2. \qquad (2.108)$$

In order to determine c we use that $\hat{a}\hat{a}^\dagger - \hat{a}^\dagger\hat{a} = 1$ according to (2.89). Since $\hat{a}^\dagger\hat{a}|n\rangle = n|n\rangle$, it follows that

$$\langle n|\hat{a}\hat{a}^\dagger|n\rangle = (n+1). \qquad (2.109)$$

As c has been chosen to be real, we may conclude from (2.108) and (2.109) that $c = (n+1)^{\frac{1}{2}}$. Thus we have established the relation

$$\hat{a}^\dagger|n\rangle = \sqrt{n+1}|n+1\rangle. \qquad (2.110)$$

In an analogous fashion we may determine the constant c' in the relation $\hat{a}|n\rangle = c'|n-1\rangle$ to be $c' = \sqrt{n}$ by using that $\langle n|\hat{a}^\dagger \hat{a}|n\rangle = n$. Thus

$$\hat{a}|n\rangle = \sqrt{n}|n-1\rangle. \tag{2.111}$$

With the help of (2.110-111) the matrices $\langle m|\hat{a}|n\rangle$ og $\langle m|\hat{a}^\dagger|n\rangle$ for $\hat{a}$ and $\hat{a}^\dagger$ respectively may now be given as

$$\hat{a}: \begin{pmatrix} 0 & \sqrt{1} & 0 & 0 & \cdots \\ 0 & 0 & \sqrt{2} & 0 & \cdots \\ 0 & 0 & 0 & \sqrt{3} & \cdots \\ 0 & 0 & 0 & 0 & \cdots \\ \vdots & \vdots & \vdots & \vdots & \ddots \end{pmatrix} \tag{2.112}$$

and

$$\hat{a}^\dagger: \begin{pmatrix} 0 & 0 & 0 & 0 & \cdots \\ \sqrt{1} & 0 & 0 & 0 & \cdots \\ 0 & \sqrt{2} & 0 & 0 & \cdots \\ 0 & 0 & \sqrt{3} & 0 & \cdots \\ \vdots & \vdots & \vdots & \vdots & \ddots \end{pmatrix}. \tag{2.113}$$

Note that the choice of equal phases, $\phi_0 = \phi_1 = \cdots$ in the state vectors (2.105), (2.106) etc. ensures that the $\hat{a}$- and $\hat{a}^\dagger$-matrices do not depend on the values of the phases. The matrix (2.113) is evidently the transposed matrix of (2.112) and vice versa. They do not satisfy the condition (2.78) and their eigenvalues are therefore not necessarily real. It is seen from (2.112), that the solution of (2.102) is given by (2.105). The matrices $\hat{a}^\dagger$ and $\hat{a}$ increase and decrease the number of energy quanta, and they are therefore named creation and annihilation operators, respectively[6]. The state (2.105) may be considered to be a vacuum state, since it contains no energy quanta according to (2.102).

From (2.112-113) and the relations (2.85-86), which express $\hat{q}$ and $\hat{p}$ in terms of $\hat{a}$ and $\hat{a}^\dagger$, the matrices for $\hat{q}$ and $\hat{p}$ may be readily found, with the result

$$\hat{q}/(\hbar/2M\omega)^{\frac{1}{2}}: \begin{pmatrix} 0 & \sqrt{1} & 0 & 0 & \cdots \\ \sqrt{1} & 0 & \sqrt{2} & 0 & \cdots \\ 0 & \sqrt{2} & 0 & \sqrt{3} & \cdots \\ 0 & 0 & \sqrt{3} & 0 & \cdots \\ \vdots & \vdots & \vdots & \vdots & \ddots \end{pmatrix} \tag{2.114}$$

---

[6] Creation and annihilation operators such as $\hat{a}^\dagger$ and $\hat{a}$ are used everywhere in quantum physics. As an example let us consider the case when the classical Hamiltonian besides $p^2/2M + Kq^2/2$ contains a term proportional to q^4. The energy matrix then contains besides (2.91) a term proportional to $(\hat{a} + \hat{a}^\dagger)^4$. Although it is not possible in the present case to bring the energy matrix into a diagonal form by a simple transformation, it may still be convenient to utilize the properties of $\hat{a}^\dagger$ and $\hat{a}$ for an approximate determination of the eigenvalues.

Quantization

and

$$\hat{p}/(M\omega\hbar/2)^{\frac{1}{2}} : \begin{pmatrix} 0 & -i\sqrt{1} & 0 & 0 & \cdots \\ i\sqrt{1} & 0 & -i\sqrt{2} & 0 & \cdots \\ 0 & i\sqrt{2} & 0 & -i\sqrt{3} & \cdots \\ 0 & 0 & i\sqrt{3} & 0 & \cdots \\ \vdots & \vdots & \vdots & \vdots & \ddots \end{pmatrix}. \quad (2.115)$$

Evidently (2.114-115) are Hermitian matrices, since the matrix for $\hat{q}$ is symmetric and real, while the matrix for $\hat{p}$ is antisymmetric and purely imaginary,

According to quantum theory the eigenvalues of the matrix associated with a physical quantity are interpreted as the possible results of a measurement of the physical quantity in question. We have seen here that the energy of a harmonic oscillator is quantized and that the size of the energy quantum is $\hbar\omega$. This is completely foreign to classical physics. It is therefore not surprising that a large body of experimental observations were needed before the step could be taken from Hamilton's equations to quantum mechanics.

Besides energy we shall need to consider other quantities such as the position and momentum of the particle. These will be discussed in the following chapter. In the development given above the position and momentum matrices are both non-diagonal. We must therefore conclude that our use of the quantum theoretical description has so far only allowed us to consider the possible results of a measurement of the energy. By forming superpositions of states corresponding to different energies it is however possible to compare the quantum mechanical description more directly with the classical one. As we shall see in Section 4.2, the classical motion (2.2) results as a certain limit, when we construct states that are superpositions of the energy eigenstates $|n\rangle$.

2.3 Problems

PROBLEM 2.1
A mathematical pendulum consists of a massless rod (length l) with a mass m at its end. The acceleration due to gravity is g. The angle of the rod relative to its vertical equilibrium position is ϕ.

a) Express the Lagrangian of the system in terms of the given constants, together with the angle ϕ and its derivative with respect to time.

b) Write down the equation of motion (2.16) for the system.

c) Determine an approximate expression for the Lagrangian valid for small departures from equilibrium, and write down the corresponding equation of motion.

d) Show that the equation of motion found in c) has solutions varying harmonically in time and express the frequency in terms of the given constants.

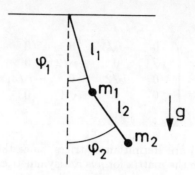

Figure 2.4: Two pendulums.

PROBLEM 2.2
Two pendulums are connected as shown in Fig. 2.4. The two rods with length l_1 and l_2 are massless, and the masses m_1 and m_2 move in the plane of the figure. The acceleration due to gravity is g.

a) Express the Lagrangian of the system in terms of the given constants and the angles ϕ_1 and ϕ_2, together with their derivatives with respect to time.

b) Write down the equations of motion (2.16) for the system.

c) Determine an approximate expression for the Lagrangian valid for small departures from equilibrium, and write down the corresponding equations of motion.

d) Show that the equations of motion found in c) have solutions varying harmonically in time and determine the corresponding frequencies in terms of the given constants.

PROBLEM 2.3
Find the eigenvalues of the following matrices

$$\begin{pmatrix} 1 & 2 \\ 1 & 2 \end{pmatrix} \tag{2.116}$$

and

$$\begin{pmatrix} 0 & -i & 0 \\ i & 0 & i \\ 0 & -i & 0 \end{pmatrix} \tag{2.117}$$

and

Quantization

$$\begin{pmatrix} 1 & 0 & 0 & 0 \\ 0 & 1 & 0 & 0 \\ 0 & 0 & 1 & 3i+1 \\ 0 & 0 & -3i+1 & 1 \end{pmatrix}. \qquad (2.118)$$

Are the matrices Hermitian?

PROBLEM 2.4
Show that the Poisson bracket $\{L_x, L_y\}$ equals L_z, where $\mathbf{L}$ is the orbital angular momentum $\mathbf{r} \times \mathbf{p}$ for a particle moving in three dimensions. Give an argument why $\{L_y, L_z\}$ equals L_x.

PROBLEM 2.5
Determine the transposed, the complex conjugate and the Hermitian conjugate matrix for each of the following two matrices A_{ij}.

1.
$$\begin{pmatrix} 1 & 2i \\ -1 & 7 \end{pmatrix}$$

2.
$$\begin{pmatrix} 0 & i & 0 & i \\ -i & 0 & i & 0 \\ 0 & -i & 0 & i \\ -i & 0 & -i & 0 \end{pmatrix}$$

PROBLEM 2.6
Show by induction that

$$(\hat{a}^\dagger)^n \hat{a}^n = \hat{N}(\hat{N}-1)(\hat{N}-2)\cdots(\hat{N}-n+1), \qquad (2.119)$$

where n is a positive integer and $\hat{N} = \hat{a}^\dagger \hat{a}$. Use this to prove that the eigenvalues λ of $\hat{N}$ satisfy the inequality

$$\lambda(\lambda-1)(\lambda-2)\cdots(\lambda-n+1) \geq 0. \qquad (2.120)$$

This inequality may only be satisfied for arbitrary (positive) n, if λ is a non-negative integer.

PROBLEM 2.7
Determine the bra vector $\langle n|$ corresponding to each of the following two ket vectors $|n\rangle$.

1.

$$\begin{pmatrix} 0 \\ i \end{pmatrix}$$

2.

$$\begin{pmatrix} e^{i\pi/4}/\sqrt{3} \\ 1/\sqrt{3} \\ e^{-i\pi/4}/\sqrt{3} \end{pmatrix}$$

PROBLEM 2.8
Find $\langle m|n\rangle$ and $|m\rangle\langle n|$ for each of the following pairs of states $|m\rangle$ and $|n\rangle$, where $|m\rangle\langle n|$ is the matrix with elements $(|m\rangle\langle n|)_{ij} = |m\rangle_i \langle n|_j$.

1.

$$\begin{pmatrix} 1 \\ 0 \end{pmatrix} \begin{pmatrix} 0 \\ 1 \end{pmatrix}$$

2.

$$\begin{pmatrix} i \\ 0 \end{pmatrix} \begin{pmatrix} i \\ 0 \end{pmatrix}$$

3.

$$\begin{pmatrix} 1/\sqrt{3} \\ 1/\sqrt{3} \\ 1/\sqrt{3} \end{pmatrix} \begin{pmatrix} i/\sqrt{2} \\ 0 \\ 1/\sqrt{2} \end{pmatrix}$$

PROBLEM 2.9
Find $\sum_n |n\rangle\langle n|$ for each of the following sets of states $|n\rangle$.

1.

$$\begin{pmatrix} 1 \\ 0 \end{pmatrix} \begin{pmatrix} 0 \\ 1 \end{pmatrix}$$

2.

$$\begin{pmatrix} i/\sqrt{2} \\ -i/\sqrt{2} \end{pmatrix} \begin{pmatrix} 1/\sqrt{2} \\ 1/\sqrt{2} \end{pmatrix}$$

3.
$$\begin{pmatrix} 1/\sqrt{3} \\ 1/\sqrt{3} \\ 1/\sqrt{3} \end{pmatrix} \begin{pmatrix} 1/\sqrt{6} \\ 1/\sqrt{6} \\ -2/\sqrt{6} \end{pmatrix} \begin{pmatrix} -1/\sqrt{2} \\ 1/\sqrt{2} \\ 0 \end{pmatrix}$$

PROBLEM 2.10
Determine AB and BA together with the sum $AB + BA$ and the difference $AB - BA$ for each of the following pairs of matrices A, B.

1.
$$\begin{pmatrix} i & 0 \\ 0 & -i \end{pmatrix} \begin{pmatrix} 0 & 1 \\ -1 & 0 \end{pmatrix}$$

2.
$$\begin{pmatrix} 0 & 1 \\ -1 & 0 \end{pmatrix} \begin{pmatrix} 0 & i \\ i & 0 \end{pmatrix}$$

3.
$$\begin{pmatrix} 0 & i \\ i & 0 \end{pmatrix} \begin{pmatrix} i & 0 \\ 0 & -i \end{pmatrix}$$

3 BASIC PRINCIPLES

In the previous chapter we used the example of a harmonic oscillator to introduce the concept of a quantum state and determined the possible results of a measurement of the energy of the harmonic oscillator. Each state $|n\rangle$ was labelled by a quantum number n, which specified the result $E_n = (n + \frac{1}{2})\hbar\omega$ of a measurement of the energy in the state $|n\rangle$.

The present chapter deals with the general principles of quantum mechanics. We shall identify states with elements in a Hilbert space, and associate observables such as momentum and energy with linear operators on the Hilbert space. The basic postulates of quantum theory are formulated. The justification of these postulates can only be obtained by comparison with experiment, in the same sense as Newton's equations are established by, for example, their success in accounting for the planetary orbits.

We shall continue to use the harmonic oscillator as a concrete example, but the concepts of states and probability amplitudes, which form the topic of the present chapter, are fundamental to all of quantum theory and its applications, whether they concern solids, atoms, molecules, nuclei or elementary particles. The quantum number n for the harmonic oscillator should in general be replaced by a collection of quantum numbers[1] associated with physical quantities which may be measured simultaneously.

3.1 Linear operators

The starting point for the formulation of quantum mechanics with the use of operators is the mathematical theory of Hilbert spaces. We shall therefore summarize the most important properties of a Hilbert space and mention some elementary results regarding linear operators.

A Hilbert space $\mathcal{H}$ is a complex vector space, in which an inner product $\langle \cdot | \cdot \rangle$ is defined as a mapping of $\mathcal{H} \times \mathcal{H}$ on the set of complex numbers. The inner product[2] should satisfy the following four conditions (f, g and h denote elements in the vector space),

$$(i) \qquad \langle f|f\rangle \geq 0, \qquad (3.1)$$

where the equality holds if and only if $f = 0$,

$$(ii) \qquad \langle f|g + h\rangle = \langle f|g\rangle + \langle f|h\rangle, \qquad (3.2)$$

---

[1] The number of different quantum numbers equals the number of classical degrees of freedom for the system under consideration, not counting possible internal non-classical degrees of freedom such as spin.

[2] The present notation for the inner product is used widely in physics. In the mathematical literature one uses $\langle \cdot, \cdot \rangle$ or $(\cdot, \cdot)$. Note that $\langle \cdot | \cdot \rangle$ is linear in the second variable and conjugate linear in the first, as is customary in physics.

Basic principles

and
$$(iii) \quad \langle f|\alpha g\rangle = \alpha\langle f|g\rangle, \quad (3.3)$$
where α denotes an arbitrary complex number, and
$$(iv) \quad \langle f|g\rangle = \langle g|f\rangle^*. \quad (3.4)$$
In addition $\mathcal{H}$ must be complete.

The properties listed above makes it possible to introduce a norm $||f||$ by the definition
$$||f|| = \langle f|f\rangle^{\frac{1}{2}}. \quad (3.5)$$
The existence of such a norm allows us to impose conditions analogous to the orthonormalization relations (2.107) for the harmonic oscillator.

Linear operators on Hilbert spaces are defined as linear mappings of one Hilbert space $\mathcal{H}$ on another $\mathcal{H}'$, which in our case is identical to $\mathcal{H}$. In this section we shall be particularly concerned with Hermitian operators $\hat{L}$ which satisfy[3]
$$\langle f|\hat{L}g\rangle = \langle \hat{L}f|g\rangle. \quad (3.6)$$
We shall now show that the eigenvalues of a Hermitian operator are real. Let $\hat{L}$ denote a Hermitian operator with eigenvalues λ. In order to prove that the eigenvalue equation
$$\hat{L}f = \lambda f \quad (3.7)$$
only has solutions for real λ, we form the inner product of both sides of (3.7) with f, resulting in
$$\langle f|\hat{L}f\rangle = \lambda\langle f|f\rangle. \quad (3.8)$$
According to (3.4) we have
$$\langle \hat{L}f|f\rangle = \lambda^*\langle f|f\rangle. \quad (3.9)$$

Because of the hermiticity (3.6) the expressions (3.8) and (3.9) are identical, and since $\langle f|f\rangle$ only vanishes for $f = 0$, it follows that λ must equal λ^*, implying that the eigenvalues are real. For a given eigenvalue λ it is possible that there exists more than one linearly-independent solution to (3.7). Under these circumstances the eigenvalue is said to be *degenerate*, and the degree of degeneracy equals the number of linearly-independent solutions.

If λ_1 and λ_2 denote two different eigenvalues associated with the Hermitian operator $\hat{L}$, corresponding to
$$\hat{L}f_1 = \lambda_1 f_1 \quad (3.10)$$
and
$$\hat{L}f_2 = \lambda_2 f_2, \quad (3.11)$$

---
[3] In mathematics such operators are called symmetric.

it follows that f_1 and f_2 are orthogonal in the sense that

$$\langle f_1|f_2\rangle = 0. \tag{3.12}$$

This may be shown by using that $\hat{L}$ is Hermitian, $\langle f_1|\hat{L}f_2\rangle = \langle \hat{L}f_1|f_2\rangle$, which according to (3.10) and (3.11) implies that $\lambda_2\langle f_1|f_2\rangle = \lambda_1\langle f_1|f_2\rangle$ or

$$(\lambda_2 - \lambda_1)\langle f_1|f_2\rangle = 0. \tag{3.13}$$

If λ_1 differs from λ_2, then according to (3.13) we have $\langle f_1|f_2\rangle = 0$. When $\lambda_1 = \lambda_2$ the eigenfunctions associated with the degenerate eigenvalue may be chosen to be orthonormal.

We shall assume in general that the eigenfunctions f_i associated with a Hermitian operator form a complete set, implying that an arbitrary function may be expanded in terms of them. Thus in the following the set of functions f_i denotes a complete orthonormal basis in the sense that

$$\langle f_i|f_j\rangle = \delta_{ij}, \tag{3.14}$$

while an arbitrary element ψ may be written as

$$\psi = \sum_i c_i f_i. \tag{3.15}$$

The coefficients c_i are determined by forming the inner product of f_i with ψ and using (3.14) with the result

$$c_i = \langle f_i|\psi\rangle. \tag{3.16}$$

If ψ is normalized, $||\psi|| = 1$, the sum of the absolute squares of the expansion coefficients c_i must equal 1,

$$\sum_i |c_i|^2 = 1. \tag{3.17}$$

An important theorem is valid for operators that commute with each other: *If two linear operators $\hat{L}_1$ and $\hat{L}_2$ commute,*

$$[\hat{L}_1, \hat{L}_2] = 0, \tag{3.18}$$

then it is possible to find simultaneous eigenfunctions for them. The theorem is shown most easily for the case when the eigenvalues are non-degenerate, but the following proof for the non-degenerate case may easily be generalized to the case where the eigenvalues are degenerate (see the *Comment* below).

If f is an eigenfunction of the operator $\hat{L}_1$,

$$\hat{L}_1 f = \lambda_1 f, \tag{3.19}$$

Basic principles

then it follows from (3.18) that

$$\hat{L}_1\hat{L}_2 f = \hat{L}_2\hat{L}_1 f = \lambda_1 \hat{L}_2 f. \tag{3.20}$$

According to our assumptions there is only one linearly-independent solution to (3.19) for a given value of λ_1. As a consequence $\hat{L}_2 f$ must be proportional to f,

$$\hat{L}_2 f = \lambda_2 f, \tag{3.21}$$

which shows that f is an eigenfunction of the operator $\hat{L}_2$ as well. Conversely, we may prove that the operators commute, if they have all their eigenfunctions in common. The proof goes as follows: Let f be a simultaneous eigenfunction of $\hat{L}_1$ and $\hat{L}_2$. According to our assumption, $\hat{L}_1\hat{L}_2 f = \lambda_1\lambda_2 f$ and $\hat{L}_2\hat{L}_1 f = \lambda_2\lambda_1 f$. Since an arbitrary function ψ belonging to the Hilbert space may be expanded in terms of the complete set of eigenfunctions of $\hat{L}_1$ and $\hat{L}_2$, we conclude that $(\hat{L}_1\hat{L}_2 - \hat{L}_2\hat{L}_1)\psi = 0$ and therefore $[\hat{L}_1, \hat{L}_2] = 0$, since ψ is arbitrary.

> *Comment.* If the eigenvalues are degenerate, the proof given above may be generalized as follows: Let f_i be one of the s linearly-independent eigenfunctions of $\hat{L}_1$ belonging to the eigenvalue λ_1 ($i = 1, \ldots, s$). According to our assumption
>
> $$\hat{L}_1 f_i = \lambda_1 f_i$$
>
> and
>
> $$\hat{L}_1\hat{L}_2 f_i = \hat{L}_2\hat{L}_1 f_i = \lambda_1 \hat{L}_2 f_i,$$
>
> corresponding to (3.19) and (3.20) above. We may now conclude that $\hat{L}_2 f_i$ may be expressed as a linear combination of the s functions f_i, and we therefore seek to determine a linear combination ψ of the s functions $f_1, \ldots, f_s$, $\psi = \sum_{j=1}^{s} c_j f_j$, in such a manner that
>
> $$\hat{L}_2 \psi = \lambda_2 \psi = \lambda_2 \sum_{j=1}^{s} c_j f_j.$$
>
> Since the s functions f_j are chosen to be orthonormal, we can find the solutions to this eigenvalue equation by setting the determinant of the matrix A_{kj} given by
>
> $$A_{kj} = \langle f_k|\hat{L}_2 f_j\rangle - \lambda_2 \delta_{kj}$$
>
> equal to zero. In this way we have shown that it is possible to find simultaneous eigenfunctions. The converse, that $\hat{L}_1$ and $\hat{L}_2$ commute if they have all their eigenfunctions in common, is proved as in the non-degenerate case.

In the operator formulation of quantum mechanics, physical quantities such as energy, angular momentum etc. are associated with definite linear operators. The eigenvalues of these linear operators are the possible results of a measurement of the corresponding physical quantity. If ψ represents a state in which the energy has a well-defined value E, the energy is found by solving the eigenvalue equation

$$\hat{H}\psi = E\psi, \qquad (3.22)$$

which corresponds to (2.94) in the matrix formulation. We also see that the matrix formulation may be derived from (3.22) by the choice of a definite basis such as (3.15), since (3.22) then results in

$$\sum_j c_j \langle f_i | \hat{H} f_j \rangle = E c_i \qquad (3.23)$$

when use is made of (3.14), corresponding to the identification (cf. (2.94))

$$|E\rangle_i = c_i; \quad \hat{H}_{ij} = \langle f_i | \hat{H} f_j \rangle. \qquad (3.24)$$

> **Comment.** The proof of (3.23) is obtained by inserting (3.15) in (3.22), with the index of summation i being named j instead. By forming the inner product of both sides of (3.22) with f_i we obtain
>
> $$\sum_j \langle f_i | \hat{H} c_j f_j \rangle = \sum_j \langle f_i | E c_j f_j \rangle, \qquad (3.25)$$
>
> which reduces to (3.23), when use is made of the orthonormality conditions (3.14).

The quantity $\langle f_i | \hat{H} f_j \rangle$ is often written in the more symmetric form $\langle f_i | \hat{H} | f_j \rangle$. Since i labels the eigenfunctions in a unique manner, the quantity $\langle f_i | \hat{H} f_j \rangle$ may equivalently be written in the form $\langle i | \hat{H} | j \rangle$ or $\langle i | H | j \rangle$.

We mentioned in the previous chapter how the choice of the operator (2.75) for the momentum of a particle agrees with the basic commutation relation (2.71). Correspondingly, when a particle moves in three dimensions its momentum operator $\hat{\mathbf{p}}$ is

$$\hat{\mathbf{p}} = \frac{\hbar}{i} \nabla = \frac{\hbar}{i} \left(\frac{\partial}{\partial x}, \frac{\partial}{\partial y}, \frac{\partial}{\partial z} \right), \qquad (3.26)$$

in accordance with the commutation relation $[x_i, p_j] = i\hbar \delta_{ij}$. The energy operator, which is named the Hamilton operator or the Hamiltonian, is therefore

$$\hat{H} = \frac{\hat{\mathbf{p}}^2}{2m} + V(\mathbf{r}) = -\frac{\hbar^2}{2m} \nabla^2 + V(\mathbf{r}) \qquad (3.27)$$

for a particle with mass m moving in an external field given by the potential energy $V(\mathbf{r})$. The solution ψ to the eigenvalue equation (3.22) is a function of

Basic principles

r and is called the wave function of the particle. The Hamiltonian is thus constructed by replacing **p** in the classical Hamiltonian with $\hbar \nabla / i$. The resulting eigenvalue equation (3.22) is the time-independent Schrödinger equation.

3.1.1 A simple example: The harmonic oscillator

The eigenfunctions of the Hamilton operator for the one-dimensional harmonic oscillator span a Hilbert space. Since the classical Hamiltonian is

$$H = \frac{p^2}{2M} + \frac{1}{2}Kx^2, \qquad (3.28)$$

the Hamilton operator is given by

$$\hat{H} = -\frac{\hbar^2}{2M}\frac{d^2}{dx^2} + \frac{1}{2}Kx^2. \qquad (3.29)$$

The eigenvalue equation (3.22) for the energy is then

$$-\frac{\hbar^2}{2M}\frac{d^2\psi}{dx^2} + \frac{1}{2}Kx^2\psi = E\psi. \qquad (3.30)$$

We shall require that ψ is normalizable and impose the normalization condition

$$\int_{-\infty}^{\infty} dx |\psi|^2 = 1. \qquad (3.31)$$

The inner product is in the present case[4] defined by an integral over the position variable x

$$\langle f | g \rangle = \int_{-\infty}^{\infty} dx f^*(x) g(x). \qquad (3.32)$$

In Problem 3.6 it is shown that (3.32) is an inner product satisfying the conditions (3.1)-(3.4).

The eigenvalue problem for a harmonic oscillator may be solved by the power series method, which is used commonly for the solution of differential equations. In the present section we shall only discuss some simple solutions to the eigenvalue equation (3.30), while the general solution obtained by the use of the power series method is the subject of Problem 3.1.

---

[4] If the particle moves in two dimensions, x and y, the inner product involves an integration over the variables x and y. In general, for a system described by s generalized coordinates q_i, the number of integrations in the inner product equals s. For the two electrons in the He atom the inner product thus involves $2 \cdot 3 = 6$ integrations over the spatial coordinates of the two electrons, cf. Section 8.2.

First we bring (3.30) into dimensionless form by introducing the variable $\xi = x/a$, where $a = \sqrt{\hbar/M\omega}$ and $\omega = \sqrt{K/M}$. The differential equation (3.30) then becomes

$$u'' + \lambda u - \xi^2 u = 0, \tag{3.33}$$

where u is the desired solution and ' denotes the derivative with respect to ξ. The dimensionless constant λ is given by $\lambda = 2E/\hbar\omega$, where E is the desired energy eigenvalue.

We first examine the solutions of the differential equation for large $|\xi|$. If we insert into (3.33) a function of the form

$$u(\xi) = \xi^p e^{\pm \xi^2/2}, \tag{3.34}$$

where p is an arbitrary exponent, the differential equation (3.33) is satisfied, provided we retain only the leading terms proportional to $\xi^2 u$.

The result (3.34), in conjunction with the requirement that the wave function is normalizable, makes it natural to seek solutions of the type

$$u = H(\xi) e^{-\xi^2/2}, \tag{3.35}$$

where $H(\xi)$ is a polynomial in ξ. By inserting (3.35) in (3.33) and carrying out the differentiations, it is seen that $H(\xi)$ must satisfy the equation

$$H'' - 2\xi H' + (\lambda - 1)H = 0. \tag{3.36}$$

It is readily verified that (3.36) is satisfied when H is a constant, provided that λ equals 1. The constant value of H may be chosen to be 1, since the wave function in any case must be normalized. We thus have the solution

$$H = 1, \ \lambda = 1, \tag{3.37}$$

which corresponds to the energy $\hbar\omega/2$ since $\lambda = 2E/\hbar\omega$. Furthermore it is seen that (3.36) has the solution $H = \xi$, provided λ equals 3,

$$H = \xi, \ \lambda = 3, \tag{3.38}$$

corresponding to the energy $3\hbar\omega/2$. By guessing functions of the simplest possible kind we have thus determined solutions associated with the lowest and second-lowest energy level of the harmonic oscillator, cf. (2.103).

If we try in a similar way to find a solution $H(\xi)$, which is proportional to ξ^2, we discover that the equation cannot be satisfied, since the second derivative of ξ^2 is a constant, while the second and third terms in (3.36) make contributions proportional to ξ^2, regardless of the value of λ. However, a solution of the form $H = \xi^2 + c$, where c is a constant, satisfies (3.36), if $-4 + (\lambda - 1) = 0$ and $2 + (\lambda - 1)c = 0$, corresponding to the solution

$$H = \xi^2 - \frac{1}{2}, \ \lambda = 5. \tag{3.39}$$

Basic principles

The energy eigenvalue associated with this solution is thus $5\hbar\omega/2$. It is evident that the solution belongs to the third-lowest energy level for the harmonic oscillator[5].

In Problem 3.1 we determine all normalizable eigenfunctions, which are given by

$$u_n = N_n e^{-\xi^2/2} H_n(\xi) \qquad (3.40)$$

with associated eigenvalues

$$E_n = (n + \frac{1}{2})\hbar\omega, \quad n = 0, 1, 2 \cdots. \qquad (3.41)$$

The function H_n is a Hermite polynomial which is defined by

$$H_n(\xi) = (-1)^n e^{\xi^2} \frac{d^n}{d\xi^n} e^{-\xi^2}. \qquad (3.42)$$

The constant N_n is given by

$$N_n = \frac{1}{\sqrt{a\sqrt{\pi}2^n n!}}, \qquad (3.43)$$

which ensures that the normalization condition

$$||u_n||^2 = \langle u_n | u_n \rangle = \int_{-\infty}^{\infty} dx |u_n(x)|^2 = 1 \qquad (3.44)$$

is satisfied. The normalized eigenfunctions belonging to the eight lowest energy eigenvalues are given in Table 3.1 and plotted in Fig. 3.1.

The Hermite polynomials satisfy the recursion relations

$$H_n' = 2nH_{n-1} \qquad (3.45)$$

and

$$H_{n+1} - 2\xi H_n + 2nH_{n-1} = 0. \qquad (3.46)$$

These recursion relations may be proved by expressing ξ and $d/d\xi$ in terms of the operators $\hat{a}$ and $\hat{a}^\dagger$, using that $\hat{a}|n\rangle$ is proportional to $|n-1\rangle$, while $\hat{a}^\dagger|n\rangle$ is proportional to $|n+1\rangle$ (Problem 3.4).

We have demonstrated above that H_0, H_1 and H_2 satisfy (3.36). The explicit expressions for these polynomials are $H_0 = 1$, $H_1 = 2\xi$ and $H_2 = 4\xi^2 - 2$, cf. (3.37)-(3.39).

---

[5] Rather than guessing the solutions we could have used the method of Section 1.4.3 and introduced the operator $\hat{a}^\dagger = (\xi - d/d\xi)/\sqrt{2}$ corresponding to (2.84). By acting with this operator n times on the function $\exp(-\xi^2/2)$, which is proportional to the ground-state wave function, we obtain a solution corresponding to the eigenvalue $(n + 1/2)\hbar\omega$. In particular, (3.38) and (3.39) correspond to $n = 1$ and $n = 2$.

$$u_0 = N_0 e^{-\xi^2/2}, \qquad N_0 = \frac{1}{\pi^{1/4}\sqrt{a}}$$

$$u_1 = N_1 2\xi e^{-\xi^2/2}, \qquad N_1 = \frac{1}{\pi^{1/4}\sqrt{2a}}$$

$$u_2 = N_2(4\xi^2 - 2)e^{-\xi^2/2}, \qquad N_2 = \frac{1}{\pi^{1/4}2\sqrt{2a}}$$

$$u_3 = N_3(8\xi^3 - 12\xi)e^{-\xi^2/2}, \qquad N_3 = \frac{1}{\pi^{1/4}4\sqrt{3a}}$$

$$u_4 = N_4(16\xi^4 - 48\xi^2 + 12)e^{-\xi^2/2}, \qquad N_4 = \frac{1}{\pi^{1/4}8\sqrt{6a}}$$

$$u_5 = N_5(32\xi^5 - 160\xi^3 + 120\xi)e^{-\xi^2/2}, \qquad N_5 = \frac{1}{\pi^{1/4}16\sqrt{15a}}$$

$$u_6 = N_6(64\xi^6 - 480\xi^4 + 720\xi^2 - 120)e^{-\xi^2/2}, \qquad N_6 = \frac{1}{\pi^{1/4}96\sqrt{5a}}$$

$$u_7 = N_7(128\xi^7 - 1344\xi^5 + 3360\xi^3 - 1680\xi)e^{-\xi^2/2}, \qquad N_7 = \frac{1}{\pi^{1/4}96\sqrt{70a}}$$

$$a = \sqrt{\frac{\hbar}{M\omega}}, \qquad \omega = \sqrt{\frac{K}{M}}$$

$$H = \frac{p^2}{2M} + \frac{1}{2}Kx^2$$

Table 3.1: Normalized energy eigenfunctions corresponding to the eight lowest energy eigenvalues ($n = 0, 1 \cdots, 7$) for the one-dimensional harmonic oscillator with $\xi = x/a$.

Basic principles 81

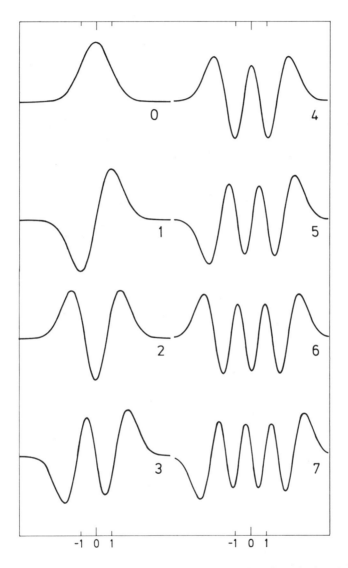

Figure 3.1: Normalized energy eigenfunctions associated with the eight lowest eigenvalues. The curves are drawn on the same horizontal and vertical scale as functions of the dimensionless variable $\xi = x/a$.

For the ground state the normalized wave function is

$$u_0(x) = (a\sqrt{\pi})^{-1/2} e^{-x^2/2a^2}, \tag{3.47}$$

where a is the characteristic length given by

$$a = \sqrt{\frac{\hbar}{M\omega}}. \tag{3.48}$$

It may be shown that the functions u_n form a complete basis in Hilbert space, since an arbitrary square integrable function $\psi(x)$ which is defined on the real axis may be expanded in terms of the eigenfunctions u_n,

$$\psi(x) = \sum_n a_n u_n. \tag{3.49}$$

With the use of this basis it is possible to turn the eigenvalue equation (3.22) into a matrix equation of the kind examined in Chapter 2, with

$$\hat{H} : H_{mn} = \int_{-\infty}^{\infty} dx\, u_m^* H u_n, \tag{3.50}$$

since the Schrödinger equation (3.30) assumes the form

$$\sum_n H_{mn} a_n = E a_m, \tag{3.51}$$

cf. the *Comment* following (3.26). Similarly, the matrices for $\hat{p}$ and $\hat{x}$ are given by

$$\hat{p} : p_{mn} = \int_{-\infty}^{\infty} dx\, u_m^* \frac{\hbar}{i} \frac{du_n}{dx}; \quad \hat{x} : x_{mn} = \int_{-\infty}^{\infty} dx\, u_m^* x u_n. \tag{3.52}$$

We have considered above a Hilbert space consisting of square integrable functions spanned by the eigenfunctions of the energy operator. However, it should be noted that the eigenfunctions for the momentum operator (3.26) are plane waves $\exp(i\mathbf{p}\cdot\mathbf{r}/\hbar)$ with corresponding eigenvalues $\mathbf{p}$. These eigenfunctions are not square integrable, since their absolute square equals 1 everywhere which yields infinity when integrated over all space. The corresponding eigenvalues form a continuum. The eigenfunctions of the momentum operator do not belong to the Hilbert space since they do not satisfy the requirements (3.1)-(3.4). The same applies to the eigenfunctions of the position operator. A satisfactory mathematical discussion of these operators requires the use of the theory of distributions. In practice one often treats the momentum operator on an equal footing with operators having a discrete spectrum. The reason is that one may consider the system in question to be enclosed in a finite volume.

Basic principles

When one uses periodic boundary conditions (cf. Chapter 4) the eigenvalues of the momentum operator become discrete and the states are normalizable in the usual sense. Alternatively one may use the delta function introduced by Dirac to express the normalization condition for such states, as discussed below. When we denote a momentum eigenstate by the symbol $|\mathbf{p}\rangle$ and employ the usual theory of Hilbert space, it is with the understanding that this does not normally lead to difficulties, even though the mathematical foundation of such a procedure may be incomplete.

3.2 The postulates of quantum mechanics

We have now established the necessary background for formulating quantum mechanics in terms of a small set of fundamental postulates. As we shall see, these postulates implicitly define the concept of a quantum mechanical state. In classical mechanics the definition of the state of a physical system appears so obvious, that it hardly needs mentioning. Classically the state of a particle is given by specifying its position and velocity. In quantum mechanics the definition of a state is much more subtle, reflecting the inadequacy of our classical intuition. The following three postulates may be considered to be the most fundamental:

I. *Each state of a physical system corresponds to an element, in the present context called a state vector, in a Hilbert space $\mathcal{H}$, and vice versa. The length of the state vector $|\nu\rangle$ is unity, $\langle \nu|\nu\rangle = 1$. Elements that only differ by a phase factor $\exp i\phi$ represent the same state of the physical system.*

> *Comment.* Let us assume that the physical system is a particle moving in one dimension in a harmonic oscillator potential. A physical state for this particle is the state of lowest energy. We denote the corresponding state vector by $|0\rangle$. Similarly the state vector associated with the second-lowest energy may be denoted by $|1\rangle$. It follows from the first postulate given above that $a|0\rangle + b|1\rangle$ is also a possible state vector corresponding to a physical state of the system, provided a and b satisfy the normalization requirement $|a|^2 + |b|^2 = 1$.
> It is necessary to take into account the normalization difficulties mentioned at the end of Section 3.1.1, in order to be able to describe properties like the momentum of a freely moving particle, but this may be accomplished by the use of the Dirac delta function (cf. Section 3.3 below) or the introduction of a finite, as opposed to an infinite, volume.

The second postulate relates the state vectors in the Hilbert space to the probability of obtaining different experimental results.

II. *Given that the state of a physical system corresponds to the state vector $|\nu\rangle$, the probability that we observe the system in a state corresponding to the state vector $|\alpha\rangle$ is $|\langle\alpha|\nu\rangle|^2$.*

> *Comment.* To observe the system in a state corresponding to the state vector $|\alpha\rangle$ means to carry out the measurements needed to decide whether the system is in the state $|\alpha\rangle$. In practice this may involve measurements of several different physical quantities. For the state $a|0\rangle + b|1\rangle$ discussed above the probability p of observing the system in a state corresponding to the state vector $|1\rangle$ is thus $p = |a\langle 1|0\rangle + b\langle 1|1\rangle|^2$. Since $\langle 1|0\rangle = 0$ (compare (3.12)), the resulting probability is $p = |b|^2$. If the quantity α characterizing the state $|\alpha\rangle$ refers to a continuum, $|\langle \alpha|\nu\rangle|^2$ becomes a probability density, $|\langle \alpha|\nu\rangle|^2 d\alpha$ giving the probability of obtaining a result of the measurement between α and $\alpha + d\alpha$.

The third postulate concerns *physical observables*. By a physical observable one means a measurable quantity such as energy, position, momentum or angular momentum. Other, less familiar, examples are spin and parity, which will be discussed in subsequent chapters. The following postulate relates physical observables to operators on a Hilbert space.

III. *Every observable corresponds to a linear, Hermitian operator on a Hilbert space. The possible results of a measurement are the eigenvalues of the corresponding operator.*

> *Comment.* The probabilities of different measurement results are determined by the state of the system according to postulate II above. In particular, if the state vector of the system is an eigenvector of the operator associated with a particular observable, the result of the measurement of this observable is given with certainty by the eigenvalue belonging to the state vector. In the example of the harmonic oscillator considered above, the energy is associated with a linear Hermitian operator, and the probability of measuring the lowest energy is $|a|^2$, given that the system is in the state corresponding to the state vector $a|0\rangle + b|1\rangle$.

In the previous chapter we quantized the harmonic oscillator by replacing the classical Poisson brackets by the quantum mechanical commutator divided by $i\hbar$. In the present context this constitutes our fourth postulate:

IV. *The operators associated with a coordinate q_i and its canonically-conjugate momentum p_i satisfy the commutation rules $[\hat{q}_i, \hat{p}_j] = i\hbar\delta_{ij}$, where δ_{ij} equals 1, if $i = j$, and zero otherwise.*

The last postulate concerns the development in time of a physical system. We denote the state vector of the system at time t by $|\psi(t)\rangle$. At a given instant of time the state vector $|\psi(t)\rangle$ contains all information that may be known about the system. The state vector $|\psi(t)\rangle$ determines the state at time $t + dt$ through an equation for the development in time.

Basic principles

V. *The development in time of the state vector is given by the first-order differential equation* $i\hbar(d/dt)|\psi(t)\rangle = \hat{H}|\psi(t)\rangle$, *where* $\hat{H}$ *is a linear Hermitian operator. In the non-relativistic limit* $\hat{H}$ *is the operator corresponding to the classical Hamiltonian.*

Comment. As a consequence of the fifth postulate, the state $|\psi\rangle = a|0\rangle + b|1\rangle$ for the harmonic oscillator develops in time according to

$$|\psi(t)\rangle = ae^{-iE_0 t/\hbar}|0\rangle + be^{-iE_1 t/\hbar}|1\rangle,$$

where E_0 is the lowest and E_1 the second-lowest energy. This may be verified by inserting a solution of the form $c_0(t)|0\rangle + c_1(t)|1\rangle$ into the equation for the development in time,

$$i\hbar\frac{d}{dt}(c_0(t)|0\rangle + c_1(t)|1\rangle) = \hat{H}(c_0(t)|0\rangle + c_1(t)|1\rangle),$$

and taking the inner product of this equation with $|0\rangle$ and $|1\rangle$. This results in

$$i\hbar\frac{dc_0}{dt} = E_0 c_0(t)$$

and

$$i\hbar\frac{dc_1}{dt} = E_1 c_1(t),$$

since $\langle 0|0\rangle = \langle 1|1\rangle = 1$ and $\langle 1|0\rangle = \langle 0|1\rangle = 0$. The constants E_0 and E_1 are the energies associated with the states $|0\rangle$ and $|1\rangle$. After use of the initial conditions $c_0(0) = a$ and $c_1(0) = b$ at $t = 0$, we obtain the result given above for the development in time of $|\psi(t)\rangle$.

3.3 Probability amplitudes

In formulating the second postulate of the previous section we introduced the quantity $\langle\alpha|\nu\rangle$ which is named a probability amplitude. According to the second postulate the absolute square of the probability amplitude determines the probability of a definite outcome of an experiment.

In the following we shall discuss the concept of probability amplitudes by starting from the energy eigenstates $|n\rangle$ for the harmonic oscillator. It follows from (2.110) that $|n+1\rangle = (n+1)^{-1/2}\hat{a}^\dagger|n\rangle$. The normalized energy eigenstates of the harmonic oscillator may therefore be written in the form

$$|n\rangle = (n!)^{-1/2}(\hat{a}^\dagger)^n|0\rangle. \tag{3.53}$$

If the harmonic oscillator is known to be in the state $|n\rangle$, the result of a measurement of its energy is certain, namely $(n+1/2)\hbar\omega$. However, the energy of the harmonic oscillator is not the only physical quantity which is observable.

The position and momentum of the particle are represented by non-diagonal matrices in the basis formed by the energy eigenstates (3.53), since neither the position operator $\hat{q}$ nor the momentum operator $\hat{p}$ commute with the energy operator. Neither the position nor the momentum has therefore a definite value in the state $|n\rangle$.

If a measurement of the position of the particle is known to give a definite result x, the state of the system is denoted by $|x\rangle$. On the assumption that the energy eigenstates form a complete set it is possible to expand the state $|x\rangle$ in terms of the energy eigenstates,

$$|x\rangle = \sum_{n=0}^{\infty} \langle n|x\rangle |n\rangle, \qquad (3.54)$$

corresponding to the expansion (3.15)-(3.16). The expression for the expansion coefficients given in (3.54) may be verified by taking the inner product of both sides of the equation with the state $|m\rangle$ and subsequently using the orthonormality condition

$$\langle n|m\rangle = \delta_{nm}. \qquad (3.55)$$

Alternatively we introduce the operator

$$\hat{1} = \sum_{n=0}^{\infty} |n\rangle \langle n|, \qquad (3.56)$$

which is a unit operator, since

$$\sum_{n=0}^{\infty} |n\rangle \langle n|n'\rangle = |n'\rangle \qquad (3.57)$$

because of the orthonormality condition (3.55), and write $|x\rangle$ as

$$|x\rangle = \hat{1}|x\rangle = \sum_{n=0}^{\infty} |n\rangle \langle n|x\rangle, \qquad (3.58)$$

which is identical to (3.54).

The expansion coefficients $\langle n|x\rangle$ are called probability amplitudes. As stated in Section 3.2 it is a basic postulate of quantum theory that the probability $P(x)dx$ of obtaining a result between x and $x + dx$ by measuring the position of the particle, given that the system is in the state $|n\rangle$, is equal to the absolute square of the probability amplitude, $|\langle x|n\rangle|^2$, times dx or

$$P(x)dx = |\langle x|n\rangle|^2 dx. \qquad (3.59)$$

Basic principles 87

Since the probability is 1 for measuring the particle somewhere, the normalization condition
$$\int_{-\infty}^{\infty} dx |\langle x|n\rangle|^2 = 1 \tag{3.60}$$
must be satisfied. The probability amplitude $\langle x|n\rangle$ is identical to Schrödinger's wave function $u_n(x)$ and obeys the same normalization condition, cf. (1.94) and (3.60). In Section 3.3.3 below we demonstrate for a particle moving in one dimension that the probability amplitude $\langle x|\psi(t)\rangle$ is the wave function $\psi(x,t)$ satisfying the time-dependent Schrödinger equation.

We have discussed the concept of probability amplitudes with a concrete example in mind, but our interpretation of the expansion coefficients is of general validity, regardless of which physical quantities we are considering. We could for instance have replaced the eigenvalues x of the position operator $\hat{q}$ with p, the eigenvalues of the momentum operator $\hat{p}$, and interpreted the expansion coefficients $\langle n|p\rangle$ as probability amplitudes. The transition from states labelled by n to states labelled by x or to states labelled by p may be considered to be a transformation from one basis in the vector space to another.

3.3.1 Expectation value

According to the second postulate the probability of observing a definite value of a physical quantity A, given that the system is in the state $|\nu\rangle$, is given by the absolute square of a probability amplitude. Let us assume that the operator associated with A only possesses discrete eigenvalues α. The probability $P(\alpha)$ for observing the eigenvalue α is then
$$P(\alpha) = |\langle \alpha|\nu\rangle|^2. \tag{3.61}$$
If the eigenvalues form a continuum, then $P(\alpha)$ is a probability density in the sense that
$$P(\alpha)d\alpha$$
yields the probability for observing a value between α and $\alpha + d\alpha$.

The sum of the probabilities must be 1, which means that
$$\sum_{\alpha} P(\alpha) = 1, \tag{3.62}$$
in the discrete case, while the corresponding condition for the continuum case is
$$\int d\alpha P(\alpha) = 1. \tag{3.63}$$

The mean value (often called the expectation value) of the physical quantity A in the state $|\nu\rangle$ may now be written as
$$<A> = \sum_{\alpha} P(\alpha)\alpha = \sum_{\alpha} |\langle \alpha|\nu\rangle|^2 \alpha \tag{3.64}$$

in the case where the eigenvalues α are discrete. When the eigenvalue spectrum is partly (or wholly) continuous, the sum over α includes an integration over the continuous part of the spectrum, corresponding to the substitution $\sum_\alpha \to \int d\alpha$.

It is convenient to give the expression for the mean value of A another form by using that the unit operator $\hat{1}$ in general may be written as a sum over states

$$\hat{1} = \sum_\alpha |\alpha\rangle\langle\alpha|, \qquad (3.65)$$

where the sum runs over all states with discrete eigenvalues (supplemented by an integration over a possible continuum), cf. Problem 2.9. With the help of (3.65) we may now express the mean value $<A>$ in terms of $\langle\nu|A|\nu\rangle$,

$$<A> = \sum_\alpha |\langle\alpha|\nu\rangle|^2 \alpha = \sum_\alpha \langle\nu|\alpha\rangle\langle\alpha|\nu\rangle \alpha = \langle\nu|A|\nu\rangle. \qquad (3.66)$$

The last equality is obtained by using that A is Hermitian, $\alpha\langle\alpha|\nu\rangle = \langle\alpha|A|\nu\rangle$, in connection with (3.65). In Example 3 below, the use of the expression (3.66) for the mean value of a physical quantity will be illustrated for the case of the two-dimensional harmonic oscillator.

The operator

$$\int_{-\infty}^{\infty} dx |x\rangle\langle x|$$

is also a unit operator. The states $|x\rangle$ satisfy the orthonormality conditions

$$\langle x'|x\rangle = \delta(x' - x), \qquad (3.67)$$

where $\delta(x)$ is the delta function[6]. The relation (3.67) generalizes (3.14) to the case where the eigenvalue spectrum is continuous. The delta function satisfies the relations

$$\int_{-\infty}^{\infty} dx \delta(x) = 1 \qquad (3.68)$$

and

$$\int_{-\infty}^{\infty} dx f(x)\delta(x) = f(0), \qquad (3.69)$$

where $f(x)$ is a continuous function of x. The delta function may be considered to be the limiting function of a series of functions, which may be given by

$$\delta(x) = \lim_{x_0 \to 0} (x_0\sqrt{\pi})^{-1} e^{-x^2/x_0^2}, \qquad (3.70)$$

---

[6] The delta function is a highly singular function which does not belong to the Hilbert space. It has a meaning only when a subsequent integration over its argument is carried out. Its use in quantum mechanics may be justified with the help of distribution theory. The delta function is an even function, $\delta(x) = \delta(-x)$ with an odd derivative, $\delta'(x) = -\delta'(-x)$. Further useful relations are given in Appendix C.

Basic principles

where x_0 tends to zero through positive values. It should be noted that the integral of the Gaussian function (3.70) equals 1, regardless of the value of x_0. The eigenstate $|x\rangle$ may thus be represented by the probability amplitude $\langle x'|x\rangle$, which is a delta function of $x' - x$, implying that $|x\rangle$ is a state in which a measurement of the position of the particle with certainty yields the result x.

The momentum operator for motion in one dimension is given by

$$\hat{p} = \frac{\hbar}{i}\frac{d}{dx}. \tag{3.71}$$

Let ψ_1 and ψ_2 be square integrable functions which vanish at infinity. It is readily verified by doing integration by parts, that $\hat{p}$ is Hermitian,

$$\int_{-\infty}^{\infty} dx (\frac{\hbar}{i}\frac{d}{dx}\psi_1)^* \psi_2 = \int_{-\infty}^{\infty} dx \psi_1^* \frac{\hbar}{i}\frac{d}{dx}\psi_2. \tag{3.72}$$

Likewise it is verified that the operator d^2/dx^2 is Hermitian. The operator d/dx is anti-Hermitian, since its Hermitian conjugate is $-d/dx$.

The eigenfunctions of $\hat{p}$ are proportional to $\exp(ipx/\hbar)$ with associated eigenvalues p. We shall now expand an arbitrary wave function $\psi(x)$ in terms of these eigenfunctions. By use of Fourier's integral theorem (see (1) and (2) in Appendix C) we obtain the desired expansion

$$\psi(x) = \frac{1}{\sqrt{2\pi\hbar}}\int_{-\infty}^{\infty} dp f(p) e^{ipx/\hbar}, \tag{3.73}$$

where

$$f(p) = \frac{1}{\sqrt{2\pi\hbar}}\int_{-\infty}^{\infty} dx \psi(x) e^{-ipx/\hbar}. \tag{3.74}$$

Let us determine $f(p)$ in the case where ψ is the ground-state wave function u_0 given by (3.47) for the harmonic oscillator. Since we have to integrate over x, it is convenient to use the identity

$$-x^2/2a^2 - ipx/\hbar = -(x + ia^2 p/\hbar)^2/2a^2 - a^2(p/\hbar)^2/2 \tag{3.75}$$

and carry out the integral in (3.74) in terms of the variable $x' = x + ia^2 p/\hbar$. The result is

$$f(p) = (a/\sqrt{\pi}\hbar)^{1/2} e^{-p^2 a^2/2\hbar^2}, \tag{3.76}$$

which shows that the Fourier transform of the Gaussian (3.47) is itself a Gaussian in the variable p. This is a special feature of a Gaussian. In general the functional form of the Fourier transform is different from that of the original function. With $u_0(x)$ denoting the wave function of the ground state of the harmonic oscillator, we have thus shown that

$$u_0(x) = \frac{1}{\sqrt{2\pi\hbar}}\int_{-\infty}^{\infty} dp f(p) e^{ipx/\hbar}, \tag{3.77}$$

where $f(p)$ is given by (3.76). The absolute square $|f(p)|^2$ of the expansion coefficients satisfies the normalization condition

$$\int_{-\infty}^{\infty} dp |f(p)|^2 = 1, \tag{3.78}$$

consistent with the fact that the absolute square of f is the probability density on the p-axis for observing a definite value of the momentum.

3.3.2 The uncertainty relations

The Fourier transform $f(p)$ of the wave function $\psi(x)$ determines the expectation value of $\hat{p} = (\hbar/i)d/dx$ according to

$$<p> = \int_{-\infty}^{\infty} dp |f(p)|^2 p, \tag{3.79}$$

since $|f|^2 dp$ is the probability of finding the particle with momentum between p and $p+dp$. It is often convenient to express the mean value directly in terms of the wave function $\psi(x)$. According to our general result (3.66) we have

$$<p> = \int_{-\infty}^{\infty} dx \psi^* (\frac{\hbar}{i}\frac{d}{dx})\psi, \tag{3.80}$$

Similarly the expectation value of the square of the momentum is given by

$$<p^2> = \int_{-\infty}^{\infty} dx \psi^* (-\hbar^2 \frac{d^2}{dx^2})\psi, \tag{3.81}$$

while the expectation value of the energy is

$$<H> = \int_{-\infty}^{\infty} dx \psi^* \hat{H} \psi, \tag{3.82}$$

where $\hat{H}$ is the operator $-(\hbar^2/2m)d^2/dx^2 + V(x)$.

The square of the expectation value of a physical quantity A is in general different from the expectation value of the square of A, unless the state considered happens to be an eigenstate for the operator associated with the physical quantity A. The root-mean-square deviation from the mean, Δ, is introduced by the definition

$$\Delta^2(A) = <(A-<A>)^2> = <A^2> - <A>^2, \tag{3.83}$$

where $<A>$ denotes the mean (or expectation) value of the physical quantity A in the state under consideration. It is seen from (3.83) that Δ^2 is always

Basic principles

positive or zero. Note that Δ^2 vanishes if the state is an eigenstate for the operator associated with A.

In order to illustrate these concepts we consider a particle in the ground state (3.47) of the harmonic oscillator. It is immediately clear that $<x>=0$, since the square of the wave function is an even function of x. The mean-square deviation $\Delta^2(x)$ is thus identical to $<x^2>$, which is found to be $a^2/2$ for the state considered. Similarly one obtains $\Delta^2(p) =<p^2> = \hbar^2/2a^2$ by using (3.81) with ψ given by (3.47). It is seen that the product $\Delta x \Delta p$ is independent of a and given by

$$\Delta(x)\Delta(p) = \frac{\hbar}{2}. \tag{3.84}$$

This is an example of the Heisenberg uncertainty relations. For a general state ψ the product $\Delta(x)\Delta(p)$ may be shown (Problem 3.13) to be greater than or equal to $\hbar/2$ as a consequence of the commutation rule $[\hat{x}, \hat{p}] = i\hbar$.

EXAMPLE 3. QUANTIZATION OF THE TWO-DIMENSIONAL HARMONIC OSCILLATOR

The two-dimensional harmonic oscillator is well suited for illustrating the concepts of quantum mechanics. In this example we shall only consider the isotropic harmonic oscillator, for which the force constants associated with the motion in the x- and y-directions both are equal to K. When the mass of the particle is denoted by M, the classical angular frequency becomes $\omega = \sqrt{K/M}$. The Hamiltonian H is[7]

$$H = \frac{p_x^2}{2M} + \frac{1}{2}Kx^2 + \frac{p_y^2}{2M} + \frac{1}{2}Ky^2. \tag{3.85}$$

We introduce dimensionless variables x_1, x_2 and p_1, p_2 by the definitions

$$x = x_1\sqrt{\frac{\hbar}{M\omega}}, \quad y = x_2\sqrt{\frac{\hbar}{M\omega}}, \tag{3.86}$$

and

$$p_x = p_1\sqrt{\hbar M\omega}, \quad p_y = p_2\sqrt{\hbar M\omega}. \tag{3.87}$$

In terms of these variables the Hamiltonian becomes

$$H = \hbar\omega\frac{1}{2}(x_1^2 + x_2^2 + p_1^2 + p_2^2). \tag{3.88}$$

The creation and annihilation operators are defined in the usual manner, cf. (2.85-86),

$$x_j = \frac{1}{\sqrt{2}}(a_j + a_j^\dagger), \quad j = 1, 2 \tag{3.89}$$

---

[7] For simplicity we leave out the 'hats' on the operators in this example.

and
$$p_j = \frac{1}{\sqrt{2}} i(a_j^\dagger - a_j), \quad j = 1, 2. \tag{3.90}$$

Note that operators with different indices j commute, $[a_1^\dagger, a_2] = 0$, since $[x_1, p_2] = [x_2, p_1] = 0$.

The angular momentum L of the particle is perpendicular to the xy-plane. In terms of the variables x, y and p_x, p_y it is

$$L = x p_y - y p_x. \tag{3.91}$$

Now we introduce the dimensionless variables x_1, x_2 and p_1, p_2 in L, with the result

$$L = \hbar(x_1 p_2 - x_2 p_1). \tag{3.92}$$

By use of the transformation (3.89)-(3.90) we get

$$L = i\hbar(a_2^\dagger a_1 - a_1^\dagger a_2). \tag{3.93}$$

The transformation of the Hamiltonian is analogous to the one-dimensional case, cf. (2.91),

$$H = H_1 + H_2, \tag{3.94}$$

where

$$H_j = \hbar\omega(a_j^\dagger a_j + \frac{1}{2}). \tag{3.95}$$

The operator L does not commute with H_1 or H_2, since

$$[a_2^\dagger a_1 - a_1^\dagger a_2, a_1^\dagger a_1] = a_1 a_2^\dagger + a_1^\dagger a_2 \tag{3.96}$$

and

$$[a_2^\dagger a_1 - a_1^\dagger a_2, a_2^\dagger a_2] = -a_1 a_2^\dagger - a_1^\dagger a_2. \tag{3.97}$$

We note, however, that L commutes with $H = H_1 + H_2$ since the sum of the right hand sides in (3.96) and (3.97) is zero. This is in agreement with the vanishing of the classical Poisson bracket $\{L, H\}$ for the classical motion (cf. Example 2).

As shown in Example 2, the classical motion is characterized by two other constants of motion besides L and H, namely M and N given by (2.66-67). According to (3.89) and (3.90) the corresponding operators are

$$M = \hbar(a_1^\dagger a_1 - a_2^\dagger a_2) \tag{3.98}$$

and

$$N = \hbar(a_1 a_2^\dagger + a_2 a_1^\dagger). \tag{3.99}$$

From these expressions for H, L, M and N it follows with the use of

$$[a_i, a_j^\dagger] = \delta_{ij}, \quad [a_i, a_j] = 0, \tag{3.100}$$

that L, M and N commute with H. By contrast, the operators L, M and N do not commute between themselves. As an example we determine the commutator $[L, N]$. This is given by

$$i\hbar^2[a_2^\dagger a_1 - a_1^\dagger a_2, a_1 a_2^\dagger + a_2 a_1^\dagger] = -i 2\hbar M. \tag{3.101}$$

Basic principles

One finds in this manner that L, M and N satisfy the same commutation relations as the Pauli matrices (cf. (2.77)) and therefore cannot simultaneously be written in diagonal form.

The different states of the oscillator are given by the set of numbers (n_1, n_2), which determine the energy eigenvalues for each of the two independent oscillators,

$$E_{n_1} = (n_1 + \frac{1}{2})\hbar\omega, \quad E_{n_2} = (n_2 + \frac{1}{2})\hbar\omega. \tag{3.102}$$

The total energy is
$$E = E_{n_1} + E_{n_2}. \tag{3.103}$$

A state (eigenvector) corresponding to the quantum numbers (n_1, n_2) is denoted by $|n_1, n_2\rangle$. It may be thought of as a column vector which has all components equal to zero everywhere except the one associated with the set of numbers (n_1, n_2), the latter being 1 if the state is normalized (cf. (2.97)). When each component of the column vector is associated with a set of numbers (n_1, n_2) it is natural to proceed according to increasing total energy E. We therefore start with $(0,0)$ and continue with $(1,0)$, $(0,1)$ and further with $(2,0)$, $(1,1)$ and $(0,2)$, etc. Note that a given value of E in general corresponds to several linearly-independent state vectors. The energy levels are degenerate, and the degree of degeneracy is the number of linearly-independent state vectors associated with the given energy level. The ground state is non-degenerate, the second-lowest level is doubly degenerate, the third-lowest level triply degenerate, etc. The matrix for the total energy is diagonal in this basis, with values

$$\hbar\omega, 2\hbar\omega, 2\hbar\omega, 3\hbar\omega, 3\hbar\omega, 3\hbar\omega, \text{etc.}$$

along the diagonal. The matrix for M is also diagonal in this basis, with values

$$0, \hbar, -\hbar, 2\hbar, 0, -2\hbar, \text{ etc.},$$

its eigenvalues being $(n_1 - n_2)\hbar$.

Neither L nor N is diagonal, since for instance $a_1 a_2^\dagger$ moves an energy quantum from the x- to the y-oscillator according to $a_1 a_2^\dagger |1, 0\rangle = |0, 1\rangle$. As a result

$$L|1,0\rangle = i\hbar|0,1\rangle, \quad L|0,1\rangle = -i\hbar|1,0\rangle, \tag{3.104}$$

which shows that the L-matrix in the subspace associated with the energy value $E = 2\hbar\omega$ is $\hbar\sigma_y$, where σ_y is the Pauli matrix, cf. (2.76). In a similar way one finds that the N-matrix associated with this subspace is $\hbar\sigma_x$.

Let us ask the following question: Given that the particle is in the state labelled by the quantum numbers (n_1, n_2), what is the probability for observing a definite value of the angular momentum L? What is the mean value (the expectation value) of L, and what is the root-mean-square deviation? To answer these questions we use the relation (3.66) and expand the state vector $|n_1, n_2\rangle$ in the complete set of basis vectors $|n, m\rangle$ which are simultaneous eigenvectors for the total energy and the angular momentum,

$$H|n,m\rangle = (n+1)\hbar\omega|n,m\rangle, \quad L|n,m\rangle = m\hbar|n,m\rangle. \tag{3.105}$$

This yields
$$|n_1, n_2\rangle = \sum_{n,m} \langle n, m|n_1, n_2\rangle |n, m\rangle. \tag{3.106}$$

The probability P for observing the energy $(n+1)\hbar\omega$ and the angular momentum $m\hbar$ is therefore

$$P = |\langle n,m|n_1,n_2\rangle|^2. \qquad (3.107)$$

The expectation value of L in the state specified by $(n_1.n_2)$ is

$$<L> = \sum_{nm} |\langle n,m|n_1,n_2\rangle|^2 m\hbar =$$

$$\sum_{nm} \langle n_1,n_2|n,m\rangle m\hbar \langle n,m|n_1,n_2\rangle = \langle n_1,n_2|L|n_1,n_2\rangle. \qquad (3.108)$$

These manipulations show, in accordance with the general result (3.66), why the mean value in the state specified by (n_1, n_2) is given by $\langle n_1, n_2|L|n_1, n_2\rangle$. The mean value is given as a weighted sum of eigenvalues, each eigenvalue being weighted by the probability for observing this particular value in the state considered. Conversely, starting from the matrix element $\langle n_1, n_2|L|n_1, n_2\rangle$, one may insert the unit operator

$$1 = \sum_{nm} |n,m\rangle\langle n,m| \qquad (3.109)$$

between L and $|n_1, n_2\rangle$ and observe that the matrix element is identical to the weighted sum of eigenvalues.

As a concrete example let us consider the state vector $|1,0\rangle$, corresponding to $n_1 = 1, n_2 = 0$. Every probability amplitude $\langle n,m|1,0\rangle$ equals zero, unless $n = n_1 + n_2 = 1$, since state vectors associated with different energies are orthogonal. The amplitudes $\langle 1,m|1,0\rangle$, where $m = \pm 1$, are however different from zero. The matrix for L, when expressed in a basis consisting of the two states $|1,0\rangle$ and $|0,1\rangle$ (corresponding to $(n_1, n_2) = (1,0)$ and $(n_1, n_2) = (0,1)$ respectively), is given by $\hbar\sigma_y$, cf. (2.76), since

$$\langle 1,0|L|1,0\rangle = \langle 0,1|L|0,1\rangle = 0, \text{ and } \langle 1,0|L|0,1\rangle = -\langle 0,1|L|1,0\rangle = -i\hbar,$$

according to (3.104). The eigenvalues of the matrix σ_y are ± 1 and the linear combinations $(|1,0\rangle \pm i|0,1\rangle)/\sqrt{2}$ are therefore the normalized eigenstates $|1,\pm 1\rangle$ for L associated with the eigenvalues ± 1. As a consequence $\langle 1,1|1,0\rangle = \langle 1,-1|1,0\rangle = 1/\sqrt{2}$. We conclude that the probability for observing the particle with angular momentum $\hbar$ is $1/2$, while there is an equal probability $(1/2)$ for observing the particle with angular momentum $-\hbar$, given that the system is in a state corresponding to $n_1 = 1, n_2 = 0$. The expectation value of L in this state is therefore zero, while the mean square deviation $\Delta^2(L)$ is seen to be $\hbar^2$.

The transition from using the states $|n_1, n_2\rangle$ as a basis to using $|n, m\rangle$ may be viewed as a generalized rotation of the basis vectors in Hilbert space. To illustrate this we shall write down the matrices of the different physical quantities considered in this example for the two choices of basis.

Basic principles

1. **Basis:** $|n_1, n_2\rangle$

The basis vectors are

$$|0,0\rangle : \begin{pmatrix} 1 \\ 0 \\ 0 \\ 0 \\ 0 \\ 0 \\ \vdots \end{pmatrix} \quad |1,0\rangle : \begin{pmatrix} 0 \\ 1 \\ 0 \\ 0 \\ 0 \\ 0 \\ \vdots \end{pmatrix} \quad |0,1\rangle : \begin{pmatrix} 0 \\ 0 \\ 1 \\ 0 \\ 0 \\ 0 \\ \vdots \end{pmatrix}$$

$$|2,0\rangle : \begin{pmatrix} 0 \\ 0 \\ 0 \\ 1 \\ 0 \\ 0 \\ \vdots \end{pmatrix} \quad |1,1\rangle : \begin{pmatrix} 0 \\ 0 \\ 0 \\ 0 \\ 1 \\ 0 \\ \vdots \end{pmatrix} \quad |0,2\rangle : \begin{pmatrix} 0 \\ 0 \\ 0 \\ 0 \\ 0 \\ 1 \\ \vdots \end{pmatrix}.$$

The matrix for $H = H_1 + H_2$ is then given by

$$H/\hbar\omega : \begin{pmatrix} 1 & 0 & 0 & 0 & 0 & 0 & \cdots \\ 0 & 2 & 0 & 0 & 0 & 0 & \cdots \\ 0 & 0 & 2 & 0 & 0 & 0 & \cdots \\ 0 & 0 & 0 & 3 & 0 & 0 & \cdots \\ 0 & 0 & 0 & 0 & 3 & 0 & \cdots \\ 0 & 0 & 0 & 0 & 0 & 3 & \cdots \\ \vdots & \vdots & \vdots & \vdots & \vdots & & \ddots \end{pmatrix}$$

since the diagonal elements of H are $(n_1 + n_2)\hbar\omega$, while the matrix for $M = H_1 - H_2$ is given by

$$M/\hbar : \begin{pmatrix} 0 & 0 & 0 & 0 & 0 & 0 & \cdots \\ 0 & 1 & 0 & 0 & 0 & 0 & \cdots \\ 0 & 0 & -1 & 0 & 0 & 0 & \cdots \\ 0 & 0 & 0 & 2 & 0 & 0 & \cdots \\ 0 & 0 & 0 & 0 & 0 & 0 & \cdots \\ 0 & 0 & 0 & 0 & 0 & -2 & \cdots \\ \vdots & \vdots & \vdots & \vdots & \vdots & & \ddots \end{pmatrix}$$

since the diagonal elements of M are $(n_1 - n_2)\hbar$.

The matrices of the two operators L and N are *not* diagonal. The L-matrix is given by

$$L/\hbar : \begin{pmatrix} 0 & 0 & 0 & 0 & 0 & 0 & \cdots \\ 0 & 0 & -i & 0 & 0 & 0 & \cdots \\ 0 & i & 0 & 0 & 0 & 0 & \cdots \\ 0 & 0 & 0 & 0 & -i\sqrt{2} & 0 & \cdots \\ 0 & 0 & 0 & i\sqrt{2} & 0 & -i\sqrt{2} & \cdots \\ 0 & 0 & 0 & 0 & i\sqrt{2} & 0 & \cdots \\ \vdots & \vdots & \vdots & \vdots & \vdots & \vdots & \ddots \end{pmatrix}$$

The eigenvalues of L are the same as those of M, but the matrices are different since we have used the basis in which M is diagonal. Finally, the N-matrix is

$$N/\hbar : \begin{pmatrix} 0 & 0 & 0 & 0 & 0 & 0 & \cdots \\ 0 & 0 & 1 & 0 & 0 & 0 & \cdots \\ 0 & 1 & 0 & 0 & 0 & 0 & \cdots \\ 0 & 0 & 0 & 0 & \sqrt{2} & 0 & \cdots \\ 0 & 0 & 0 & \sqrt{2} & 0 & \sqrt{2} & \cdots \\ 0 & 0 & 0 & 0 & \sqrt{2} & 0 & \cdots \\ \vdots & \vdots & \vdots & \vdots & \vdots & \vdots & \ddots \end{pmatrix}.$$

Again, the eigenvalues are the same as those of M, but the matrix is non-diagonal.

2. **Basis:** $|n, m\rangle$

We might have chosen from the outset to work in the basis where H and L are diagonal. For reasons of symmetry L must have the same eigenvalues as those of M (just as the three Pauli matrices σ_x, σ_y and σ_z have the same eigenvalues ± 1). The L-matrix in the basis $|n, m\rangle$ is therefore

$$L/\hbar : \begin{pmatrix} 0 & 0 & 0 & 0 & 0 & 0 & \cdots \\ 0 & 1 & 0 & 0 & 0 & 0 & \cdots \\ 0 & 0 & -1 & 0 & 0 & 0 & \cdots \\ 0 & 0 & 0 & 2 & 0 & 0 & \cdots \\ 0 & 0 & 0 & 0 & 0 & 0 & \cdots \\ 0 & 0 & 0 & 0 & 0 & -2 & \cdots \\ \vdots & \vdots & \vdots & \vdots & \vdots & \vdots & \ddots \end{pmatrix}.$$

In the $|n, m\rangle$-basis H is

$$H/\hbar\omega : \begin{pmatrix} 1 & 0 & 0 & 0 & 0 & 0 & \cdots \\ 0 & 2 & 0 & 0 & 0 & 0 & \cdots \\ 0 & 0 & 2 & 0 & 0 & 0 & \cdots \\ 0 & 0 & 0 & 3 & 0 & 0 & \cdots \\ 0 & 0 & 0 & 0 & 3 & 0 & \cdots \\ 0 & 0 & 0 & 0 & 0 & 3 & \cdots \\ \vdots & \vdots & \vdots & \vdots & \vdots & \vdots & \ddots \end{pmatrix}$$

Basic principles

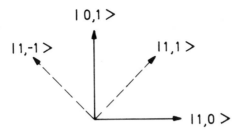

Figure 3.2: Rotation in Hilbert space (symbolic).

while N and M are given by

$$
N/\hbar : \begin{pmatrix}
0 & 0 & 0 & 0 & 0 & 0 & \cdots \\
0 & 0 & -i & 0 & 0 & 0 & \cdots \\
0 & i & 0 & 0 & 0 & 0 & \cdots \\
0 & 0 & 0 & 0 & -i\sqrt{2} & 0 & \cdots \\
0 & 0 & 0 & i\sqrt{2} & 0 & -i\sqrt{2} & \cdots \\
0 & 0 & 0 & 0 & i\sqrt{2} & 0 & \cdots \\
\vdots & \vdots & \vdots & \vdots & \vdots & \vdots & \ddots
\end{pmatrix}
$$

and

$$
M/\hbar : \begin{pmatrix}
0 & 0 & 0 & 0 & 0 & 0 & \cdots \\
0 & 0 & 1 & 0 & 0 & 0 & \cdots \\
0 & 1 & 0 & 0 & 0 & 0 & \cdots \\
0 & 0 & 0 & 0 & \sqrt{2} & 0 & \cdots \\
0 & 0 & 0 & \sqrt{2} & 0 & \sqrt{2} & \cdots \\
0 & 0 & 0 & 0 & \sqrt{2} & 0 & \cdots \\
\vdots & \vdots & \vdots & \vdots & \vdots & \vdots & \ddots
\end{pmatrix}.
$$

In Fig. 3.2 we have illustrated the change of basis as a rotation of the axes of the system of coordinates.

3.3.3 Probability amplitudes and the Schrödinger equation

We conclude the present section by showing how the Schrödinger equation is derived from the postulate of Section 3.2 concerning the development in time of the state vector $|\psi(t)\rangle$,

$$i\hbar \frac{d}{dt}|\psi(t)\rangle = \hat{H}|\psi(t)\rangle. \tag{3.110}$$

For simplicity we consider the motion in one dimension of a single particle of mass m, but the derivation may readily be generalized to motion in three dimensions and to systems containing several particles. The classical Hamiltonian of the particle is thus assumed to be given by

$$H = \frac{p^2}{2m} + V(x). \tag{3.111}$$

In Section 3.3 we identified Schrödinger's wave function with a probability amplitude. In the present case of motion of a single particle in one dimension, the wave function $\psi(x,t)$ is equal to the probability amplitude $\langle x|\psi(t)\rangle$,

$$\psi(x,t) = \langle x|\psi(t)\rangle. \tag{3.112}$$

The equation of motion for $\psi(x,t)$ is obtained from (3.110) by taken the inner product of each side with the state vector $|x\rangle$ and inserting the unit operator

$$\hat{1} = \int_{-\infty}^{\infty} dx' |x'\rangle\langle x'| \tag{3.113}$$

on the right hand side. The result is

$$i\hbar \frac{\partial \psi(x,t)}{\partial t} = \int_{-\infty}^{\infty} dx' \langle x|\hat{H}|x'\rangle \langle x'|\psi(t)\rangle$$
$$= \int_{-\infty}^{\infty} dx' \langle x|\frac{\hat{p}^2}{2m}|x'\rangle \langle x'|\psi(t)\rangle + \int_{-\infty}^{\infty} dx' \langle x|V|x'\rangle \langle x'|\psi(t)\rangle. \tag{3.114}$$

Since

$$\langle x|V|x'\rangle = V(x)\delta(x-x') \tag{3.115}$$

the last of the two terms in (3.114) equals $V(x)\psi(x,t)$. The kinetic energy term is rewritten by using that $\langle x|p\rangle$ is a wave function describing a particle with definite momentum p,

$$\langle x|p\rangle = \frac{1}{\sqrt{2\pi\hbar}} e^{ipx/\hbar}. \tag{3.116}$$

By inserting the unit operator

$$\hat{1} = \int_{-\infty}^{\infty} dp' |p'\rangle\langle p'| \tag{3.117}$$

to the right and to the left of the operator $\hat{p}^2$, we obtain

$$i\hbar \frac{\partial \psi(x,t)}{\partial t} - V(x)\psi(x,t) =$$
$$\int_{-\infty}^{\infty} dp' \int_{-\infty}^{\infty} dp'' \int_{-\infty}^{\infty} dx' \langle x|p'\rangle \langle p'|\frac{\hat{p}^2}{2m}|p''\rangle \langle p''|x'\rangle \langle x'|\psi(t)\rangle =$$
$$\frac{1}{2\pi\hbar} \int_{-\infty}^{\infty} dp' \int_{-\infty}^{\infty} dp'' \delta(p'-p'') \frac{p'^2}{2m} \int_{-\infty}^{\infty} dx' e^{ip'x/\hbar} e^{-ip''x'/\hbar} \psi(x',t). \tag{3.118}$$

Basic Principles

We now complete the integral over p'' and write (3.118) in the form

$$i\hbar\frac{\partial\psi(x,t)}{\partial t} - V(x)\psi(x,t) =$$
$$-\frac{\hbar^2}{2m}\frac{1}{2\pi\hbar}\int_{-\infty}^{\infty}dp'\int_{-\infty}^{\infty}dx'\frac{\partial^2}{\partial x'^2}e^{ip'(x-x')/\hbar}\psi(x',t). \quad (3.119)$$

Integrating by parts with respect to x', assuming that the contributions from the limits $x = \pm\infty$ vanish, and using the identity (cf. App. C)

$$\int_{-\infty}^{\infty}dp'e^{i(x-x')p'/\hbar} = 2\pi\hbar\delta(x-x') \quad (3.120)$$

to complete the integration over x', we arrive at the Schrödinger equation

$$i\hbar\frac{\partial\psi(x,t)}{\partial t} = -\frac{\hbar^2}{2m}\frac{\partial^2}{\partial x^2}\psi(x,t) + V(x)\psi(x,t). \quad (3.121)$$

We have thus derived the Schrödinger equation from (3.110), which describes the development in time of the state vector. We could equally well have started out by postulating the Schrödinger equation (3.121), but the postulate (3.110) is more general.

3.4 Problems

PROBLEM 3.1
The eigenvalue problem for the harmonic oscillator may be treated by a polynomial method, which is used commonly for solving differential equations.
As discussed in Section 3.1.1 we seek solutions of the type

$$u = H(\xi)e^{-\xi^2/2}, \quad (3.122)$$

where H is a polynomial in ξ.
a) Show that H must satisfy the equation

$$H'' - 2\xi H' + (\lambda - 1)H = 0. \quad (3.123)$$

The equation (3.123) is solved by writing H as a series expansion

$$H(\xi) = \xi^s\sum_{\nu=0}^{\infty}a_\nu\xi^\nu, \quad (3.124)$$

where s is greater than or equal to zero, and $a_0 \neq 0$. According to (3.123) the series expansion of the left hand side of the equation must equal zero, implying that the coefficient of each power of ξ in the expansion of the left hand side is equal to zero.

b) Use this condition to show that $a_{\nu+2}/a_\nu \to 2/\nu$ for $\nu \to \infty$, unless the series breaks off for some value of ν, and show that this ratio also appears in the series expansion of $\xi^p e^{\xi^2}$, where p is an arbitrary exponent. Use this to prove that the series (3.124) must terminate at a finite value of ν, if the wave function is to be square integrable.

c) Show that the series (3.124) terminates, if

$$\lambda = 2s + 2\nu + 1, \tag{3.125}$$

and that the condition $a_0 \neq 0$ implies that s is either zero or 1. It therefore follows from (3.125) that λ must be an odd integer, $\lambda = 2n + 1$, in order that $a_{n-s+2}, a_{n-s+4}, \cdots$ are equal to zero. Consequently $n - s$ must be even since the series (3.124) contains terms with even ν. As the series of terms with odd ν does not terminate as a result of the condition (3.125), a_1 must be set equal to zero in order that the coefficients associated with the odd values of ν vanish.

d) Use the result from c) to prove that the energy eigenvalues are given by (2.103) and determine the normalized wave functions corresponding to $n = 0, 1$ and 2. The quantum number n is the largest value of $s + \nu$ in the series for H. The corresponding polynomial H given by (3.124) is denoted by H_n. It is named a Hermite polynomial. Note that H_n is either an even or an odd function of ξ, depending on whether n is even or odd.

Problem 3.2

Use the operators $\hat{a}$ and $\hat{a}^\dagger$ in (2.83-84) to show that $\Delta(x)\Delta(p)$ for the harmonic oscillator equals $(n + \frac{1}{2})\hbar$ in the state $|n\rangle$.

Problem 3.3

Determine for the two-dimensional isotropic harmonic oscillator the mean value and the root-mean-square of the angular momentum in the state $|0, 2\rangle$ corresponding to $n_1 = 0, n_2 = 2$.

Problem 3.4

Prove the recursion relations (3.45) and (3.46) by writing ξ and $d/d\xi$ in terms of the operators $\hat{a}$ and $\hat{a}^\dagger$ and using that $\hat{a}|n\rangle$ is proportional to $|n-1\rangle$, while $\hat{a}^\dagger|n\rangle$ is proportional to $|n+1\rangle$.

Problem 3.5

A particle of mass M moves in a three-dimensional harmonic oscillator potential given by $V(x, y, z) = K(x^2 + y^2 + z^2)/2$.

Basic principles

a) Show that the Hamiltonian may be written in the form

$$\hat{H} = \frac{3}{2}\hbar\omega + \sum_{i=1,2,3} \hbar\omega \hat{a}_i^\dagger \hat{a}_i \qquad (3.126)$$

and write down the commutation relations between the creation and annihilation operators $\hat{a}_i^\dagger$ and $\hat{a}_j$.

b) Determine the spectrum and the degree of degeneracy for the three lowest energy levels.

c) Show that the Schrödinger equation separates in the cartesian coordinates and sketch how the method of Problem 3.1 may be used to determine the energy levels and the corresponding wave functions. Determine the wave functions associated with the two lowest energy levels.

PROBLEM 3.6
Show that (3.32) satisfies the four conditions (3.1)-(3.4).

PROBLEM 3.7
A particle with mass M moves in the potential given by

$$V(x) = \frac{1}{2}Kx^2 + Ax^4, \qquad (3.127)$$

where A is a positive constant.

a) Use (2.110-111) for expressing $(\hat{a} + \hat{a}^\dagger)^2|0\rangle$ as a superposition of states $|n\rangle$.

b) Determine with the help of the result from a) the mean value of Ax^4 in the state $|0\rangle$. Compare with the result of the dimensional analysis, Problem 1.12.

c) Show that you obtain the same result for the mean value of x^4 by weighting x^4 with the probability density for the ground-state wave function and integrating the expression over all x.

PROBLEM 3.8
Determine the Fourier transform of the function

$$\psi(x) = e^{iqx} \text{ for } |x| < a, \; \psi(x) = 0 \text{ for } |x| > a, \qquad (3.128)$$

as well as the function

$$\psi(x) = \frac{a}{a^2 + x^2}, \qquad (3.129)$$

where a is a positive constant (use a table of integrals).

Problem 3.9

A particle moves in one dimension. At a certain time it is described by the wave function

$$\psi(x) = \sqrt{\frac{2a^3}{\pi}} \frac{1}{a^2 + x^2} e^{ikx}. \qquad (3.130)$$

a) Find the momentum probability density for this state and use it to determine the mean value $<p>$ and the mean square deviation $\Delta^2(p)$ (use a table of integrals).

b) Determine the product $\Delta(x)\Delta(p)$ and compare with the ground state for a harmonic oscillator.

Problem 3.10

The Hamiltonian for a two-dimensional harmonic oscillator is

$$H = -\frac{\hbar^2}{2m}\left(\frac{\partial^2}{\partial x^2} + \frac{\partial^2}{\partial y^2}\right) + \frac{1}{2}(K_1 x^2 + K_2 y^2), \qquad (3.131)$$

where K_1 and K_2 are positive constants.

a) Show that the solutions ψ of the Schrödinger equation

$$H\psi = E\psi \qquad (3.132)$$

may be separated according to

$$\psi = u_{n_1}(x) u_{n_2}(y) \qquad (3.133)$$

where n_1 and n_2 are positive integers. The function $u_n(x)$ denotes the solution of the Schrödinger equation for a one-dimensional harmonic oscillator.

b) Show by inserting the separated solution (3.133) in (3.132) that the energy eigenvalues are given by

$$E = \hbar\omega_1\left(n_1 + \frac{1}{2}\right) + \hbar\omega_2\left(n_2 + \frac{1}{2}\right), \qquad (3.134)$$

where $n_1, n_2 = 0, 1, 2 \cdots$, and express the frequencies ω_1 and ω_2 in terms of the constants of the problem.

Problem 3.11

The Hamiltonian of a particle moving in two dimensions is

$$H = -\frac{\hbar^2}{2m}\left(\frac{\partial^2}{\partial x^2} + \frac{\partial^2}{\partial y^2}\right) + \frac{1}{2}K(x^2 + y^2) + Axy, \qquad (3.135)$$

where K is a positive constant, while the constant A may be positive or negative.

Basic principles

a) Use a change of variables to show that the Hamiltonian may be written as the sum of the Hamiltonians for two independent oscillators, provided that $|A| < K$. Show that the energy eigenvalues under these circumstances are given by

$$E = \hbar\omega_1(n_1 + \frac{1}{2}) + \hbar\omega_2(n_2 + \frac{1}{2}), \qquad (3.136)$$

where $n_1, n_2 = 0, 1, 2 \cdots$. Express the frequencies ω_1 and ω_2 in terms of the given constants and compare with Problem 3.10.

Problem 3.12
Show that the functions $u = \exp(\xi^2/2)$ and $u = \xi \exp(\xi^2/2)$ are solutions to (3.33) and determine the corresponding values of λ. Do these values have physical significance?

Problem 3.13
In this problem we shall show that the root-mean-square product $\Delta(A)\Delta(B)$ associated with the Hermitian operators $\hat{A}$ and $\hat{B}$ cannot be less than $\hbar/2$ if $\hat{A}$ and $\hat{B}$ satisfy the commutation rule

$$[\hat{A}, \hat{B}] = i\hbar. \qquad (3.137)$$

An example of the relation (3.137) is the commutation rule between position and momentum, cf. (2.71).

a) Show that the inequality $\langle h|h\rangle \geq 0$ with

$$h = f - g\frac{\langle g|f\rangle}{\langle g|g\rangle} \qquad (3.138)$$

implies that

$$\langle f|f\rangle\langle g|g\rangle \geq |\langle f|g\rangle|^2. \qquad (3.139)$$

We define the operators $\hat{\alpha}$ and $\hat{\beta}$ by

$$\hat{\alpha}\psi = (\hat{A} - <A>)\psi \text{ and } \hat{\beta}\psi = (\hat{B} - <B>)\psi, \qquad (3.140)$$

where ψ is an arbitrary wave function.

b) Identify f with $\hat{\alpha}\psi$ and g with $\hat{\beta}\psi$, and use (3.139) to show that

$$\Delta(A)\Delta(B) \geq |\langle\psi|\hat{\alpha}\hat{\beta}\psi\rangle|. \qquad (3.141)$$

c) Use the identity

$$\hat{\alpha}\hat{\beta} = \frac{1}{2}(\hat{\alpha}\hat{\beta} + \hat{\beta}\hat{\alpha}) + \frac{1}{2}(\hat{\alpha}\hat{\beta} - \hat{\beta}\hat{\alpha}) \qquad (3.142)$$

to deduce that
$$\Delta(A)\Delta(B) \geq \frac{\hbar}{2} \tag{3.143}$$
from the inequality (3.141) (hint: Note that the operator $(\hat{\alpha}\hat{\beta} - \hat{\beta}\hat{\alpha})$ is an anti-Hermitian operator since it is (-1) times its Hermitian conjugate, and use that $\langle\psi|(\hat{\alpha}\hat{\beta} - \hat{\beta}\hat{\alpha})\psi\rangle$ as a consequence is purely imaginary).

d) Discuss how the inequality (3.143) is changed if the right hand side of (3.137) is replaced by an arbitrary anti-Hermitian operator $\hat{C}$.

Problem 3.14
Indicate which of the following operators $\hat{O}$ are linear,
$$\hat{O}f = f(x) + x, \quad \hat{O}f = -d^2 f/dx^2, \quad \hat{O}f = (df/dx)^2,$$
where f is a function of x.

Problem 3.15
In the following we introduce an inner product by the definition
$$\langle f_i | f_j \rangle = \int_0^\pi \sin\theta d\theta \int_0^{2\pi} d\phi f_i^*(\theta, \phi) f_j(\theta, \phi),$$
with θ being the polar angle and ϕ the azimuthal angle in a polar coordinate-system. We seek to determine matrices $A_{ij} = \langle f_i | A f_j \rangle$ associated with different operators A. Later on we shall encounter these operators in the context of the theory of angular momentum (Chapter 7).

1. Let us consider the subspace spanned by the three functions f_1, f_2 and f_3, which are proportional to
$$\sin\theta e^{i\phi}, \quad \cos\theta, \quad \sin\theta e^{-i\phi},$$
respectively. Determine normalization constants such that
$$\langle f_i | f_j \rangle = \delta_{ij}.$$

2. Find the matrices associated with each of the two operators
$$-i\frac{\partial}{\partial\phi}, \quad -\frac{\partial^2}{\partial\phi^2}.$$

3. Determine the matrices associated with each of the two operators A and B given by
$$A = e^{i\phi}\left(\frac{\partial}{\partial\theta} + i\cot\theta\frac{\partial}{\partial\phi}\right), \quad B = e^{-i\phi}\left(-\frac{\partial}{\partial\theta} + i\cot\theta\frac{\partial}{\partial\phi}\right).$$

4. Use the results from 3. to determine the matrices associated with each of the two operators
$$\frac{1}{2}(A+B); \quad \frac{1}{2i}(A-B)$$
and determine the eigenvalues of these matrices.

5. Determine the commutation relations between the matrices found in 2. and 4. above.

4 THE SCHRÖDINGER EQUATION

In the following we discuss some properties of the time-dependent Schrödinger equation, which is the fundamental equation of motion in the quantum theory, corresponding to Newton's equations in classical mechanics. Like Newton's equations the validity of the Schrödinger equation is limited to situations where relativistic effects may be neglected.

We shall see how the time-dependent Schrödinger equation is reduced to the eigenvalue equation for the energy under stationary conditions. The eigenstates that are solutions to this equation are called stationary states, since the associated probability densities are independent of time. By forming superpositions of stationary states we are able to describe time-dependent phenomena and compare with the result of solving the classical equations of motion. Although a given state may not be an eigenstate for the energy operator, it may be an eigenstate for other operators associated with, say, momentum or angular momentum. As we shall see in Example 4 below, the state of a system may also be characterized by being an eigenstate for a non-Hermitian operator. The special feature of the eigenstates of the energy operator is the simplicity of their development in time, as demonstrated below.

We consider a single particle with mass m moving in the potential $V(\mathbf{r})$. As shown in Section 3.3.3 of the previous chapter, the time-dependent Schrödinger equation has the form

$$-\frac{\hbar^2}{2m}\nabla^2\psi(\mathbf{r},t) + V(\mathbf{r})\psi(\mathbf{r},t) = i\hbar\frac{\partial\psi(\mathbf{r},t)}{\partial t}, \tag{4.1}$$

where the left-hand side is $\hat{H}\psi$ with $\hat{H}$ being the Hamiltonian.

In general the wave function $\psi(\mathbf{r},t)$ is not necessarily an eigenstate of the Hamiltonian and may therefore not be labelled by an energy eigenvalue. Its physical interpretation is that of a probability amplitude. Thus

$$|\psi(\mathbf{r},t)|^2 d\mathbf{r} \tag{4.2}$$

is the probability that a measurement of the position of the particle yields a result in the volume element $d\mathbf{r}(=dxdydz)$ at $\mathbf{r}$.

Since the potential energy $V(\mathbf{r})$ is independent of time, it is possible to find solutions to (4.1) which separate according to

$$\psi = u_E(\mathbf{r})e^{-iEt/\hbar}. \tag{4.3}$$

Here $u_E(\mathbf{r})$ denotes the spatial part of the wave function, which must satisfy $\hat{H}u_E = Eu_E$ in order that (4.3) is a solution to (4.1). The state (4.3) is called a stationary state, since the absolute value of ψ is independent of time. In a state characterized by a definite energy the absolute square of the probability amplitude is therefore independent of time, implying that the probability of the

The Schrödinger equation

result of a measurement of the corresponding physical quantity is independent of time. This is in harmony with classical physics, in so far as the energy is a constant of motion for a system characterized by a Hamiltonian, which has no explicit dependence on time, cf. (2.51-52).

The general solution of (4.1) is

$$\psi = \sum_E c_E u_E(\mathbf{r}) e^{-iEt/\hbar}, \tag{4.4}$$

where $\hat{H} u_E = E u_E$, and the constants c_E may be determined from the knowledge of the wave function at time $t = 0$. We have assumed that the energy eigenstates form a complete basis. If some of the energy eigenstates belong to a continuum of eigenvalues, the summation in (4.4) includes an integration over this continuum.

The time-dependent Schrödinger equation (4.1) applies to the motion of a single particle, but it may readily be generalized to the case where a system is specified by a large number of generalized coordinates. As an example we may take the conduction electrons (approximately 10^{22}) contained in one cubic centimetre of copper (Chapter 9). In this case the Hamiltonian is the sum of the kinetic energy of the conduction electrons and the potential energy due to their mutual repulsion together with their interaction with the positively charged ions. The ability in such a case to write down the Schrödinger equation in analogy with (4.1) does not, of course, imply that one is able to solve it.

It is important when solving the Schrödinger equation to take into account the possible existence of symmetries. Some of these are discussed in Chapter 8. Often it is necessary to add terms without a classical analogue to the Hamiltonian in the Schrödinger equation. An example of this is given by the coupling between the orbital angular momentum of a particle and its spin, which is a relativistic effect (Example 7). It should also be mentioned that the Hamilton operator may not be uniquely determined by the knowledge of the classical Hamiltonian. The quantization is based on the Poisson brackets (2.44) for the generalized coordinates and momenta, the left hand side being replaced by the commutator for the relevant operators and the right hand side by $i\hbar \delta_{ij}$. Now, if the classical Hamiltonian contains products of generalized coordinates and generalized momenta, the transition to the Hamilton operator is not sufficiently defined if the relevant operators do not commute. Under such circumstances one writes the Hamiltonian in a symmetric form in the classical variables before replacing these by operators, corresponding to the substitution $ab \to (\hat{a}\hat{b} + \hat{b}\hat{a})/2$.

The general form of the time-dependent Schrödinger equation is thus

$$\hat{H} \psi = i\hbar \frac{\partial \psi}{\partial t}, \tag{4.5}$$

where the Hamiltonian $\hat{H}$ of the system under consideration is derived from the classical Hamiltonian as described above, with the possible addition of relativistic correction terms.

The relativistic equation for a single electron, known as the Dirac equation, is a linear equation like the Schrödinger equation, but the wave function has in this case four components, which means that (4.5) is replaced by a 4×4 matrix differential equation, cf. (1.140). In the limit where the relativistic effects may be treated as small correction terms, it is sufficient to take into account two of the four components of the wave function. This introduces the column vectors which will be used in Chapter 7 to represent the electron spin, cf. (2.79-80).

4.1 Ehrenfest's theorem

Quantum mechanics reduces to classical mechanics in an appropriate limit. To elucidate the nature of this limit we shall derive an expression for the time derivative of an arbitrary physical quantity. It follows from the Schrödinger equation

$$\hat{H}\psi = i\hbar \frac{\partial \psi}{\partial t} \tag{4.6}$$

by complex conjugation that

$$\hat{H}\psi^* = -i\hbar \frac{\partial \psi^*}{\partial t}. \tag{4.7}$$

The time derivative of the expectation value $<A> = \langle \psi | \hat{A} \psi \rangle$ of an arbitrary physical quantity A is

$$\frac{d<A>}{dt} = \langle \psi | \hat{A} \frac{\partial \psi}{\partial t} \rangle + \langle \frac{\partial \psi}{\partial t} | \hat{A} \psi \rangle + \langle \psi | \frac{\partial \hat{A}}{\partial t} \psi \rangle. \tag{4.8}$$

Here we have taken into account that the operator $\hat{A}$ may depend explicitly on time. By inserting (4.6-7) in (4.8) and using that $\hat{H}$ is Hermitian we obtain the result

$$\frac{d<A>}{dt} = \frac{1}{i\hbar} < [\hat{A}, \hat{H}] > + < \frac{\partial \hat{A}}{\partial t} >. \tag{4.9}$$

This is *Ehrenfest's theorem*. We shall use (4.9) to demonstrate how the mean values of the momentum and the position of a particle in a suitable limit obey the classical equations of motion.

For simplicity we consider a particle with mass m which moves in one dimension under the influence of the potential energy $V(x)$. Since $[\hat{x}, \hat{H}] = i\hbar \hat{p}/m$ and $[\hat{p}, \hat{H}] = -i\hbar dV/dx$, we conclude from (4.9) that

$$\frac{d<x>}{dt} = \frac{<p>}{m}, \tag{4.10}$$

The Schrödinger equation 109

and
$$\frac{d<p>}{dt} = -<\frac{dV}{dx}>. \qquad (4.11)$$

If the potential energy is nearly constant in the part of space where the wave function is substantially different from zero, then we are allowed to make the replacement $<dV/dx> \simeq dV(<x>)/d<x>$, which shows that $<x>$ and $<p>$ under this condition obey classical equations of motion.

4.2 The superposition principle

The Schrödinger equation (4.5) is a linear differential equation. Therefore, if ψ_1 and ψ_2 are solutions of the equation, so is the superposition $\psi_1 + \psi_2$. Such a superposition principle is fundamental to quantum mechanics and applies even when the physical system may not be characterized by a wave function obeying the Schrödinger equation. In its general form the superposition principle states the following: If $|n_1\rangle$ and $|n_2\rangle$ both are possible states of the system, with n_1 and n_2 denoting a set of quantum numbers, then the superpositions $c_1|n_1\rangle + c_2|n_2\rangle$ also represent possible states of the system, provided the coefficients c_1 and c_2 satisfy the normalization condition $|c_1|^2 + |c_2|^2 = 1$.

Let us consider a superposition of two states ψ_{n_1} and ψ_{n_2}, both of which are eigenstates for the energy of a one-dimensional harmonic oscillator with eigenvalues E_{n_1} and E_{n_2}. We assume that $n_1 \neq n_2$, which ensures that the two energies are different. The absolute square of the superposition $\psi = \psi_{n_1} + \psi_{n_2}$ therefore varies in time according to

$$|\psi|^2 = |\psi_{n_1}|^2 + |\psi_{n_2}|^2 + 2u_{n_1}u_{n_2}\cos(E_{n_1} - E_{n_2})t/\hbar. \qquad (4.12)$$

Here $\psi_n = u_n \exp(-iE_n t/\hbar)$, and $u_n = u_n(x)$ has been chosen to be real. The probability density is thus periodic in time, with the period determined by the energy difference $(E_{n_1} - E_{n_2})$. The period is the longest, equal to $2\pi/\omega$ where ω is the classical angular frequency, when the difference between the two energies is $\hbar\omega$, corresponding to $n_1 = n_2 \pm 1$.

EXAMPLE 4. THE CLASSICAL LIMIT FOR A HARMONIC OSCILLATOR.

In this example we shall construct a *coherent state* from a superposition of energy eigenstates $|n\rangle$ for a harmonic oscillator. Such states play an important role in quantum optics. The superposition is formed in such a manner that the absolute square of the wave function ψ is a Gaussian of constant width at all times, while the position of the maximum oscillates in time with the classical angular frequency. The coherent state forms a minimum wave packet in the sense that it satisfies the uncertainty relation (3.84) at any time. We shall see in the following how such a state may be constructed by requiring it to be an eigenstate of the annihilation operator $\hat{a}$.

In Chapter 2 we introduced the operators $\hat{a}$ and $\hat{a}^\dagger$, the sum of which is proportional to $\hat{q}$ according to (2.83-84),

$$\hat{q} = \sqrt{\frac{\hbar}{2M\omega}}(\hat{a} + \hat{a}^\dagger). \tag{4.13}$$

Let us assume that the system is in a state $|\beta\rangle$ which is an eigenstate of $\hat{a}$ with eigenvalue β. The mean value $\langle\beta|\hat{q}|\beta\rangle$ of the position of the particle is then given by $(\beta + \beta^*)\sqrt{\hbar/2M\omega}$, since $\langle\beta|\hat{a}^\dagger = \langle\beta|\beta^*$. The eigenvalue β may be complex, since $\hat{a}$ is not a Hermitian operator. We write the state $|\beta\rangle$ as a superposition

$$|\beta\rangle = \sum_{n=0}^{\infty} c_n |n\rangle, \tag{4.14}$$

and seek to determine the expansion coefficients c_n by requiring that $|\beta\rangle$ is an eigenstate for $\hat{a}$. In order that the superposition (4.14) may satisfy the Schrödinger equation, the time dependence of the coefficients c_n must be given by

$$c_n \propto \exp(-iE_n t/\hbar). \tag{4.15}$$

Furthermore we note that $\hat{a}$ according to (2.111) has two effects when acting on $|n\rangle$: i) It changes the quantum number from n to $n - 1$, and ii) it multiplies the state by $\sqrt{n}$. We therefore require that the coefficients satisfy the recursion relation

$$c_n = c_{n-1} \frac{\beta}{\sqrt{n}}, \tag{4.16}$$

whereby $c_n \hat{a}|n\rangle = c_{n-1}\beta|n-1\rangle$, so that $|\beta\rangle$ satisfies the eigenvalue equation

$$\hat{a}|\beta\rangle = \beta|\beta\rangle. \tag{4.17}$$

The superposition (4.14) is a solution to the time-dependent Schrödinger equation (4.5), if the time dependence of β is given by

$$\beta = |\beta|e^{-i\omega t + i\phi}, \tag{4.18}$$

since $\hbar\omega$ is the energy difference between the n'th and $(n-1)$'th energy level. In (4.18) we have also introduced an arbitrary phase ϕ such that β at time $t = 0$ is given by

$$\beta = |\beta|e^{i\phi}. \tag{4.19}$$

As we shall see below the two constants $|\beta|$ and ϕ may be related to the solution (2.2) of the classical equation of motion. By using (4.17) and its Hermitian conjugate it is seen that the mean value of $\hat{H}$ is

$$\langle\beta|\hat{H}|\beta\rangle = \hbar\omega(|\beta|^2 + \frac{1}{2}). \tag{4.20}$$

Evidently $|\beta|^2$ is the mean number of energy quanta in the state $|\beta\rangle$.

The Schrödinger equation

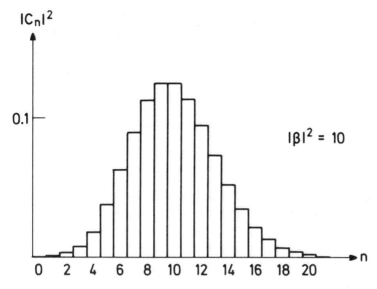

Figure 4.1: Poisson distribution.

The quantity $|c_0|^2$ may be determined from the recursion relation (4.16) and the normalization condition

$$\sum_{n=0}^{\infty} |c_n|^2 = 1. \tag{4.21}$$

According to (4.16), $|c_n|^2 = |c_0|^2 |\beta|^{2n}/n!$. By inserting this in (4.21) we obtain

$$|c_0|^{-2} = \sum_{n=0}^{\infty} \frac{|\beta|^{2n}}{n!} = e^{|\beta|^2}. \tag{4.22}$$

The absolute square of the expansion coefficient c_n may therefore be written as

$$|c_n|^2 = e^{-|\beta|^2} \frac{|\beta|^{2n}}{n!}. \tag{4.23}$$

The distribution (4.23) is called a Poisson distribution. It is shown in Fig. 4.1 for the case when the mean number of energy quanta is 10.

According to (4.13) and (4.18) the mean value $<q> = \langle \beta|\hat{q}|\beta \rangle$ of the position of the particle in the state $|\beta\rangle$ is therefore

$$<q> = |\beta|\sqrt{\frac{2\hbar}{M\omega}} \cos(\omega t - \phi). \tag{4.24}$$

In the classical limit, $|\beta| \gg 1$, the dimensionless quantity $|\beta|\sqrt{2}$ thus denotes the maximum departure of the oscillator from its equilibrium position, in units of the characteristic length $\sqrt{\hbar/M\omega}$, cf. (2.2). By forming the expectation value of the operator $\hat{p}$, which is proportional to $i(\hat{a} - \hat{a}^\dagger)$, we obtain

$$<p> = M\frac{d<q>}{dt} \tag{4.25}$$

corresponding to Hamilton's equation (2.56). Similarly, we get

$$\frac{d<p>}{dt} = -K<q> \tag{4.26}$$

which corresponds to (2.57).

It is also possible to find higher moments of the distribution (4.23) besides the mean value of the position and the momentum of the particle. The mean-square deviation $\Delta^2(H)$ is given by

$$\Delta^2(H) = <H^2> - <H>^2, \tag{4.27}$$

according to the general definition (3.83). We shall determine the value of (4.27) in two different ways. The first method is based directly on the use of the Poisson distribution (4.23). The second and more elegant one utilizes the commutation relations between $\hat{a}$ and $\hat{a}^\dagger$.

Let us first use the Poisson distribution (4.23). By changing the summation index from n to $\nu = n - 1$ and using (4.16) it is seen that

$$\sum_{n=0}^{\infty} |c_n|^2 n = \sum_{\nu=0}^{\infty} |c_\nu|^2 |\beta|^2 = |\beta|^2. \tag{4.28}$$

Similarly one finds

$$\sum_{n=0}^{\infty} |c_n|^2 n(n-1) = |\beta|^4. \tag{4.29}$$

With the use of (4.28-29) we may now determine the mean value of the square of the Hamiltonian,

$$<H^2> = (\hbar\omega)^2 \sum_{n=0}^{\infty} |c_n|^2 (n+\frac{1}{2})^2 = (\hbar\omega)^2(|\beta|^4 + 2|\beta|^2 + \frac{1}{4}). \tag{4.30}$$

The square of the mean value (4.20) is $<H>^2 = (\hbar\omega)^2(|\beta|^4 + |\beta|^2 + 1/4)$, which implies that

$$<H^2> = <H>^2 + (\hbar\omega)^2 |\beta|^2. \tag{4.31}$$

From (4.31) it follows that the ratio of the root-mean-square deviation to the mean value is given by

$$\frac{\Delta(H)}{<H>} = \frac{|\beta|}{|\beta|^2 + \frac{1}{2}}, \tag{4.32}$$

The Schrödinger equation

which shows that the ratio is negligible when the mean value of the energy is much larger than $\hbar\omega$.

Instead of using the Poisson distribution we may derive the result (4.30) for the mean value of $\hat{H}^2$ and the expression (4.32) for the root-mean-square deviation by using the commutation relations for the creation and annihilation operators together with (4.17). It follows from the commutation relations (2.89) that

$$(\hat{a}^\dagger \hat{a} + \frac{1}{2})^2 = \hat{a}^\dagger \hat{a} \hat{a}^\dagger \hat{a} + \hat{a}^\dagger \hat{a} + \frac{1}{4} = \hat{a}^\dagger \hat{a}^\dagger \hat{a} \hat{a} + 2\hat{a}^\dagger \hat{a} + \frac{1}{4}. \tag{4.33}$$

The mean value of $\hat{H}^2/(\hbar\omega)^2$ therefore equals the mean value of $(\hat{a}^\dagger\hat{a}^\dagger\hat{a}\hat{a}+2\hat{a}^\dagger\hat{a}+1/4)$, which is seen by the use of (4.17) and its Hermitian conjugate to be $(|\beta|^4+2|\beta|^2+1/4)$ in agrement with (4.30).

The mean-square deviations for position and momentum are determined in a similar way. They are found to satisfy the relation

$$\Delta(x)\Delta(p) = \frac{\hbar}{2}. \tag{4.34}$$

When the product of $\Delta(x)$ and $\Delta(p)$ is given by (4.34), the state is called a minimal wave packet. The ground state $|0\rangle$ of the oscillator is also a minimal wave packet, since it corresponds to the eigenvalue $\beta = 0$. The relation (4.34) is a special case of the Heisenberg uncertainty relations, according to which the product is always greater than or equal to $\hbar/2$. For an energy eigenstate $|n\rangle$ the product is found to be $\Delta(x)\Delta(p) = (n + \frac{1}{2})\hbar$, which only agrees with (4.34) for $n = 0$ (Problem 3.2).

While the expectation value of q in the state $|\beta\rangle$ varies in time according to (4.24), the root-mean-square product (4.34) is seen to be independent of time. As shown in Problem 4.4 the absolute square of the wave function has the same form as the absolute square of the ground-state wave function (3.47) for the harmonic oscillator. If the system is described by such a minimal wave packet at some instant of time, its subsequent development in time is very simple; the probability density oscillates with the classical frequency without changing its form. This is a special feature of the harmonic oscillator potential and does not, for instance, hold for a free particle, cf. Problem 4.1.

4.3 Boundary conditions

The time-independent Schrödinger equation is a second-order differential equation in the spatial variables. Its solution requires the boundary conditions to be specified. We have already used - in connection with the discussion of the harmonic oscillator in Section 3.1.1 - that the wave function had to vanish at infinity, $x \to \pm\infty$. This boundary condition arose from the requirement that the wave function should be normalizable, which meant that the solutions growing exponentially in the limit $x \to \pm\infty$ had to be discarded (cf. (3.34)). It is not always possible to require that the wave function vanishes at infinity. The momentum eigenstates, which are proportional to $\exp(i\mathbf{p}\cdot\mathbf{r}/\hbar)$, are examples of wave functions that are non-zero everywhere in space. This is

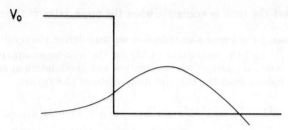

Figure 4.2: Potential step and wave function.

related to the existence of a continuum of momentum eigenvalues, while the energy eigenvalues for a harmonic oscillator are given by the discrete spectrum (2.103).

In the present section we shall discuss in more general terms the boundary conditions on the solutions of the Schrödinger equation and introduce the useful concept of the density of states.

Let us first examine the boundary conditions in the simplest possible case, when a particle of mass m moves in a piecewise constant potential $V(x)$ defined by

$$V(x) = V_0 \quad \text{for} \quad x < 0, \quad V(x) = 0 \quad \text{for} \quad x > 0, \tag{4.35}$$

where V_0 is a positive constant. This potential step is illustrated in Fig. 4.2. Since the time-independent Schrödinger equation is a second-order differential equation in x, its solution must be continuous and possess a continuous derivative for all x, including $x = 0$.

The continuity of the wave function and its derivatives with respect to its spatial variables is a general property of the Schrödinger equation. The continuity of the derivatives may be shown by integrating the equation once, while the continuity of the wave function itself follows from a second integration.

We shall consider solutions that belong to energy eigenvalues less than the height of the potential step, $E < V_0$. In the two regions they are

$$\psi = A \cos kx + B \sin kx \quad \text{for} \quad x > 0, \tag{4.36}$$

where $k^2 = 2mE/\hbar^2$ and

$$\psi = Ce^{\kappa x} \quad \text{for} \quad x < 0, \tag{4.37}$$

where $\kappa = (2m(V_0 - E))^{1/2}/\hbar$. Since the wave function should not diverge for $x \to -\infty$, we have discarded the solution proportional to $\exp(-\kappa x)$ in the region $x < 0$ (see Fig. 4.2).

The wave function must be continuous in $x = 0$, and we may therefore conclude that $A = C$. Similarly we obtain from (4.36-37) that $\kappa C = kB$ is the

The Schrödinger equation

condition ensuring the continuity of the derivative. In the limit $V_0 \gg E$, the coefficient C therefore vanishes, which means that the wave function in the region $x > 0$ is proportional to $\sin kx$. This wave function is not an eigenstate of the momentum operator, but a superposition of eigenstates according to

$$\sin kx = \frac{1}{2i}(e^{ikx} - e^{-ikx}). \tag{4.38}$$

We conclude that the wave function vanishes in the region where the potential energy is infinite.

In quantum mechanics it is often useful to imagine that the system under consideration is contained in a box represented by walls of infinitely large potential energy. An example of this arises in the description of the conduction electrons in a metal (cf. Chapter 9), where the box represents the macroscopic volume occupied by the metal. Under such circumstances it is advantageous to use periodic boundary conditions instead of the ordinary boundary condition described above, namely that the wave function should vanish at the wall. The periodic boundary conditions imply that the value of the wave function at a point situated on one wall is the same as the value at the corresponding point on the opposite wall. The advantage of using these periodic boundary conditions is that they enable one to use eigenstates of the momentum operator. This simplifies the applications of quantum mechanics to macroscopic systems such as metals or semiconductors.

In the following we discuss the motion of a particle in a one-dimensional box and compare the use of the conventional boundary conditions with the periodic conditions. As we shall see, the use of the two different boundary conditions leads to the same physical result when the box is chosen sufficiently large, so that the energy levels are closely spaced.

We shall thus determine the solutions to the Schrödinger equation for a particle with mass m moving in a potential $V(x)$ given by

$$V(x) = 0, \quad 0 < x < L; \quad V = \infty, \quad x > L \text{ and } x < 0. \tag{4.39}$$

For simplicity the potential energy has been set equal to zero in the interior of the box. The Schrödinger equation in the region $0 < x < L$ becomes

$$-\frac{d^2\psi}{dx^2} = k^2\psi, \tag{4.40}$$

where $k^2 = 2mE/\hbar^2$. The complete solution of (4.40) is

$$\psi = Ae^{ikx} + Be^{-ikx}. \tag{4.41}$$

First we assume that the wave function of the particle is zero on the 'wall' of the one-dimensional box, at $x = 0$ and $x = L$. The resulting wave function is proportional to $\sin kx$, where k must satisfy the condition

$$k = n\frac{\pi}{L}, n = 1, 2, 3 \cdots. \tag{4.42}$$

The corresponding energy eigenvalues are

$$E_n = n^2 \frac{\hbar^2 \pi^2}{2mL^2}. \tag{4.43}$$

For an electron moving in a region of size $L = 1$ cm, the difference between the n'th and the $(n-1)$'th energy level is very small. For $n \gg 1$ the difference equals $n\hbar^2\pi^2/mL^2$ which is seen to be n times 10^{-33} joule, a very small energy even when n is large. In practice it is therefore permissible to consider the energy eigenstates as a continuum and introduce a density of states as follows: The number of states Δn in the interval Δk is according to (4.42) given by $\Delta n = L\Delta k/\pi$. The number $\Gamma(E)$ of states with energy less than E is seen from (4.43) to be equal to

$$\Gamma(E) = \frac{L}{\pi}\sqrt{\frac{2mE}{\hbar^2}}. \tag{4.44}$$

The density of states $g(E)$ is defined by expressing the number of states with energy between E and $E + dE$ as $g(E)dE$. From this it follows that

$$g(E) = \frac{d\Gamma}{dE}. \tag{4.45}$$

With the use of (4.44-45) we conclude that

$$g(E) = \frac{L}{\pi}\sqrt{\frac{m}{2E\hbar^2}}. \tag{4.46}$$

In the case under consideration the density of states thus decreases with increasing energy. This is due to the restriction of the motion to one dimension. For motion in three dimensions the density of states is proportional to the square root of the energy, as we shall presently see.

We now generalize our discussion to two and three dimensions, using the same boundary condition that the wave function vanishes on the walls of the box. In two dimensions the particle moves in the region defined by $0 < x < L$ and $0 < y < L$. The wave function has the form $\sin(\pi n_x x/L)\sin(\pi n_y y/L)$, where n_x and n_y are positive integers. The number of states with energy less than E is given by the area enclosed by a circle of radius kL/π in the first quadrant of the $n_x n_y$-plane,

$$\Gamma(E) = \frac{1}{4}\pi k^2 \frac{L^2}{\pi^2}, \tag{4.47}$$

where

$$k^2 = \frac{2mE}{\hbar^2}. \tag{4.48}$$

The Schrödinger equation

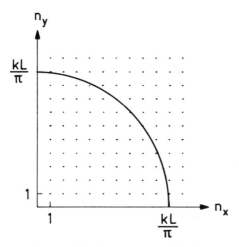

Figure 4.3: Illustration of (4.47).

The factor of 1/4 in (4.47) is due to the restriction of n_x and n_y to positive integers (Fig. 4.3), since a state obtained by replacing for instance n_x with $-n_x$ is linearly dependent on the original state (the states only differ by a sign). The density of states defined by (4.45) is obtained by differentiating (4.47) with respect to energy and is therefore independent of energy.

For motion in three dimensions we obtain

$$\Gamma(E) = \frac{1}{8}\frac{4\pi}{3}k^3\frac{L^3}{\pi^3}, \qquad (4.49)$$

where k as before is given by (4.48). It is seen from (4.49) that the density of states in the three-dimensional case increases with energy as the square root of E, since k^3 is proportional to $E^{3/2}$, yielding the density of states $g(E) = (L^3/\pi^2\hbar^3)\sqrt{m^3E/2}$.

It should be noted that we have disregarded the existence of electron spin in determining the density of states. Since the spin quantum number may assume two values (Chapter 7), the results above should be multiplied by 2 when applied to electrons or other spin-1/2 particles.

Now let us turn to the use of periodic boundary conditions. The complete solution of the Schrödinger equation for motion of a free particle in one dimension has been given in (4.41). Instead of demanding that the wave function vanishes on the 'wall' given by $x = 0$ and $x = L$ we apply the periodic boundary condition

$$\psi_{x=0} = \psi_{x=L}. \qquad (4.50)$$

Such a boundary condition might be realized physically by joining the two ends of the one-dimensional region together to form a ring. A similar construction is not possible in three dimensions. It is therefore better to consider the use of periodic boundary conditions to be an approximation which is sufficiently accurate to be of use for determining bulk properties[1]. We shall now demonstrate that the use of the periodic boundary conditions (4.50) yields the same expression for the density of states as that given in (4.46).

The complete solution of the Schrödinger equation (4.40) is a superposition of states that are eigenfunctions of the momentum operator with eigenvalues given by $\hbar k$ and $-\hbar k$, respectively. We now wish to describe the motion of the particle in terms of normalized states that are eigenfunctions of the momentum operator, corresponding to the wave functions

$$\psi = \frac{1}{\sqrt{L}} e^{ikx}. \qquad (4.51)$$

A wave function such as (4.51) cannot become zero at $x = 0$ and $x = L$. The use of the periodic boundary condition (4.50) yields

$$k = n\frac{2\pi}{L}, \, n = 0, \pm 1, \pm 2, \cdots. \qquad (4.52)$$

Note that the interval between the allowed values of k according to (4.52) is twice that given by (4.42). However, the number of states within the energy interval $\Delta \epsilon$ is the same in the two cases, since k according to (4.52) assumes both positive and negative values. When the density of states is determined from (4.52), one therefore recovers the result (4.46).

To illustrate the use of periodic boundary conditions we shall complete our discussion of the particle in the box by determining again the density of states, this time by starting from eigenstates of the momentum operator. To emphasize the role of the dimension we formulate the problem in d dimensions and consider the density of states in the cases $d = 1, 2$ and 3.

In order to find the density of states we start from the eigenstates for the momentum operator which are proportional to $\exp(i\mathbf{k} \cdot \mathbf{r})$, and label the states according to

$$|\mathbf{k}\rangle. \qquad (4.53)$$

The corresponding eigenvalues of the momentum operator are $\hbar \mathbf{k}$, where $\mathbf{k}$ is a d-dimensional vector. The use of periodic boundary conditions corresponding to (4.50) implies that each of the components of $\mathbf{k}$ satisfies a condition such as (4.52). We choose the sides of the box to be of equal length, given by L, though this is of course not necessary. In the general case the volume of the box (which in the present case is L^d) enters the result for the density of states.

---

[1] For macroscopic systems, the surface contribution to, for instance, the total energy is normally of negligible importance compared to the contribution from the bulk.

The Schrödinger equation

The normalized wave functions associated with the state vectors (4.53) are

$$\psi = \frac{1}{\sqrt{L^d}} \exp(i\mathbf{k} \cdot \mathbf{r}). \tag{4.54}$$

In the applications of quantum mechanics one often has to sum over states labelled by $\mathbf{k}$. Such a sum is changed into an integral according to the prescription

$$\sum_{\mathbf{k}} \cdots = \frac{L^d}{(2\pi)^d} \int d\mathbf{k} \cdots, \tag{4.55}$$

since the relation (4.52) holds for each component of the vector $\mathbf{k}$. When the integral (4.55) in $\mathbf{k}$-space is changed to an integral over energy, the dimension is seen to play an important role in determining the energy dependence of the density of states $g(E)$. In three dimensions we obtain

$$\frac{L^3}{(2\pi)^3} \int d\mathbf{k} \cdots = \frac{L^3}{(2\pi)^3} 4\pi \int_0^\infty dk\, k^2 \cdots = \int_0^\infty dE\, g(E) \cdots. \tag{4.56}$$

Here we have assumed for simplicity that the integrand $(\cdots)$ only depends on the length of $\mathbf{k}$, thereby allowing one to multiply the differential dk by the surface area $4\pi k^2$ when making the transition to polar coordinates, $dk_x dk_y dk_z \to 4\pi k^2 dk$. Using the connection (4.48) between E and k we find

$$g(E) = \frac{L^3 m k(E)}{2\pi^2 \hbar^2} = \sqrt{\frac{m^3 E}{2}} \frac{L^3}{\pi^2 \hbar^3}, \quad d = 3. \tag{4.57}$$

In two dimensions the density of states is constant, since $dk_x dk_y \to 2\pi k\, dk$. This results in

$$g(E) = \frac{L^2}{2\pi} \frac{m}{\hbar^2}, \quad d = 2, \tag{4.58}$$

while the density of states in one dimension is

$$g(E) = \frac{L}{2\pi} 2 \frac{m}{\hbar^2 k(E)} = \frac{L}{\pi\hbar} \sqrt{\frac{m}{2E}}, \quad d = 1. \tag{4.59}$$

in agreement with the result (4.46). Note that the transition from the differential dk to dE in one dimension requires multiplying by 2, since the one-dimensional wave vector assumes both positive and negative values, cf. (4.52). As mentioned earlier, spin has not been taken into account in these results which must be multiplied by 2, if they are applied to electrons in solids (Chapter 9).

4.4 Summary

The present section summarizes the basic elements of quantum mechanics as introduced in the present and previous chapters.

Observables

Physical observables such as energy, momentum or angular momentum are represented by Hermitian operators or matrices with eigenvalues giving the possible results of a measurement of the corresponding physical quantity.

States

A physical system is characterized by a state vector (ket), which is a simultaneous eigenvector for one or more observables represented by matrices which commute with each other. A complete characterization of the state of the system is given by the set of quantum numbers specifying the eigenvalues of the matrices. Alternatively the system may be characterized by a wave function ψ, which is a simultaneous eigenfunction for operators that commute with each other.

Superposition

Given that the states $|\nu_1\rangle$ and $|\nu_2\rangle$ both are possible states of a system, then the linear combination $c_1|\nu_1\rangle + c_2|\nu_2\rangle$ also represents a possible physical state. The coefficients c_1 and c_2 are arbitrary except for the normalization condition $|c_1|^2 + |c_2|^2 = 1$, while ν_1 and ν_2 denote a set of quantum numbers.

Probability amplitudes

A physical state $|\nu\rangle$ may be expanded in a complete set of eigenstates for a matrix (or an operator) associated with a definite physical quantity. The absolute square of the expansion coefficients (the probability amplitudes) yield the probability $P(\alpha)$ (or the probability density) for measuring the value α for the corresponding physical quantity, given that the system is in state $|\nu\rangle$, $P(\alpha) = |\langle\alpha|\nu\rangle|^2$.

Mean value

The mean value (or expectation value) of a physical quantity A in the state $|\nu\rangle$ is given by the matrix element $\langle\nu|A|\nu\rangle$. If the state of the system is specified by the wave function ψ, the mean value is the inner product of ψ and $A\psi$.

Momentum and position

When the state of a particle is given by a wave function $\psi(\mathbf{r},t)$, where $\mathbf{r}$ denotes the cartesian coordinates of the particle, the momentum operator is represented by the differential operator $(\hbar/i)\boldsymbol{\nabla}$.

The Schrödinger equation

The wave function develops in time according to the Schrödinger equation $H\psi = i\hbar\partial\psi/\partial t$, where H in the non-relativistic limit is the Hamilton operator corresponding to the classical Hamiltonian.

4.5 Problems

PROBLEM 4.1
Show for a free particle with mass M that

$$\frac{d\Delta^2(x)}{dt} = \frac{<xp+px>}{M} - 2\frac{<x><p>}{M}, \tag{4.60}$$

and

$$\frac{d^2\Delta^2(x)}{dt^2} = 2\frac{\Delta^2(p)}{M^2}. \tag{4.61}$$

Consider a free particle which at time $t = 0$ is in a state with a given value of $\Delta^2(p)$, and show from the results given above that the mean-square deviation $\Delta^2(x)$ for a free particle is proportional to t^2 for large values of t.

PROBLEM 4.2
A particle with a mass of 10 g executes harmonic motion with an amplitude of 1 cm and a period of 1 s. Determine the mean number of energy quanta.

PROBLEM 4.3
In this problem we consider a particle of mass M moving in a harmonic oscillator potential $V(x) = Kx^2/2$. As usual ω denotes the classical angular frequency.

a) Show that the wave function

$$\psi(x,t) = c_0 e^{-iE_0 t/\hbar} e^{-M\omega x^2/2\hbar} + c_1 e^{-iE_1 t/\hbar} x e^{-M\omega x^2/2\hbar}, \tag{4.62}$$

where E_n is given by (2.103), satisfies the time-dependent Schrödinger equation. Here c_0 and c_1 are constants chosen such that ψ is normalized (verify that the integral over x of $|\psi|^2$ from $-\infty$ to ∞ is independent of time).

b) Sketch the probability density at times $t = 0$ and $t = \pi/\omega$, when $c_1 = \sqrt{M\omega/\hbar}c_0$. Show that the probability density varies periodically in time and determine the period.

PROBLEM 4.4
In the present problem we first determine the ground-state wave function for the harmonic oscillator by means of the operator equation corresponding to (2.102), and subsequently the wave function describing the superposition (4.14)

(the connection between the state vector $|\beta\rangle$ and the wave function $\psi(x,t)$ is $\psi(x,t) = \langle x|\beta\rangle$ as discussed in the previous chapter).

According to (2.83) the differential operator $\hat{a}$ is

$$\hat{a} = \sqrt{\frac{M\omega}{2\hbar}}(x + \frac{\hbar}{M\omega}\frac{d}{dx}). \tag{4.63}$$

a) Verify that the ground-state wave function (3.47) satisfies the first order differential equation
$$\hat{a}u_0(x) = 0 \tag{4.64}$$
where $\hat{a}$ is given by (4.63).

b) Find the solution $\psi(x,t)$ to the first-order differential equation corresponding to (4.17)
$$(\hat{a} - \beta)\psi = 0, \tag{4.65}$$
where the time dependence originates in β as given by (4.18).

c) Show that the absolute square of the normalized solution found in b) is given by
$$|\psi(x,t)|^2 = (a\sqrt{\pi})^{-1} e^{-(x-\langle x\rangle)^2/a^2}, \tag{4.66}$$
where $<x>$ is the mean value corresponding to (4.24).

PROBLEM 4.5
A particle of mass m moves in a potential given by

$$V(x) = \frac{1}{2}Kx^2 \text{ for } x > 0, \quad V(x) = \infty \text{ for } x < 0, \tag{4.67}$$

where K is a positive constant.

a) Determine the ground-state energy of the system and sketch the ground-state wave function.

b) At time $t = 0$ the potential energy is suddenly changed into

$$V(x) = \frac{1}{2}Kx^2 \text{ for all } x. \tag{4.68}$$

The wave function at $t = 0$ is assumed to be that determined in a). What is the probability that a subsequent measurement (at time $t > 0$) results in the value $\hbar\omega/2$ for the energy of the particle? What is the probability for measuring the value $3\hbar\omega/2$?

PROBLEM 4.6
A particle with mass m moves in the xy-plane within a region defined by the inequalities
$$0 \leq x \leq L_1; \quad 0 \leq y \leq L_2. \tag{4.69}$$

Within this region the potential V is a constant, $V = 0$, while it is infinitely large outside. It is assumed that the wave function of the particle is zero at the boundaries of the region.

a) Determine the ground-state energy and the corresponding wave function.

b) Indicate on an energy axis the four lowest energy eigenvalues and the corresponding quantum numbers in the case $L_1 = 2L_2$.

c) Repeat b) in the case $L_1 = L_2$.

5 TUNNELLING

Some of the most surprising consequences of quantum mechanics appear as tunnelling effects. By tunnelling is meant the penetration of energy barriers that separate one region in space from another region which is inaccessible from a classical point of view. Tunnelling is responsible for such diverse phenomena as the α-decay of nuclei and the possibility of having an electrical current flow through a thin insulating layer between two metals. In the present chapter we first discuss bound states for motion in a potential well. Then we introduce the concept of a probability current which will allow us to discuss the penetration of simple one-dimensional barriers.

5.1 Bound states

When a particle moves in a time-independent potential there exist stationary-state solutions ψ of the Schrödinger equation, characterized by the absolute square $|\psi|^2$ being independent of time. We shall study here a simple example of motion in one dimension in a potential given by

$$V(x) = V_0 \quad \text{for} \quad |x| > \frac{a}{2}, \quad V(x) = 0 \quad \text{for} \quad |x| < \frac{a}{2}, \qquad (5.1)$$

where V_0 is a positive constant (Fig. 5.1). The potential (5.1) is called a potential well.

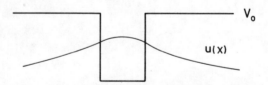

Figure 5.1: Potential well and wave function.

The time-dependent Schrödinger equation is

$$\hat{H}\psi = -\frac{\hbar^2}{2m}\frac{\partial^2 \psi(x,t)}{\partial x^2} + V(x)\psi(x,t) = i\hbar\frac{\partial \psi}{\partial t}. \qquad (5.2)$$

Since the potential $V(x)$ is independent of time, it is possible to find stationary-state solutions of the Schrödinger equation of the form

$$\psi(x,t) = u(x)e^{-iEt/\hbar} \qquad (5.3)$$

Tunnelling

in accordance with (4.3). By inserting (5.3) into (5.2) and dividing by the common factor $\exp(-iEt/\hbar)$ we obtain the time-independent Schrödinger equation in the form $\hat{H}u = Eu$ or

$$-\frac{\hbar^2}{2m}\frac{d^2u}{dx^2} + V_0 u = Eu \quad \text{for} \quad |x| > \frac{a}{2}, \tag{5.4}$$

and

$$-\frac{\hbar^2}{2m}\frac{d^2u}{dx^2} = Eu \quad \text{for} \quad |x| < \frac{a}{2}. \tag{5.5}$$

According to the Schrödinger equation, the second derivative of u is discontinuous at $x = \pm a/2$, since the potential energy has a discontinuity at these two points. The function u itself and its derivative are continuous. We introduce the quantities k and κ by the definitions

$$k = \frac{\sqrt{2mE}}{\hbar}, \quad \kappa = \frac{\sqrt{2m(V_0 - E)}}{\hbar}. \tag{5.6}$$

Rather than attempting to find the general solution of the Schrödinger equation we shall here examine the bound states. A bound state is characterized by a wave function which decreases exponentially at infinity, $x \to \pm\infty$. In the present case the two linearly-independent solutions to (5.4) are given by $\exp(\pm \kappa x)$. When κ is real, one of these two solutions decreases exponentially at infinity[1]. Such a solution is square integrable, in the sense that the integral $\int_{-\infty}^{\infty} dx |\psi|^2$ exists. As a necessary condition for the existence of bound states we therefore have $0 \leq E < V_0$. The bound-state solutions of the Schrödinger equation in the two regions $x > a/2$ and $x < -a/2$ are thus given by

$$u = Ae^{-\kappa x} \quad \text{for} \quad x > \frac{a}{2} \tag{5.7}$$

and

$$u = Be^{\kappa x} \quad \text{for} \quad x < -\frac{a}{2}, \tag{5.8}$$

where A and B are arbitrary constants.

In the region $-a/2 < x < a/2$ the general solution may be written as a linear combination of $\sin kx$ and $\cos kx$. Let us first investigate whether there exist even solutions that are bound states. We therefore choose

$$u = C \cos kx \tag{5.9}$$

---

[1] If κ is purely imaginary, $\kappa = i|\kappa|$ corresponding to $E > V_0$, the solutions are not square integrable, since the absolute value of $\exp(\pm i|\kappa|x)$ is 1. For $E > V_0$ there exist two linearly-independent solutions for each value of E in the interval $V_0 < E < \infty$.

as the solution in this region. Both the solution $u(x)$ and its derivative du/dx must be continuous in $x = a/2$ and $x = -a/2$. The continuity of u at $x = a/2$ yields

$$C\cos(ka/2) = Ae^{-\kappa a/2} \tag{5.10}$$

while the continuity of du/dx results in

$$-kC\sin(ka/2) = -\kappa A e^{-\kappa a/2} \tag{5.11}$$

When (5.11) is divided by (5.10) we obtain

$$k\frac{\sin(ka/2)}{\cos(ka/2)} = \kappa. \tag{5.12}$$

The number of solutions of this equation depends on the magnitude of the dimensionless parameter $ma^2V_0/\hbar^2$. The associated energies may for instance be obtained graphically by intersecting the family of curves $\tan ka/2$ with κ/k (Problem 5.2).

The energy of the lowest-lying bound state may be determined by squaring (5.12) with the result

$$E = V_0 - E\frac{\sin^2(a\sqrt{2mE}/2\hbar)}{\cos^2(a\sqrt{2mE}/2\hbar)}, \tag{5.13}$$

It is always possible to find a solution of this equation, no matter how small V_0 is. For $V_0 \ll \hbar^2/ma^2$ we may solve (5.13) by iteration. To first order in V_0 the solution of (5.13) is given by $E = V_0$. After inserting this first-order result on the right hand side of (5.13) we obtain the energy E to second order in V_0,

$$E \simeq V_0(1 - \frac{ma^2V_0}{2\hbar^2}). \tag{5.14}$$

The numerical solution of (5.13) is shown in Fig. 5.2. We note that a bound state exists for any value of V_0. In Fig. 5.1 we have illustrated a wave function corresponding to the value $E/V_0 = 0.770$ in the case $V_0ma^2/\hbar^2 = 0.649$.

By examining the odd solutions in a similar fashion, one finds that bound states associated with odd solutions only exist if the parameter V_0 exceeds a certain critical value (Problem 5.2).

For energies E larger than V_0 the states are not square integrable. The wave functions do not decrease exponentially, and the possible energy values form a continuum, extending from V_0 to infinity.

5.2 Probability current

It is a consequence of the Schrödinger equation for a single particle that the probability density satisfies a continuity equation, which relates the divergence

Tunnelling

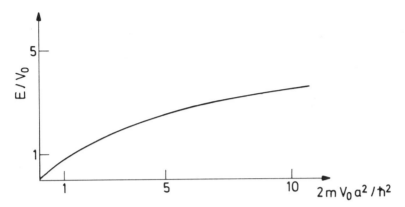

Figure 5.2: Ground state energy for a potential well.

of a probability current density to the time derivative of the probability density. To show this we introduce the probability current density $\mathbf{j}(\mathbf{r},t)$ by the definition

$$\mathbf{j}(\mathbf{r},t) = \frac{\hbar}{2mi}(\psi^*\boldsymbol{\nabla}\psi - \psi\boldsymbol{\nabla}\psi^*). \tag{5.15}$$

We shall now verify that this probability current density satisfies the continuity equation

$$\frac{\partial(\psi^*\psi)}{\partial t} + \boldsymbol{\nabla}\cdot\mathbf{j} = 0, \tag{5.16}$$

as a consequence of the Schrödinger equation for the motion of a particle with mass m in the potential $V(\mathbf{r})$. The continuity equation (5.16) is verified by inserting the expression (5.15) into the probability current density in (5.16). The left hand side then becomes

$$\frac{\partial(\psi^*\psi)}{\partial t} + \boldsymbol{\nabla}\cdot\mathbf{j} = \psi^*(\frac{\partial\psi}{\partial t} + \frac{\hbar}{2mi}\nabla^2\psi) + \psi(\frac{\partial\psi^*}{\partial t} - \frac{\hbar}{2mi}\nabla^2\psi^*), \tag{5.17}$$

which is zero because of the Schrödinger equation (4.6) and its complex conjugate (4.7), the Hamiltonian being $\hat{H} = \hat{p}^2/2m + V(\mathbf{r})$.

It follows from the continuity equation (5.16), that the integral

$$\int d\mathbf{r}|\psi|^2$$

over all space of the absolute square of ψ is a constant in time, provided the wave function ψ vanishes sufficiently rapidly at infinity. This may be seen by integrating (5.16) over a volume V, turning the volume integral of $\boldsymbol{\nabla}\cdot\mathbf{j}$ into

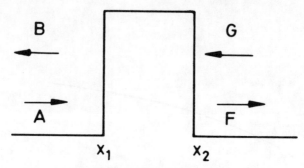

Figure 5.3: Potential barrier.

a surface integral of the normal component of **j** according to the divergence theorem of Gauss,

$$\int_V d\mathbf{r}\, \boldsymbol{\nabla} \cdot \mathbf{j} = \int_S d\mathbf{S} \cdot \mathbf{j}.$$

The integral over the surface S vanishes when the surface recedes to infinity, provided the wave function is zero or approaches zero sufficiently rapidly at infinity.

In the following we shall be interested in one-dimensional motion only and therefore consider stationary states of the form

$$\psi = Ae^{ikx}, \qquad (5.18)$$

where A is a constant. The probability current density associated with the wave function (5.18) is $\mathbf{j} = (j, 0, 0)$, where

$$j = |A|^2 \frac{\hbar k}{m}. \qquad (5.19)$$

Note that the probability current density associated with a real wave function is zero.

5.3 Barrier transmission

We shall use a state of the form (5.18) with k positive, corresponding to a particle moving towards a barrier in the positive direction of the x-axis. We shall see that quantum mechanics allows a particle to penetrate a barrier which is impenetrable from a classical point of view. In order to simplify the mathematical treatment of this tunnelling problem we shall assume that the potential V is given by

$$V(x) = V_0 \quad \text{for} \quad x_1 \leq x \leq x_2, \quad V(x) = 0 \quad \text{otherwise,} \qquad (5.20)$$

Tunnelling

where V_0 is a positive constant (Fig. 5.3). To facilitate the generalization to the case of several barriers we have introduced the two coordinates x_1 and x_2.

The transmission coefficient of the barrier is obtained from the general solution of the time-independent Schrödinger equation $\hat{H}\psi = E\psi$. This solution is

$$\psi(x) = Ae^{ikx} + Be^{-ikx} \quad \text{for} \quad x < x_1 \tag{5.21}$$

in the region to the left of the barrier (see Fig. 5.3), where the energy eigenvalue E is related to the wave number k through $E = \hbar^2 k^2/2m$. In the region to the right of the barrier the solution is

$$\psi(x) = Fe^{ikx} + Ge^{-ikx} \quad \text{for} \quad x > x_2. \tag{5.22}$$

In the barrier itself, for $x_1 < x < x_2$, it is

$$\psi = Ce^{-\kappa x} + De^{\kappa x}, \tag{5.23}$$

where $\kappa^2 = 2m(V_0 - E)/\hbar^2$. The six constants of integration are related by the conditions that the wave function and its derivative are continuous.

In the following we shall only consider the case in which κ is real, corresponding to energies E less than the barrier height V_0. The calculations may readily be taken over to the case where E is greater than V_0, the only change being that κ becomes purely imaginary (cf. Problem 5.3).

Each of the wave functions (5.21)-(5.23) are seen to satisfy the Schrödinger equation in the corresponding region. In addition we must ensure that the boundary conditions are satisfied, since the wave function and its derivative should be continuous at $x = x_1$ and $x = x_2$. This yields

$$Ae^{ikx_1} + Be^{-ikx_1} = Ce^{-\kappa x_1} + De^{\kappa x_1}, \tag{5.24}$$

and

$$ikAe^{ikx_1} - ikBe^{-ikx_1} = -\kappa Ce^{-\kappa x_1} + \kappa De^{\kappa x_1} \tag{5.25}$$

from the matching of the wave function and its derivative at $x = x_1$. At $x = x_2$ the matching conditions give

$$Ce^{-\kappa x_2} + De^{\kappa x_2} = Fe^{ikx_2} + Ge^{-ikx_2} \tag{5.26}$$

and

$$-\kappa Ce^{-\kappa x_2} + \kappa De^{\kappa x_2} = ikFe^{ikx_2} - ikGe^{-ikx_2}. \tag{5.27}$$

We now eliminate C and D from the system of equations (5.24)-(5.27) and obtain

$$\begin{pmatrix} A \\ B \end{pmatrix} = t(x_1, x_2) \begin{pmatrix} F \\ G \end{pmatrix} \tag{5.28}$$

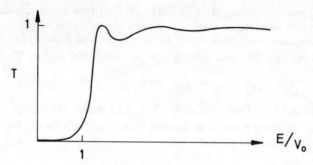

Figure 5.4: Transmission coefficient as a function of energy for a barrier with $V_0 a^2 m/\hbar^2 = 12.5$.

where $t(x_1, x_2)$ is a 2×2 matrix,

$$t = \begin{pmatrix} c_{11} & c_{12} \\ c_{21} & c_{22} \end{pmatrix}. \tag{5.29}$$

The elements in the matrix are given by

$$c_{11} = c_{22}^* = \frac{1}{4} e^{ik(x_2-x_1)}[(1+i\frac{\kappa}{k})(1-i\frac{k}{\kappa})e^{\kappa(x_2-x_1)} + (1-i\frac{\kappa}{k})(1+i\frac{k}{\kappa})e^{-\kappa(x_2-x_1)}] \tag{5.30}$$

and

$$c_{12} = c_{21}^* = \frac{1}{4} e^{-ik(x_2+x_1)}[(1+i\frac{\kappa}{k})(1+i\frac{k}{\kappa})e^{\kappa(x_2-x_1)} + (1-i\frac{\kappa}{k})(1-i\frac{k}{\kappa})e^{-\kappa(x_2-x_1)}]. \tag{5.31}$$

The equations (5.28) and (5.29) relate the wave function to the right of the barrier to the wave function to the left of the barrier. Let us now examine the particular case in which the wave function to the right of the barrier corresponds to a particle moving towards the right, which implies that $G = 0$. This describes the experimental situation in which a particle current is sent towards the barrier from the left with one part of the current being reflected and another part transmitted. It follows from the Schrödinger equation, in agreement with the continuity equation (5.16), that the current approaching from the left (which is proportional to $|A|^2$) is equal to the sum of the transmitted current (which is proportional to $|F|^2$) and the reflected current (which is proportional to $|B|^2$).

Since we are interested in obtaining the transmission coefficient T, defined by

$$T = \frac{|F|^2}{|A|^2}, \tag{5.32}$$

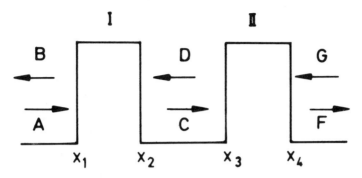

Figure 5.5: Quantum well.

and the reflection coefficient

$$R = \frac{|B|^2}{|A|^2}, \qquad (5.33)$$

the ratios F/A and B/A are determined from (5.28) and (5.29). It will turn out, as expected, that $T = 1 - R$ in agreement with the continuity equation (5.16).

When G is set equal to zero, the ratio F/A is seen from (5.28) to be

$$\frac{F}{A} = \frac{1}{c_{11}}. \qquad (5.34)$$

With the help of (5.30) we are now able to find the transmission coefficient for a barrier of thickness $a = x_2 - x_1$, yielding

$$c_{11} = e^{ika}(\cosh \kappa a + i\frac{1}{2}(\frac{\kappa}{k} - \frac{k}{\kappa})\sinh \kappa a), \qquad (5.35)$$

which results in the transmission coefficient

$$T = |c_{11}|^{-2} = \frac{1}{1 + (\kappa^2 + k^2)^2(\sinh \kappa a)^2/4\kappa^2 k^2}. \qquad (5.36)$$

The energy dependence of the transmission coefficient is shown in Fig. 5.4 (see also Problem 5.3).

EXAMPLE 5. QUANTUM WELLS

As an application of the result (5.30)-(5.31) we shall determine the transmission coefficient for a particle current moving through two barriers (Fig. 5.5). Such barrier configurations are generally named quantum wells. In practice they may be realized by using the method of molecular beam epitaxy to build a semiconducting material from elements such as Ga and As in the third and fifth group of the periodic system. With the help of this method it is possible in a controlled manner to build the material by putting one atomic layer on top of another. By changing the composition of the layers from GaAs to GaAlAs for every tenth layer one may produce wells of a typical width a equal to 5 nm. The height of the energy barrier V_0 may in practice be a few tenths of an electron volt, in the example mentioned 0.3 eV. Since the effective mass m entering the tunnelling process is considerably less than the electron mass m_e (for GaAs $m = 0.07 m_e$), the dimensionless quantity $V_0 m a^2 / \hbar^2$ becomes 7 with the parameters given above (the explanation of the reason for the difference between m and m_e requires a consideration of the effect of the periodic potential arising from the atoms in the crystalline lattice, cf. Chapter 9). This allows one experimentally to study resonances in the tunnelling through two barriers.

We assume that two identical barriers are separated by the distance l, the distance between their midpoints being $(l + a)$. The coordinates of the second barrier are x_3 and x_4. Thus we have $x_4 - x_3 = a$ and $x_3 - x_2 = l$ (see Fig. 5.5). The matrix elements c_{ij} for the other barrier corresponding to (5.30)-(5.31) are then given by the same expressions, except that x_1 is replaced by x_3 and x_2 by x_4 in (5.30)-(5.31). Note that the factor

$$e^{ik(x_3 - x_2)} = e^{ikl} \tag{5.37}$$

will be significant for the resonance phenomena in the tunnelling.

The transmission coefficient for the two barriers considered as a whole is obtained by multiplying the matrix t^{II} associated with barrier II with the matrix t^I associated with barrier I, since

$$\begin{pmatrix} A \\ B \end{pmatrix} = t^I(x_1, x_2) \begin{pmatrix} C \\ D \end{pmatrix} = t^I(x_1, x_2) t^{II}(x_3, x_4) \begin{pmatrix} F \\ G \end{pmatrix}. \tag{5.38}$$

This yields the transmission coefficient

$$T = \frac{1}{|c_{11}^{tot}|^2}, \tag{5.39}$$

where

$$c_{11}^{tot} = c_{11}^I c_{11}^{II} + c_{12}^I c_{21}^{II}, \tag{5.40}$$

as a result of the multiplication of the two matrices.

To simplify the calculation let us first consider the case where $k = \kappa$, corresponding to the condition that the energy E equals one half of the barrier height. We then obtain

$$c_{11}^I c_{11}^{II} = (e^{ika} \cosh \kappa a)^2 \tag{5.41}$$

and

$$c_{12}^I c_{21}^{II} = e^{ik(2l + 2a)} (\sinh \kappa a)^2. \tag{5.42}$$

We may now combine (5.39)-(5.42) with the result

$$T = \frac{1}{|\cosh^2 \kappa a + e^{i2kl} \sinh^2 \kappa a|^2}. \tag{5.43}$$

For $l = 0$ the transmission coefficient (5.43) becomes

$$T = \frac{1}{\cosh^2 2\kappa a}. \tag{5.44}$$

Tunnelling

This is seen to agree with the value of (5.36) for $k = \kappa$ when a is replaced by $2a$, corresponding to a barrier of double the width. When l is finite, the value of the transmission coefficient depends on the magnitude of kl. Specifically, $T = 1$ when $kl = (n + 1/2)\pi$, where n is an integer. This is called perfect transmission.

In the general case, when κ differs from k, the transmission coefficient (5.39) becomes

$$T = \frac{1}{|(\cosh(\kappa a) + i(\kappa^2 - k^2)\sinh(\kappa a)/2\kappa k)^2 + (\kappa^2 + k^2)^2 e^{i2kl} \sinh^2(\kappa a)/4\kappa^2 k^2|^2}. \quad (5.45)$$

An investigation of (5.45) shows (Problem 5.4), that it is possible to achieve perfect transmission for the double barrier (unlike the case of a single barrier) for definite values of the energy $E < V_0$.

The example given above illustrates that it may be easier to penetrate two barriers than one. In Chapter 9 we shall treat the case where (infinitely) many barriers are brought together to form a periodic potential. It will be shown that it is possible in this case to find stationary-state wave functions of the form of a plane wave e^{ikx} multiplied by a periodic function $u(x)$, where the periodicity of $u(x)$ is the same as that of the potential. The possible energy values form a band separated by intervals, the energy gaps, in which no stationary states may be found. This band structure is of fundamental importance for the understanding of the properties of metals, semiconductors and insulators.

5.4 The golden rule

The preceding sections of this chapter have demonstrated how a particle of energy E smaller than the height V_0 of a potential barrier may penetrate the barrier. This tunnelling is a pure quantum phenomenon with no analogue in classical mechanics. At the same time there is a non-zero probability that the particle is reflected, even though its energy is greater than the height of the potential barrier.

In this section we shall use the tunnelling phenomenon as the starting point for considering transitions between states which in a first approximation may be considered to be stationary. This allows a description of a variety of physical phenomena such as γ-decay of nuclei, light emission from atoms, and electrical resistivity in a metal.

If the energy is large compared to the barrier height, $E \gg V_0$, we may consider the barrier to be a perturbation, that is a small disturbance[2] of the free motion of the particle. The perturbation gives rise to transitions between free-particle states $|p\rangle$ and $|-p\rangle$, where $|p\rangle$ denotes a state of the form (5.18)

---

[2] The concept of a perturbation is used in classical physics as well, for instance in connection with the calculation of the orbits of the planets around the sun, when the small gravitational effects of other planets are taken into account.

with $p = \hbar k$. If k is positive, this state corresponds to motion along the x-axis, cf. (5.19).

We shall now show that the reflection coefficient in the limit $E \gg V_0$ is proportional to the absolute square of the matrix element $\langle -p|V|p\rangle$, where $V(x)$ is the potential energy, which is given by (5.20) for a rectangular barrier. By comparing this approximate result for the reflection coefficient with the exact one we are able also to investigate the validity of our approximation. Our main purpose in carrying out an approximate determination of the probability for the transition from $|p\rangle$ to $|-p\rangle$ is to illustrate the use of time-dependent perturbation theory (Section 5.4.2). With the help of this perturbation theory we shall prove the so-called 'golden rule' for the transition probability[3]. The golden rule may also be applied to three-dimensional scattering problems (cf. Problem 5.6), in which case it yields results that are identical with those obtained from the *Born approximation* introduced by Max Born in 1926.

5.4.1 Reflection from a barrier

We shall start from the result (5.36) for the transmission coefficient, which may be readily generalized to the case $E \geq V_0$ by replacing κ with $i\sqrt{2m(E-V_0)}/\hbar$ (cf. Problem 5.3). Since $\sinh(ix) = i\sin x$, the transmission coefficient for $E \gg V_0$ becomes

$$T = [1 + \frac{V_0^2}{4E(E-V_0)}\sin^2(a\sqrt{2m(E-V_0)}/\hbar)]^{-1} \simeq 1 - \frac{V_0^2}{4E^2}\sin^2(a\sqrt{2mE}/\hbar). \tag{5.46}$$

Here we have used that $(1+\epsilon)^{-1} \simeq 1 - \epsilon$, when $\epsilon \ll 1$ ($E \gg V_0$). As above V_0 denotes the height of the barrier, while a is the width. In the following we transform this approximate expression for the reflection coefficient $R = 1 - T$ and express R in terms of the absolute square of the matrix element $\langle -k|V|k\rangle$ given by

$$\langle -k|V|k\rangle = \frac{1}{(\sqrt{L})^2}\int_{-\infty}^{\infty}dx(e^{i(-k)x})^*V(x)e^{ikx}. \tag{5.47}$$

Here $V(x)$ is the potential energy given by

$$V(x) = V_0 \quad \text{for} \quad |x| < \frac{a}{2}, \quad V(x) = 0 \quad \text{for} \quad |x| > \frac{a}{2}, \tag{5.48}$$

and we have used free-particle states of the form (5.18) with a normalization constant $A = 1/\sqrt{L}$. The length L of the one-dimensional box in which the system is enclosed is assumed to be very much greater than the barrier width

---

[3] The name 'golden rule' was used by Enrico Fermi, and (5.63) is therefore often called Fermi's golden rule, although it is due to Dirac, Proc. Roy. Soc. **A112**, 661, 1926 and Proc. Roy. Soc. **A114**, 243, 1927.

a. By inserting the potential energy (5.48) in (5.47) we get

$$\langle -k|V|k\rangle = \frac{V_0}{i2kL}(e^{ika} - e^{-ika}) = \frac{V_0}{kL}\sin ka. \tag{5.49}$$

The reflection coefficient R, which according to (5.46) is given by

$$R = \frac{V_0^2}{4E^2}\sin^2(a\sqrt{2mE}/\hbar), \tag{5.50}$$

for $E \gg V_0$, may then be transformed to

$$R = \frac{1}{\hbar k/mL}\frac{2\pi}{\hbar}\frac{mL}{2\pi\hbar^2 k}|\langle -k|V|k\rangle|^2, \tag{5.51}$$

since $E = \hbar^2 k^2/2m$. The motivation for writing R in the form (5.51) is the following: We wish to express the reflection coefficient R in terms of a transition probability per unit time, w, defined by $w = JR$, where J is the flux. In the one-dimensional case under consideration the flux is identical to the current density j given by the expression (5.19), that is $J = \hbar k/mL$. From (5.51) it follows that w is given by

$$w = \frac{2\pi}{\hbar}\frac{mL}{2\pi\hbar^2 k}|\langle -k|V|k\rangle|^2 = \frac{2\pi}{\hbar}g_f|\langle -k|V|k\rangle|^2. \tag{5.52}$$

We have introduced the density of states associated with the final state, g_f, which in one dimension is one half of (4.59), since the particle in its final state ($|-k\rangle$) moves along the negative x-axis, yielding $g_f = g/2$. The perturbation thus causes a transition from $|k\rangle$ to $|-k\rangle$, which are states of equal energy[4].

Let us recapitulate: In (5.46) we have written an exact expression in an approximate form which is only valid when $E \gg V_0$. Following that, we introduced the transition probability per unit time, w, in order to express the reflection coefficient in terms of w and the flux $J = \hbar k/mL$, according to $R = w/J$ (note that the dimension of w is inverse time). The result for w may then be written as $2\pi/\hbar$ times the density of states g_f times the absolute square of the matrix element $\langle -k|V|k\rangle$, which connects the initial state $|k\rangle$ with the final state $|-k\rangle$. Note that the matrix element, except for an overall constant, equals the Fourier transform of the potential $V(x)$ with respect to the change $2k$ of the particle wave vector. In this form the result is general, as long as we may treat the perturbation as a small quantity. It may be applied to both time-independent and time-dependent perturbations, provided that it makes sense to introduce a density of final states g_f. The existence of a continuum of final states is thus a necessary condition for the validity of the golden rule.

---

[4] In other cases of interest, such as the absorption of light in atoms, the transition occurs between states of different energy, due to the presence of a perturbation oscillating in time.

5.4.2 Time-dependent perturbation theory

Let us assume that a system is described by the Hamiltonian

$$\hat{H} = \hat{H}_0 + \lambda \hat{H}'(t), \tag{5.53}$$

where $\hat{H}_0$ is a time-independent Hamilton operator with eigenstates ψ_k and eigenvalues E_k that are assumed to be known,

$$\hat{H}_0 \psi_k = E_k \psi_k. \tag{5.54}$$

By k is meant a set of quantum numbers[5] characterizing the eigenvalues and eigenstates of $\hat{H}_0$. The set of states ψ_k is assumed to be complete, allowing an arbitrary state vector to be expanded in terms of it. We denote by $\hat{H}'$ the part of the Hamiltonian to be treated as a perturbation. To be able to keep track of the order in the perturbation expansion we have multiplied $\hat{H}'$ with the dimensionless parameter λ, which may either be set equal to unity in the final result or divided out as below.

The Schrödinger equation for the system is

$$i\hbar \frac{\partial \Psi}{\partial t} = (\hat{H}_0 + \lambda \hat{H}') \Psi. \tag{5.55}$$

We now expand the solution Ψ in terms of the complete set ψ_k,

$$\Psi = \sum_l c_l(t) \psi_l e^{-iE_l t/\hbar}. \tag{5.56}$$

Here we have separated out the time-dependence $\exp(-iE_l t/\hbar)$ associated with the unperturbed states. The coefficients c_l are thus in general functions of time, except when λ is zero.

By inserting the state (5.56) in the Schrödinger equation (5.55) and taking the inner product of each side with the state ψ_k, one gets a system of coupled differential equations for $c_k(t)$,

$$i\hbar \frac{dc_k(t)}{dt} = \sum_l \langle k | \lambda \hat{H}' | l \rangle e^{-i(E_l - E_k)t/\hbar} c_l(t). \tag{5.57}$$

We have used (5.54) together with the orthonormalization condition $\langle \psi_k | \psi_l \rangle = \delta_{kl}$ and multiplied each side of the equation with $\exp(iE_k t/\hbar)$. As usual the matrix element $\langle k | \hat{H}' | l \rangle$ denotes the inner product $\langle \psi_k | \hat{H}' \psi_l \rangle$.

---

[5] For a three-dimensional harmonic ocillator, k denotes all the possible values of the set of numbers (n_x, n_y, n_z), where $n_x, n_y,$ and n_z are non-negative integers.

The coupled system of equations (5.57) is exact and just as difficult to solve as the original Schrödinger equation. However, the reformulation of the Schrödinger equation provides a suitable starting point for the perturbation treatment, since the right hand side of (5.57) is proportional to λ. The coefficients c_k may now be expanded in powers of λ,

$$c_k = c_k^{(0)} + \lambda c_k^{(1)} + \cdots. \tag{5.58}$$

Here $c_k^{(0)}$ is independent of time, the wave function

$$\Psi^{(0)} = \sum_k c_k^{(0)} \psi_k e^{-iE_k t/\hbar} \tag{5.59}$$

being the general solution of the Schrödinger equation

$$i\hbar \frac{\partial \Psi}{\partial t} = \hat{H}_0 \Psi \tag{5.60}$$

in the absence of the perturbation.

Let us assume that $\hat{H}'$ is zero for $t < 0$ and constant in time for $t > 0$. Before the perturbation is turned on, the system is known to be in a definite state ψ_m, corresponding to

$$c_k^{(0)} = \delta_{km}. \tag{5.61}$$

By keeping only terms proportional to λ and neglecting all higher order terms we obtain from (5.57) that

$$i\hbar \frac{dc_k^{(1)}(t)}{dt} = \sum_l \langle k|\hat{H}'|l\rangle e^{-i(E_l - E_k)t/\hbar} \delta_{lm}. \tag{5.62}$$

Due to the presence of δ_{lm} the sum over l yields only a single term, corresponding to $l = m$. By integrating over time and using that the perturbation vanishes before $t = 0$ together with the boundary condition $c_k^{(1)}(0) = 0$, we find that

$$c_k^{(1)}(t) = \frac{1}{i\hbar} \int_0^t dt' \langle k|\hat{H}'(t')|m\rangle e^{-i(E_m - E_k)t'/\hbar} = -\frac{1}{\hbar \omega_{km}} (e^{i\omega_{km}t} - 1)\langle k|\hat{H}'|m\rangle \tag{5.63}$$

with the abbreviation $\omega_{km} = (E_k - E_m)/\hbar$. In carrying out the integration over t' we have used our assumption that the matrix element $\langle k|\hat{H}'|m\rangle$ is independent of time. If $\hat{H}'$ varies harmonically or decreases exponentially in time, the integration over time may be carried out without difficulty.

The transition probability P is equal to the sum over k of the probability p_k for finding the system in the state k ($k \neq m$). The probability p_k is given by the absolute square of the inner product $\langle k|\Psi\rangle$,

$$p_k = |\langle k|\Psi\rangle|^2 = |c_k^{(1)}(t)|^2, \tag{5.64}$$

and the transition probability P is therefore

$$P = \sum_{k \neq m} p_k = \sum_k \frac{4}{\hbar^2 \omega_{km}^2} \sin^2(\omega_{km} t/2) |\langle k|\hat{H}'|m\rangle|^2. \tag{5.65}$$

It is convenient to change the summation over k to an integration by introducing the density of states $g_f(E_k)$, as the energy levels are assumed to lie sufficiently close that they may be regarded as a continuum,

$$P = \int dE_k g_f(E_k) \frac{4}{\hbar^2 \omega_{km}^2} \sin^2(\omega_{km} t/2) |\langle k|\hat{H}'|m\rangle|^2. \tag{5.66}$$

Before carrying out the energy integration we examine the function

$$F(t,\omega) = \omega^{-2} \sin^2(\omega t/2), \tag{5.67}$$

where ω denotes the integration variable ω_{km} (note that $dE_k = \hbar d\omega_{km}$). For $\omega \to 0$ we have $F = t^2/4$, while F, which is a symmetric function of ω, has its first zero at $\omega = 2\pi/t$. The area under the function is given by

$$\int_{-\infty}^{\infty} d\omega \, \omega^{-2} \sin^2(\omega t/2) = \frac{\pi}{2} t. \tag{5.68}$$

The function F thus resembles the delta function $\pi t \delta(\omega)/2$, and for sufficiently large t we must be able to approximate $F(t,\omega)$ by this function. Physically this corresponds to considering the matrix element and the density of states as functions that vary sufficiently slowly[6] with ω_{km}, that we can consider these to be constants. These quantities may therefore be taken outside the integral with a value given by the condition $E_k = E_m$.

The result for the transition probability per unit time, w, is then

$$w = \frac{P}{t} = \frac{2\pi}{\hbar} g_f(E_m) |\langle k|\hat{H}'|m\rangle|^2_{E_k = E_m}, \tag{5.69}$$

which is the *golden rule*. Note that we have indicated explicitly that the energy in the final state, E_k, should be set equal to the initial energy E_m in calculating the matrix element.

---

[6] The fact that this approximation breaks down for sufficiently small t does not need to worry us, since we are interested in determining the time-independent transition probability per unit time, $w = P/t$. The related worry that P is proportional to t and therefore in principle may be greater than 1 (a nonsensical result) disappears for the same reason.

Tunnelling

We have thus derived the golden rule for the case when the perturbation does not depend explicitly on time. By comparing (5.52) with (5.69) it is seen that (5.69) has the same form as the approximate expression (5.52) obtained by expanding the reflection coefficient for large energies.

5.5 Problems

PROBLEM 5.1
A particle with mass M moves in the potential V given by

$$V = \infty \text{ for } x < 0, \quad V = 0 \text{ for } 0 < x < a, \quad V = V_0 \text{ for } x > a, \qquad (5.70)$$

where V_0 and a are positive quantities. In this problem we wish to determine a condition which V_0 must satisfy, in order that there exists at least one bound state.

a) Find the smallest possible value, V_{0c}, which the constant V_0 may assume in order that there exists one bound state, for the case $a = 2 \cdot 10^{-15}$ m and $M = 8.4 \cdot 10^{-28}$ kg (it may be shown that the wave function belonging to the ground state has no nodes in the open interval $0 < x < \infty$).

b) Sketch the ground-state wave function and the potential in each of the two cases a) $V_0 = 1.2 V_{0c}$ and b) $V_0 = 10 V_{0c}$.

c) Which of the following three physical systems might be described by this potential, with the parameters given in a): 1) a deuterium nucleus, 2) a hydrogen atom, 3) a nitrogen molecule?

PROBLEM 5.2
Use (5.12) and the corresponding equation for the odd solutions to determine the energies of the bound states for the case $V_0 = 5\hbar^2/ma^2$ by graphical or numerical methods.

PROBLEM 5.3
Sketch how T depends on the particle energy $E = \hbar^2 k^2/2m$ for a barrier with $V_0 ma^2/\hbar^2 = 4$. Indicate how (5.36) is modified when E is larger than V_0, and show that the transmission coefficient in this case may equal 1 for certain values of the energy E.

PROBLEM 5.4
Verify (5.45) and examine how the transmission coefficient varies as a function of the parameter kl.

PROBLEM 5.5
Use the golden rule to find the transmission coefficient for the barrier

$$V(x) = V_0 e^{-x^2/a^2}, \qquad (5.71)$$

where V_0 is a positive constant. Is perfect transmission ($T = 1$) possible? Use the golden rule to state a condition for obtaining perfect transmission for an arbitrary barrier. Does the result change when the sign of the potential is changed?

Problem 5.6

The present problem treats scattering in three dimensions. A plane wave is incident on a scattering center described by the potential $V(\mathbf{r})$. We imagine that the system is contained within a cube of side L (volume L^3). We shall study the transition $\mathbf{k} \to \mathbf{k}'$, where $\mathbf{k}$ is the wave vector for the incident plane wave (4.54), and seek to determine the transition probability per unit time from $\mathbf{k}$ to the element of solid angle $d\Omega$ around $\mathbf{k}'$.

a) Use periodic boundary conditions to show that the density of states associated with the solid angle element $d\Omega$ is

$$g(E) = \frac{L^3}{(2\pi)^3} \frac{\sqrt{2m^3 E}}{\hbar^3} d\Omega, \qquad (5.72)$$

where $E = \hbar^2 k^2/2m = \hbar^2 k'^2/2m$.

We shall now determine the transition probability in terms of the potential. The quantity of physical interest is the differential cross-section $d\sigma$, which is defined as the area obtained by dividing the transition probability per unit time with the flux, i. e. the number of particles per unit time incident on a unit area.

b) Show that the flux associated with the wave function of the incident particle is $\hbar k/mL^3$. Show furthermore that the differential cross-section for scattering from $\mathbf{k}$ to an element of solid angle $d\Omega$ around $\mathbf{k}'$ is

$$d\sigma = \frac{m^2}{4\pi^2 \hbar^4} |\tilde{V}(\mathbf{k} - \mathbf{k}')|^2 d\Omega \qquad (5.73)$$

where $\tilde{V}(\mathbf{q})$ is the Fourier coefficient (cf. Appendix C)

$$\tilde{V}(\mathbf{q}) = \int d\mathbf{r} V(\mathbf{r}) \exp(-i\mathbf{q} \cdot \mathbf{r}). \qquad (5.74)$$

The integration in (5.74) extends over all space, since we allow the size of the box to tend towards infinity.

c) Show that the cross-section (5.73) only depends on $q = |\mathbf{k} - \mathbf{k}'|$, provided the potential is spherically symmetric. Let θ denote the angle between $\mathbf{k}$ and $\mathbf{k}'$, with $k = k'$. Show that

$$q = 2k \sin(\theta/2). \qquad (5.75)$$

Tunnelling

The total scattering cross-section σ is obtained by integrating the differential cross-section over solid angle

$$\sigma = \frac{m^2}{4\pi^2\hbar^4} \int d\Omega |\tilde{V}(\mathbf{k} - \mathbf{k}')|^2. \tag{5.76}$$

The integration is carried out conveniently by taking the polar axis to be along $\mathbf{k}$, whereby $d\Omega = d(\cos\theta)d\phi$.

d) Find $d\sigma$ and σ for the potential

$$V(r) = \frac{Ze_0^2}{r} e^{-r/a}, \tag{5.77}$$

which is a screened Coulomb potential (in nuclear physics the Yukawa potential). Examine $d\sigma$ for $a \to \infty$ and compare with the Rutherford scattering formula from classical physics. Show that the total cross-section diverges for $a \to \infty$. Indicate the dependence of σ on the energy $E = \hbar^2 k^2/2m$ for large values of E. Under which circumstances may our approximate treatment of the scattering problem (use of the golden rule) be expected to be valid?

e) Determine $d\sigma$ and σ for the potential

$$V(\mathbf{r}) = A\delta(\mathbf{r}), \tag{5.78}$$

where A is a constant, and compare with the result for the screened Coulomb potential.

6 ELECTRON IN A MAGNETIC FIELD

The quantization of the motion of a charged particle in a homogeneous magnetic field forms the topic of the present chapter. First we shall see how the magnetic field enters the classical Hamiltonian for a charged particle, this being the starting point for the quantization of its motion. The resulting Hamilton operator contains a term which has the same form as the Hamilton operator of the one-dimensional harmonic oscillator. This allows us to determine the energy eigenvalues by using the results of Chapter 2. It turns out that the energy quantum involved in the motion is the classical cyclotron frequency times the Planck constant. We discuss the significance of this quantization for the Hall effect in an electron gas which is only able to move in two dimensions, perpendicular to the direction of the magnetic field.

6.1 The classical Hamiltonian

An electron with charge $-e$ moving with velocity $\mathbf{v}$ in a magnetic field $\mathbf{B}$ experiences a force perpendicular to both the velocity and the magnetic field. The force $\mathbf{F}$ is proportional to the charge of the electron and the magnitude of the velocity perpendicular to the magnetic field. This *Lorentz force* is given by

$$\mathbf{F} = -e\mathbf{v} \times \mathbf{B}. \qquad (6.1)$$

Since the force is perpendicular to the magnetic field, it cannot change the velocity component in the direction of the magnetic field. It also follows that the speed of the particle must be constant in time, since the magnetic field does not do any work, the velocity being perpendicular to the Lorentz force. If the direction of the magnetic field is taken to be the z-axis, the velocity component v_z is thus a constant of the motion. Likewise $v_x^2 + v_y^2$ must be a constant of the motion. In the following we shall show that the classical motion of the particle is a circular orbit in the plane perpendicular to the magnetic field.

When the force on the electron is given by (6.1), Newton's second law becomes

$$m\frac{d\mathbf{v}}{dt} = -e\mathbf{v} \times \mathbf{B}, \qquad (6.2)$$

where the mass of the electron is denoted by m. If the direction of the magnetic field is taken to be the z-axis, the equations of motion (6.2) become

$$m\dot{v}_x = -eBv_y, \quad m\dot{v}_y = eBv_x, \quad m\dot{v}_z = 0. \qquad (6.3)$$

It follows from (6.3) that the velocity v_z along the direction of the magnetic field is unchanged. Likewise $v_x^2 + v_y^2$ is seen to be independent of time, since $v_x\dot{v}_x + v_y\dot{v}_y = 0$. By eliminating v_y from the first two equations in (6.3) we arrive at

$$\ddot{v}_x + \omega_c^2 v_x = 0, \qquad (6.4)$$

where
$$\omega_c = \frac{eB}{m} \tag{6.5}$$
is the cyclotron frequency. It is seen from (6.4) that
$$v_x = v_0 \cos \omega_c t \tag{6.6}$$
is a solution satisfying the boundary condition $v_x = v_0$ at $t = 0$. Furthermore $v_y = v_0 \sin \omega_c t$ is seen to be a solution for the y-component of the velocity with the property that $v_y = 0$ at the same time, $t = 0$.

Taking the expression (6.1) for the Lorentz force as our point of departure we shall now verify that the Lagrangian for a particle with mass m and charge $-e$ in a static magnetic field $\mathbf{B} = \mathbf{B}(\mathbf{r})$ is given by
$$L = \frac{1}{2} m \dot{\mathbf{r}}^2 - e\mathbf{A} \cdot \dot{\mathbf{r}}. \tag{6.7}$$
Here $\mathbf{A}$ denotes the vector potential, which determines the magnetic field $\mathbf{B}$ according to
$$\mathbf{B} = \nabla \times \mathbf{A}. \tag{6.8}$$
Our aim is to show that the Hamilton equations based on (6.7-8) result in the expression (6.1) for the force on a charged particle.

The relation (6.8) does not determine the vector potential uniquely for a given magnetic field $\mathbf{B}$. The two vector potentials
$$\mathbf{A} : B(0, x, 0) \tag{6.9}$$
and
$$\mathbf{A}' : \frac{B}{2}(-y, x, 0) \tag{6.10}$$
both yield the same magnetic field directed along the z-axis,
$$\mathbf{B} : (0, 0, B). \tag{6.11}$$
A change of vector potential $\mathbf{A}$, which leaves the magnetic field $\mathbf{B}$ unaffected - corresponding to the transition from (6.9) til (6.10) - is called a gauge transformation. Such a gauge transformation is accomplished by adding the gradient of a scalar function of $\mathbf{r}$ to $\mathbf{A}$, in the present case the gradient of the function $-Bxy/2$.

The generalized momentum $\mathbf{p}$ is introduced by the definition (2.31),
$$\mathbf{p} = \frac{\partial L}{\partial \dot{\mathbf{r}}}. \tag{6.12}$$
By inserting (6.7) in (6.12) we see that the generalized momentum is not simply $m\dot{\mathbf{r}}$, but rather
$$\mathbf{p} = m\dot{\mathbf{r}} - e\mathbf{A}. \tag{6.13}$$

Let us determine the derivative $\dot{\mathbf{p}}$ of the momentum with respect to time. According to (2.34)

$$\dot{\mathbf{p}} = \frac{\partial L}{\partial \mathbf{r}}. \tag{6.14}$$

Denoting $\dot{\mathbf{r}}$ by $\mathbf{v}$ one obtains by using the rules of vector analysis[1]

$$\dot{\mathbf{p}} = -e\nabla(\mathbf{A} \cdot \mathbf{v}) = -e\mathbf{v} \times (\nabla \times \mathbf{A}) - e(\mathbf{v} \cdot \nabla)\mathbf{A}. \tag{6.15}$$

Since $\mathbf{r}$ and $\mathbf{v}$ are independent variables, the right hand side of (6.15) only contains derivatives of $\mathbf{A}$.

The magnetic field $\mathbf{B}$ is assumed to be independent of time, which implies that $\mathbf{A}$ cannot depend explicitly on time. The time derivative of $\mathbf{A}$ is therefore given by

$$\dot{\mathbf{A}} = \frac{\partial \mathbf{A}}{\partial t} + (\mathbf{v} \cdot \nabla)\mathbf{A} = (\mathbf{v} \cdot \nabla)\mathbf{A}. \tag{6.16}$$

From (6.13) and (6.15-16) it follows that

$$m\dot{\mathbf{v}} = \dot{\mathbf{p}} + e\dot{\mathbf{A}} = -e\mathbf{v} \times \mathbf{B} \tag{6.17}$$

in agreement with (6.1).

The classical Hamiltonian for a particle with charge $-e$ in a magnetic field is therefore

$$H = \mathbf{p} \cdot \dot{\mathbf{r}} - L = \frac{1}{2m}(\mathbf{p} + e\mathbf{A})^2. \tag{6.18}$$

The result (6.18) is the starting point for the quantum mechanical treatment of the motion of an electron in a magnetic field. In the following section we shall use this Hamiltonian with $\mathbf{p} = \hbar\nabla/i$ to determine the energy eigenvalues in a magnetic field.

6.2 Quantization

According to the quantization procedures discussed in Chapter 2 the Hamilton operator describing the motion of an electron in a homogeneous magnetic field is obtained from the classical Hamiltonian by replacing the generalized momentum $\mathbf{p}$ with the operator $\hbar\nabla/i$. This results in a Hamiltonian which resembles that of a particle in a harmonic oscillator potential. We shall therefore introduce creation and annihilation operators that satisfy commutation relations analogous to those of Chapter 2. Alternatively, the energy eigenvalues and their degree of degeneracy may be determined by solving the Schrödinger equation. As we shall see, the degree of degeneracy is given by the magnetic flux divided by the magnetic flux quantum h/e (neglecting the spin of the electron).

---

[1] In general one has

$$\nabla(\mathbf{A} \cdot \mathbf{B}) = \mathbf{A} \times (\nabla \times \mathbf{B}) + \mathbf{B} \times (\nabla \times \mathbf{A}) + (\mathbf{A} \cdot \nabla)\mathbf{B} + (\mathbf{B} \cdot \nabla)\mathbf{A}.$$

6.2.1 The Hamiltonian and its eigenvalues

The Hamilton operator for an electron moving in a magnetic field is given by

$$\hat{H} = \frac{1}{2}m\hat{\mathbf{v}}^2 = \frac{1}{2m}(\hat{\mathbf{p}} + e\mathbf{A})^2 \qquad (6.19)$$

according to (6.18). It is most convenient to use the vector potential (6.9), which is called the Landau gauge. The components of the velocity operator $\hat{\mathbf{v}}$ are then given by

$$\hat{v}_x = \frac{\hbar}{im}\frac{\partial}{\partial x}; \quad \hat{v}_y = \frac{\hbar}{im}\frac{\partial}{\partial y} + \frac{eB}{m}x; \quad \hat{v}_z = \frac{\hbar}{im}\frac{\partial}{\partial z}. \qquad (6.20)$$

It may be seen from (6.20) that the x- and y-components of the velocity do not commute because of the presence of the magnetic field, since

$$[\hat{v}_x, \hat{v}_y] = \frac{\hbar\omega_c}{im}, \qquad (6.21)$$

where

$$\omega_c = \frac{eB}{m} \qquad (6.22)$$

is the classical cyclotron frequency for the electron.

In analogy with the harmonic oscillator it is convenient to introduce operators by the definition

$$\hat{v}_+ = \hat{v}_x + i\hat{v}_y \qquad (6.23)$$

and

$$\hat{v}_- = \hat{v}_x - i\hat{v}_y. \qquad (6.24)$$

These operators satisfy commutation relations similar to those of the creation and annihilation operators for the harmonic oscillator, since (6.21) implies that

$$[\hat{v}_-, \hat{v}_+] = 2\frac{\hbar\omega_c}{m}. \qquad (6.25)$$

In analogy with (2.91) the Hamiltonian can therefore be written as

$$\hat{H} = \frac{1}{2}m\hat{\mathbf{v}}^2 = \frac{1}{2}m\hat{v}_z^2 + \frac{1}{2}m\hat{v}_+\hat{v}_- + \frac{1}{2}im[\hat{v}_x, \hat{v}_y]. \qquad (6.26)$$

We define the number operator $\hat{N}$ by

$$\hat{N} = \frac{m}{2\hbar\omega_c}\hat{v}_+\hat{v}_-, \qquad (6.27)$$

which allows the Hamiltonian (6.26) to be expressed in the form

$$\hat{H} = \hbar\omega_c(\hat{N} + \frac{1}{2}) + \frac{1}{2}m\hat{v}_z^2. \quad (6.28)$$

Except for the last term in (6.28), which concerns the motion of the particle along the direction of the magnetic field, the Hamiltonian has the same appearance as that of the harmonic oscillator, since the commutation relations for $\hat{N}$ and the operators $\hat{v}_-$, $\hat{v}_+$ are identical to (2.92-93) apart from constants. It is only necessary to replace the classical angular frequency ω of the harmonic oscillator with the cyclotron frequency ω_c.

6.2.2 Degeneracy

The eigenstates for the Hamiltonian (6.28) may be given in the form

$$|\nu, k_z\rangle, \quad (6.29)$$

with periodic boundary conditions applied to the motion in the z-direction. The associated eigenvalues are

$$E_{\nu,k_z} = (\nu + \frac{1}{2})\hbar\omega_c + \frac{\hbar^2 k_z^2}{2m}. \quad (6.30)$$

As for the harmonic oscillator the quantum number ν may assume any of the values $0, 1, 2, \cdots$, while k_z on account of the periodic boundary conditions may assume the values given by $k_z L = 2\pi p$. Here p is an integer, which may be positive, negative or zero, and we have taken the volume to be a cube of side $L = V^{1/3}$.

Unlike that of the one-dimensional harmonic oscillator, the spectrum (6.30) is strongly degenerate, in the sense that there are many linearly-independent eigenstates belonging to a particular value of ν. The reason for this degeneracy is that there exists - in addition to the Hamiltonian $\hat{H}$ and the momentum component $\hat{p}_z$ - another operator, which commutes with both of these and therefore corresponds to a classical constant of the motion. When the vector potential is chosen as the Landau gauge (6.9), this operator is $\hat{p}_y$. As shown in Chapter 3 it is possible to find simultaneous eigenstates for Hermitian operators which commute with each other. We may therefore label these eigenstates with the eigenvalue k_y for $\hat{p}_y/\hbar$,

$$|\nu, k_y, k_z\rangle, \quad (6.31)$$

but it should be noted, that the eigenvalues k_y do not enter the energy eigenvalues. They determine however the degree of degeneracy, which we now turn to consider.

An alternative method of deriving the energy spectrum (6.30) is to solve the Schrödinger equation for the wave function ψ_{ν,k_y,k_z}, which depends on x, y

Electron in a magnetic field

and z. As we shall see, this also allows the degree of degeneracy of the levels to be determined. Since the operators $\hat{p}_z$ and $\hat{p}_y$ commute with the Hamiltonian, we seek a solution in the form

$$\psi_{\nu,k_y,k_z}(x,y,z) = e^{ik_y y} e^{ik_z z} f(x), \qquad (6.32)$$

which is an eigenfunction of $\hat{p}_z$ and $\hat{p}_y$. By inserting (6.32) into the Schrödinger equation one finds that the function f must satisfy

$$-\frac{\hbar^2}{2m}\frac{d^2 f}{dx^2} + \frac{1}{2m}(\hbar k_y + eBx)^2 f + \frac{\hbar^2 k_z^2}{2m} f = Ef, \qquad (6.33)$$

which is the result of dividing the Schrödinger equation by $\exp i(k_y y + k_z z)$ on both sides. Apart from the presence of the constant $\hbar^2 k_z^2/2m$, the equation (6.33) has the same form as the Schrödinger equation for a harmonic oscillator with its minimum displaced by the amount x_0 along the x-axis, where

$$x_0 = -\frac{\hbar k_y}{eB}. \qquad (6.34)$$

By comparing with the eigenvalues (2.103) for the one-dimensional harmonic oscillator it is readily seen from the Schrödinger equation (6.33), that the effective force constant is $(eB)^2/m$ and the eigenvalues those given by (6.30).

The degree of degeneracy of a level with a given ν, k_z is determined by the requirement that the minimum of the oscillator (6.34) lies somewhere within the volume considered[2]. The number of states belonging to the interval Δk_y is given by $L\Delta k_y/2\pi$, and the condition $0 < x_0 < L$ implies that

$$\Delta k_y = \frac{eBL}{\hbar}. \qquad (6.35)$$

The number N_u of linearly-independent state vectors belonging to the label ν, k_z is therefore given by[3]

$$N_u = \frac{eBL^2}{2\pi\hbar}. \qquad (6.36)$$

The result (6.36) for the degree of degeneracy N_u may be interpreted pictorially by considering h/e to be a quantum of flux. The degree of degeneracy is thus the number of flux quanta corresponding to the flux BL^2 of the magnetic field through the volume considered.

---

[2] Strictly speaking one cannot use this condition for the states corresponding to an oscillator with a minimum near (or at) the boundary of the volume considered. However, as long as the degree of degeneracy is large, the minimum is far from the surface for nearly all states considered, well in the interior of the volume in question.

[3] The existence of the electron spin is neglected here.

EXAMPLE 6. THE SYMMETRIC GAUGE.

We shall consider the Schrödinger equation for an electron in a homogeneous magnetic field described by the symmetric gauge (6.10) and determine the degree of degeneracy of the energy levels. Apart from giving an alternative derivation of the energies and the degree of degeneracy, our purpose in doing this is to show how the transition to cylindrical coordinates introduces an effective potential energy in the equation for the radial part of the wave function. Later on we shall see (Section 7.4), how a similar effective potential enters the description of the motion in a central field.

a) The use of the symmetric gauge makes it natural to solve the Schrödinger equation in cylindrical coordinates (see Appendix B) given by

$$x = \rho \cos\phi, \quad y = \rho \sin\phi. \tag{6.37}$$

The inverse relations are

$$\tan\phi = \frac{y}{x}, \quad \rho = \sqrt{x^2 + y^2}. \tag{6.38}$$

We shall express the Laplace operator ∇^2 in cylindrical coordinates by using

$$\frac{\partial}{\partial x} = \frac{\partial \rho}{\partial x}\frac{\partial}{\partial \rho} + \frac{\partial \phi}{\partial x}\frac{\partial}{\partial \phi} \tag{6.39}$$

and the corresponding expression for $\partial/\partial y$. Using (6.37-38) we get

$$\frac{\partial}{\partial x} = \cos\phi \frac{\partial}{\partial \rho} - \frac{\sin\phi}{\rho}\frac{\partial}{\partial \phi} \tag{6.40}$$

and

$$\frac{\partial}{\partial y} = \sin\phi \frac{\partial}{\partial \rho} + \frac{\cos\phi}{\rho}\frac{\partial}{\partial \phi} \tag{6.41}$$

From these relations it follows that the Laplace operator in cylindrical coordinates becomes

$$\nabla^2 = \frac{\partial^2}{\partial \rho^2} + \frac{1}{\rho}\frac{\partial}{\partial \rho} + \frac{1}{\rho^2}\frac{\partial^2}{\partial \phi^2} + \frac{\partial^2}{\partial z^2}. \tag{6.42}$$

The solution to the Schrödinger equation may be separated according to

$$\psi = e^{ik_z z} e^{im\phi} f(\rho). \tag{6.43}$$

The function $f(\rho)$ is determined by inserting (6.43) into the Schrödinger equation with the Hamiltonian given by (6.19). The equation for f then becomes

$$-\frac{\hbar^2}{2m_e}\left(\frac{1}{\rho}\frac{df}{d\rho} + \frac{d^2f}{d\rho^2} - \frac{m^2}{\rho^2}f\right) + \frac{em\hbar}{2m_e}Bf + \frac{e^2B^2}{8m_e}\rho^2 f = \left(E - \frac{\hbar^2 k_z^2}{2m_e}\right)f. \tag{6.44}$$

Here we have denoted the mass of the electron by m_e in order to avoid any confusion with the quantum number m. It is convenient to introduce the dimensionless variables $x = \rho(eB/\hbar)^{1/2}$ and $\epsilon = (E - \hbar^2 k_z^2/2m_e)/\hbar\omega_c$. We also introduce the function $g(x)$

by the definition $f = g(x)/\sqrt{x}$ in order to eliminate the first-order derivative from the equation. As a result the equation (6.44) becomes

$$-\frac{1}{2}\frac{d^2 g}{dx^2} + \frac{1}{2}(m + \frac{x^2}{4} + [m^2 - \frac{1}{4}]\frac{1}{x^2})g = \epsilon g. \qquad (6.45)$$

The problem is now analogous to motion in one dimension, except that x varies between 0 and ∞. For $m < 0$ and $|m| \gg 1$ the effective potential has its minimum at $x = x_0 = \sqrt{2|m|}$ with the minimum value 0. In the neighborhood of this minimum we may approximate the effective potential, which is the term multiplying g on the left hand side of the equation, by $(x - x_0)^2/2$ (this may be seen by differentiating the effective potential twice with respect to x and inserting $x = x_0$). The corresponding harmonic oscillator has the ground-state energy $1/2$ (in units of $\hbar\omega_c$), in agreement with (6.30). We assume that the electron is contained in a cylindrical volume. The degree of degeneracy is then determined by the condition $x_0 < R$, where R is the radius of the cylindrical volume in units of $\sqrt{\hbar/eB}$. Since the maximum value of $|m|$ belonging to the energy $1/2$ is the degree of degeneracy N_u, we obtain $N_u = \max|m| = \max x_0^2/2$, or $N_u = eBR^2/2\hbar$. We conclude that $N_u = eB\pi R^2/h$ in agreement with (6.36).

The large degree of degeneracy (6.36) has significance for the occurrence of the quantum Hall effect, which was discovered experimentally in two-dimensional electron systems by K. v. Klitzing in 1980. Before we describe this discovery we shall remind the reader of the classical Hall effect, which was first observed by E. H. Hall in 1880. When a current-carrying wire is placed in a magnetic field perpendicular to the direction of the current, one observes a potential difference across the wire, perpendicular to the direction of the current and to the magnetic field. This implies that the electric field and the current density in the wire are not parallel to each other. If the magnetic field is sufficiently strong, the electric field may be nearly perpendicular to the direction of the current.

When a current is passed through a wire in a magnetic field perpendicular to the direction of the current, the magnetic field acts to deflect the electrons in a direction perpendicular to itself and the velocity of the electrons. Since the drift velocity $\mathbf{v}_d$ points along the direction of the wire, a component of the electric field must exist perpendicular to the current direction. The perpendicular field component has a magnitude which is just sufficient to ensure that the electrons move along the direction of the wire. The perpendicular field component is named the Hall field and denoted by E_H. Its magnitude is determined by the requirement that the two forces perpendicular to the direction of the current, the Lorentz force due to the magnetic field and the force due to the perpendicular component of the electric field, cancel each other,

$$0 = -eE_H + ev_d B. \qquad (6.46)$$

The Hall field is thus proportional to both the drift velocity and the magnetic field. The current density $\mathbf{j}$ and the drift velocity $\mathbf{v}_d$ are connected by the

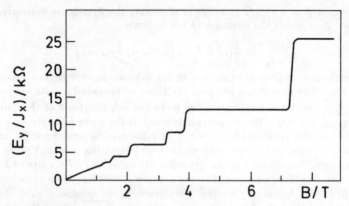

Figure 6.1: The quantum Hall effect. The figure shows the measured value of E_y/j_x as a function of the strength of the magnetic field B, measured in tesla.

relation $\mathbf{j} = -ne\mathbf{v}_d$, where n is the number density of the conduction electrons. This allows the Hall field to be written as

$$E_\mathrm{H} = R_\mathrm{H} j B, \tag{6.47}$$

where

$$R_\mathrm{H} = -\frac{1}{ne} \tag{6.48}$$

is called the Hall constant. Within the simple treatment given here the Hall constant is seen to depend solely on the charge of the electrons and their density.

In a two-dimensional system of electrons moving perpendicular to a homogeneous magnetic field the energy spectrum for a single electron is given by (6.30) except for the last term, since the value of k_z to a good approximation may be set equal to zero. According to the Pauli exclusion principle, which was introduced in Chapter 1 and will be further discussed in Chapter 8, each level belonging to a quantum number ν may contain as many electrons as the degree of degeneracy N_u given by (6.36). If a certain number, say ν_0, of the levels are completely occupied, while the others are empty, it follows that the number of electrons per unit area is given by

$$n = \nu_0 \frac{N_u}{L^2} = \nu_0 \frac{Be}{h}. \tag{6.49}$$

In practice the degree of degeneracy N_u may be about 10^8.

Having expressed the number of electrons per unit area, n, in terms of the number ν_0 of completely filled energy levels we may use the expression

(6.48) for the Hall constant to obtain the relation between the current per unit length, j_x, and the perpendicular component E_y of the electric field strength (note that the dimension of E_y is volt/m, while that of j_x is amp/m),

$$j_x = -\frac{ne}{B}E_y = -\nu_0 \frac{e^2}{h} E_y. \qquad (6.50)$$

According to (6.50) the observed ratio between j_x and E_y is an integer divided by the constant h/e^2, which is equal to 25813 ohm. Measurements of this quantum Hall effect may be used to determine the combination h/e^2 of the fundamental constants with a relative accuracy of about 10^{-7}. In Fig. 6.1 we exhibit results of measurements[4] of E_y/j_x in a GaAs-AlGaAs heterostructure at $T = 8$ mK. The simple free-electron description of the quantum Hall effect which we have given here, relates the degree of degeneracy (6.36) to the existence of the observed plateaus in E_y/j_x, but it does not explain *why* one observes plateaus. If the free-electron model is taken literally, one is led to the conclusion that the relation between E_y/j_x and B in Fig. 6.1 should be a straight line with the slope $-1/ne$, as for the classical Hall effect. The width of the plateaus may be due to the presence of imperfections, but a discussion of this issue would fall outside the framework of the present text.

6.3 Problems

PROBLEM 6.1
A particle with charge q is moving in a harmonic oscillator potential given by $V(x, y, z) = K(x^2 + y^2 + z^2)/2$ under the influence of a homogeneous magnetic field **B**. In this problem we shall use the vector potential (6.9).

a) Show that the momentum operator $\hat{\mathbf{p}}$ commutes with **A**, corresponding to $\hat{\mathbf{p}} \cdot \mathbf{A} = \mathbf{A} \cdot \hat{\mathbf{p}}$.
In the following we assume that the magnetic field is sufficiently weak that terms quadratic in B may be neglected in the Hamiltonian. The eigenstates for the Hamiltonian without magnetic field are denoted by $|n_x, n_y, n_z\rangle$ (cf. Problem 3.5). The part of the Hamiltonian which is due to the magnetic field is denoted by H'.

b) Determine the expectation value of H' for each of the states $|0, 0, 0\rangle$, $|1, 0, 0\rangle$, $|0, 1, 0\rangle$ and $|0, 0, 1\rangle$.

c) Express the Hamiltonian as a 3×3 matrix in the basis given by the orthonormal states $|1, 0, 0\rangle, |0, 1, 0\rangle$ and $|0, 0, 1\rangle$. Determine the eigenvalues of this matrix and find the corresponding eigenstates.

---

[4] K. v. Klitzing and G. Ebert, Physica B+C **117-118**, 682, 1983.

7 ANGULAR MOMENTUM

According to classical mechanics the angular momentum of a particle with respect to a given point in space is conserved, when the potential energy of the particle only depends on the distance from the particle to the given point. Such a potential is called a central field. Since the force on the particle moving in a central field is directed towards or away from the given point, it follows that the moment of the force with respect to this point must be zero, and the angular momentum with respect to the same point is therefore independent of time. The conservation of angular momentum is a consequence of the symmetry of the system, the potential energy being invariant under rotations about the given point. As we shall see, the existence of this symmetry means that the operators for each component of the angular momentum commute with the Hamiltonian. However, the individual components of the angular momentum operator do not commute with each other. This may be seen from the classical expression for the angular momentum $\mathbf{L}$,

$$\mathbf{L} = \mathbf{r} \times \mathbf{p}, \tag{7.1}$$

where $\mathbf{r}$ and $\mathbf{p}$ in quantum mechanics are represented by operators which do not commute.

The commutation relations for the components of the angular momentum will be derived in Section 7.1.2 below, starting from the classical expression (7.1). These commutation relations imply that the three components cannot take on definite values at the same time. It is however possible to find states which are simultaneous eigenstates for one of the components of the angular momentum and the square of its length. The eigenvalues for one of the components of the angular momentum turn out to be an integer or a half-integer times the Planck constant $\hbar$. The half-integer values cannot be derived from the orbital angular momentum alone. They are a consequence of the Lorentz invariance that must apply to the wave equation describing the motion of the particle. In this sense the result of using the classical expression has a greater generality than the starting point (7.1).

7.1 Quantization of angular momentum

Let us start by considering the example treated in Chapter 6 of a charged particle moving in a constant magnetic field and discuss the quantization of a single component of the orbital angular momentum. The choice of the symmetric form of the vector potential (6.10) means that the z-component of the angular momentum is a constant of the motion, since the operator belonging to L_z commutes with the Hamiltonian. This reflects the axial symmetry, the Hamiltonian being invariant with respect to rotations about the axis of the magnetic field.

Angular momentum

The Hamiltonian for an electron (charge $-e$), which moves in a magnetic field described by the vector potential **A**, is given in (6.19). Instead of using the asymmetric Landau gauge (6.9) we shall employ here the symmetric gauge given by (6.10). When this expression for the vector potential is inserted into (6.19), the result becomes

$$\hat{H} = \frac{\hat{\mathbf{p}}^2}{2m} + \frac{e}{2m}B\hat{L}_z + \frac{e^2 B^2}{8m}(\hat{x}^2 + \hat{y}^2), \tag{7.2}$$

where

$$\hat{L}_z = \hat{x}\hat{p}_y - \hat{y}\hat{p}_x \tag{7.3}$$

is the operator associated with the z-component of the angular momentum (7.1). Using the commutation relations corresponding to (2.44),

$$[\hat{q}_i, \hat{p}_j] = i\hbar\delta_{ij}, \quad i,j = 1,2,3, \tag{7.4}$$

where (q_1, q_2, q_3) stand for (x, y, z) and (p_1, p_2, p_3) for (p_x, p_y, p_z), we find

$$[\hat{L}_z, \hat{H}] = [\hat{p}_z, \hat{H}] = [\hat{L}_z, \hat{p}_z] = 0. \tag{7.5}$$

Since the three operators $\hat{H}$, $\hat{L}_z$ and $\hat{p}_z$ thus commute with each other, it is possible to find states that are simultaneous eigenstates for all three operators. We shall not consider the eigenvalues of $\hat{H}$. As shown in Example 6 in Chapter 6 they are identical to the eigenvalues (6.30), which were found using the asymmetric Landau gauge. This is to be expected, since the vector potentials (6.9) and (6.10) correspond to the same magnetic field.

We shall examine the eigenvalue equation for $\hat{L}_z$,

$$\hat{L}_z |m\rangle = m\hbar |m\rangle, \tag{7.6}$$

where we have introduced the notation $m\hbar$ for the eigenvalues of $\hat{L}_z$ and labelled the states by m. Since $\hbar$ has the same dimension as an angular momentum, m must be dimensionless.

The operator $\hat{L}_z$ is given by (7.3). It is convenient to discuss the eigenvalue equation (7.6) in terms of cylindrical coordinates (ρ, ϕ, z), which are connected to the cartesian coordinates by the transformation

$$x = \rho\cos\phi; \quad y = \rho\sin\phi; \quad z = z. \tag{7.7}$$

According to (7.3) the z-component of the angular momentum corresponds to the differential operator

$$\frac{\hbar}{i}\left(x\frac{\partial}{\partial y} - y\frac{\partial}{\partial x}\right) = \frac{\hbar}{i}\frac{\partial}{\partial \phi}, \tag{7.8}$$

where we have used the relations (6.37-41).

We shall now solve the eigenvalue equation (7.6) by considering the corresponding equation for the wave function $\psi = \psi_m(\phi) = \langle \phi | m \rangle$,

$$-i\frac{\partial}{\partial \phi}\psi_m(\phi) = m\psi_m(\phi). \tag{7.9}$$

Note that the wave function ψ_m in addition to ϕ also depends on ρ and z. This dependence does not need to be considered in the present context, since the operator associated with the z-component of the angular momentum only involves the partial derivative $\partial/\partial\phi$, cf. (7.8).

The solutions of the differential equation (7.9) are

$$\psi_m = Ce^{im\phi}, \tag{7.10}$$

where C is a constant which may depend on ρ and z. We shall now impose the additional condition that the wave function is a unique function of the spatial variables and therefore periodic in ϕ,

$$\psi_m(\phi) = \psi_m(\phi + 2\pi). \tag{7.11}$$

Such a condition appears natural from a classical point of view, but it excludes, as we shall see, the possibility of describing the spin of the electron. The periodicity condition (7.11) implies that

$$m\hbar, \quad \text{where} \quad m = 0, \pm 1, \pm 2, \cdots \tag{7.12}$$

represents the possible results of a measurement of the component of the angular momentum along the magnetic field. Note that the condition (7.11) has the same form as the periodic boundary conditions (4.50), and the result (7.12) therefore corresponds to the quantization rule (4.52).

As long as the electron is free to move anywhere in space there is no upper limit to the values of $|m|$. As we saw in Example 6 of the previous chapter, the states

$$|\nu, m, k_z\rangle, \quad \nu = 0, \ k_z = 0, \tag{7.13}$$

belong to the same energy eigenvalue $\hbar\omega_c/2$. The degree of degeneracy is therefore infinite, since m may assume any of the values (7.12). Classically speaking this corresponds to the fact that the diameter of the electron orbit can be arbitrarily large for a given energy. For a finite system the degree of degeneracy is determined by the area perpendicular to the direction of the magnetic field, cf. the discussion below (6.45).

7.1.1 Spherical harmonics

Instead of considering motion in a magnetic field we shall now assume that the particle moves in a central field, which means that its potential energy V

only depends on the length r of the radius vector. Under such conditions there exists an additional constant of the motion, the square of the length of the angular momentum, since the three operators $\hat{H}$, $\hat{\mathbf{L}}^2$ and $\hat{L}_z$ commute with each other. We shall demonstrate this important property by expressing the operators $\hat{H}$, $\hat{\mathbf{L}}^2$ and $\hat{L}_z$ in polar coordinates r, θ, ϕ defined by

$$x = r\sin\theta\cos\phi, \quad y = r\sin\theta\sin\phi, \quad z = r\cos\theta. \tag{7.14}$$

By exploiting the inverse relations $r = \sqrt{x^2 + y^2 + z^2}$ together with $\tan\phi = y/x$ and $\tan\theta = \sqrt{x^2 + y^2}/z$ we may proceed in analogy with (6.37-41) with the following result for $\hat{L}_z$,

$$\hat{L}_z = \frac{\hbar}{i}\frac{\partial}{\partial\phi}. \tag{7.15}$$

The details of this transformation are carried out in Appendix B. Note that the operator $\hat{L}_z$ only involves the angle ϕ, because x and y enter in a symmetric fashion in $\tan\theta$. The expression for $\hat{L}_z$ is thus the same in cylindrical as in spherical coordinates.

It is somewhat more involved to express the square of the length of the angular momentum in polar coordinates. Using relations corresponding to (6.40-41) we obtain, as demonstrated in Appendix B, that

$$\hat{\mathbf{L}}^2 = -\hbar^2\left[\frac{1}{\sin\theta}\frac{\partial}{\partial\theta}\left(\sin\theta\frac{\partial}{\partial\theta}\right) + \frac{1}{\sin^2\theta}\frac{\partial^2}{\partial\phi^2}\right]. \tag{7.16}$$

This expression for $\hat{\mathbf{L}}^2$ shows immediately that $\hat{\mathbf{L}}^2$ and $\hat{L}_z$ commute with each other. This commutation rule holds whether or not the particle moves in a central field. It is in fact proved most easily by using cartesian coordinates, as we shall see in the following subsection.

Since $\hat{\mathbf{L}}^2$ and $\hat{L}_z$ only depend on the variables θ and ϕ, but not on r, these operators commute with the potential energy V, which only depends on r. The two operators $\hat{\mathbf{L}}^2$ and $\hat{L}_z$ also commute with the kinetic energy. This is seen most easily by expressing these operators in cartesian coordinates, but we show it here by using polar coordinates, since we shall need the Laplace operator ∇^2 in polar coordinates for the purpose of solving the Schrödinger equation.

The kinetic energy operator is proportional to the Laplace operator ∇^2, which is given in cartesian coordinates by

$$\nabla^2 = \frac{\partial^2}{\partial x^2} + \frac{\partial^2}{\partial y^2} + \frac{\partial^2}{\partial z^2}. \tag{7.17}$$

In polar coordinates this becomes

$$\nabla^2 = \frac{1}{r^2}\frac{\partial}{\partial r}\left(r^2\frac{\partial}{\partial r}\right) + \frac{1}{r^2}\left[\frac{1}{\sin\theta}\frac{\partial}{\partial\theta}\left(\sin\theta\frac{\partial}{\partial\theta}\right) + \frac{1}{\sin^2\theta}\frac{\partial^2}{\partial\phi^2}\right], \tag{7.18}$$

as demonstrated in Appendix B. By comparison of (7.18) with (7.16) we note that the angular dependence of the kinetic energy operator is given by the square of the angular momentum. This completes the proof that the operator for the kinetic energy commutes with both $\hat{\mathbf{L}}^2$ and $\hat{L}_z$. Since the potential energy V in a central field only depends on the length r, it follows that the operator for the total energy, $\hat{H}$, also commutes with $\hat{\mathbf{L}}^2$ and $\hat{L}_z$.

The spherical harmonics, which are eigenfunctions of $\hat{\mathbf{L}}^2$ and $\hat{L}_z$, play an important role in the applications of quantum mechanics. Let us try to guess a few simple functions, which are simultaneous eigenfunctions of $\hat{L}_z$ and $\hat{\mathbf{L}}^2$ given in (7.15) and (7.16). First we observe that any function which only depends on r is an eigenfunction with the eigenvalue zero. Likewise it is clear that we only need to consider the θ- and ϕ-dependence of the eigenfunctions, since their r-dependence is not affected by the operators (7.15) and (7.16). We have already seen, that the eigenfunctions of $\hat{L}_z$ are proportional to $\exp(im\phi)$, where m is an integer. Let us first assume that $m = 0$ and investigate whether there exists an eigenfunction for $\hat{L}_z$ and $\hat{\mathbf{L}}^2$ which is proportional to z. From the expression for $\hat{\mathbf{L}}$ in cartesian coordinates it is seen that $\hat{L}_x z = (\hbar/i)y$ and $\hat{L}_y z = -(\hbar/i)x$, while $\hat{L}_z z = 0$. From this it follows that $\hat{L}_x^2 z = \hbar^2 z$ and $\hat{L}_y^2 z = \hbar^2 z$, resulting in $\hat{\mathbf{L}}^2 z = 2\hbar^2 z$. We conclude that z is a simultaneous eigenfunction for $\hat{L}_z$ and $\hat{\mathbf{L}}^2$. For symmetry reasons x and y must also be eigenfunctions of $\hat{\mathbf{L}}^2$ with the eigenvalue $2\hbar^2$, which we shall write in the form $1(1+1)\hbar^2$ in order to establish the connection with the general theory of angular momentum discussed in the following subsection. It is clear from (7.3), that neither x nor y is an eigenfunction of $\hat{L}_z$. However, we may form the linear combinations $(x \pm iy)$, which are seen to be eigenfunctions of $\hat{L}_z$ with eigenvalues $\pm \hbar$, respectively.

We have thus determined the first four spherical harmonics $Y_{lm}(\theta, \phi)$ belonging to $(l, m) = (0, 0), (1, 1), (1, 0)$ and $(1, -1)$. Except for a normalization constant, which is determined by the condition

$$\int_{-1}^{1} d(\cos \theta) \int_0^{2\pi} d\phi Y_{l'm'}^*(\theta, \phi) Y_{lm}(\theta, \phi) = \delta_{l'l}\delta_{m'm}, \qquad (7.19)$$

these functions are given by $Y_{00} \propto 1$, $Y_{11} \propto \sin\theta \exp(i\phi)$, $Y_{10} \propto \cos\theta$ and $Y_{1-1} \propto \sin\theta \exp(-i\phi)$. In a similar way one may construct the five spherical harmonics belonging to $l = 2$ by starting from the functions $xy, yz, zx, x^2 - y^2$ and $3z^2 - r^2$ (these functions have been chosen such that they are orthogonal to Y_{00}). It is shown in Sect. 7.1.2 below that the eigenvalues for $\hat{\mathbf{L}}^2$ generally are given by $l(l+1)\hbar^2$, where l is a positive integer or zero. Table 7.1 lists the spherical harmonics belonging to $l = 0, 1, 2$ and 3, normalized according to (7.19).

Angular momentum

If we consider a quantum mechanical state in which the square of the angular momentum has a definite value it is natural to expect that this implies the existence of an upper limit on the numerical value of the eigenvalues of $\hat{L}_z$, since classically speaking the magnitude of one of the components of the angular momentum cannot exceed the length of the angular momentum vector. The existence of such an upper limit will be shown in the following section, where we determine the eigenfunctions and eigenvalues of $\hat{\mathbf{L}}^2$ and $\hat{L}_z$, starting from the commutation relations between the components of the angular momentum.

7.1.2 Eigenstates of $\hat{\mathbf{L}}^2$ and $\hat{L}_z$

In the following we shall consider the quantization of angular momentum from a different point of view. Our starting point is now the commutation relations between the three components of the angular momentum, and we shall use these to determine the eigenvalues and eigenfunctions of $\hat{\mathbf{L}}^2$ and $\hat{L}_z$. The commutation relations between the components of the angular momentum are derived from (7.3) and the analogous expressions for $\hat{L}_x$ and $\hat{L}_y$ with the use of (7.4). It is seen that the commutator $[\hat{L}_x, \hat{L}_y]$ is given by

$$[\hat{L}_x, \hat{L}_y] = i\hbar \hat{L}_z. \tag{7.20}$$

The other commutation relations are obtained from (7.20) by cyclic permutation of the indices x, y and z. The commutation relations may thus be written in the form of a vector equation,

$$\hat{\mathbf{L}} \times \hat{\mathbf{L}} = i\hbar \hat{\mathbf{L}}. \tag{7.21}$$

By using (7.21) one sees that any of the components of the angular momentum, e. g. $\hat{L}_z$, commutes with the square of the angular momentum vector,

$$[\hat{\mathbf{L}}^2, \hat{L}_z] = 0. \tag{7.22}$$

The three commutation relations satisfied by each component may also be written as a vector equation

$$[\hat{\mathbf{L}}^2, \hat{\mathbf{L}}] = \mathbf{0}. \tag{7.23}$$

According to (7.23) it is possible to find simultaneous eigenstates for the square of the angular momentum and any one of its components.

In the derivation that follows we utilize that the eigenvalues of the square of the angular momentum must be either positive or zero, since the expectation value $<\mathbf{L}^2>$ of $\hat{\mathbf{L}}^2$ in an arbitrary state ψ satisfies the inequality

$$<\mathbf{L}^2> \geq 0. \tag{7.24}$$

$$Y_{00} = N_{00}, \qquad N_{00} = \frac{1}{\sqrt{4\pi}}$$

$$Y_{10} = N_{10} \cos\theta, \qquad N_{10} = \frac{\sqrt{3}}{\sqrt{4\pi}}$$

$$Y_{1\pm 1} = N_{1\pm 1} \sin\theta e^{\pm i\phi}, \qquad N_{1\pm 1} = \mp\frac{\sqrt{3}}{\sqrt{8\pi}}$$

$$Y_{20} = N_{20}(3\cos^2\theta - 1), \qquad N_{20} = \frac{\sqrt{5}}{\sqrt{16\pi}}$$

$$Y_{2\pm 1} = N_{2\pm 1} \sin\theta\cos\theta e^{\pm i\phi}, \qquad N_{2\pm 1} = \mp\frac{\sqrt{15}}{\sqrt{8\pi}}$$

$$Y_{2\pm 2} = N_{2\pm 2} \sin^2\theta e^{\pm 2i\phi}, \qquad N_{2\pm 2} = \frac{\sqrt{15}}{\sqrt{32\pi}}$$

$$Y_{30} = N_{30}(5\cos^3\theta - 3\cos\theta), \qquad N_{30} = \frac{\sqrt{7}}{\sqrt{16\pi}}$$

$$Y_{3\pm 1} = N_{3\pm 1} \sin\theta(5\cos^2\theta - 1)e^{\pm i\phi}, \qquad N_{3\pm 1} = \mp\frac{\sqrt{21}}{\sqrt{64\pi}}$$

$$Y_{3\pm 2} = N_{3\pm 2} \sin^2\theta\cos\theta e^{\pm 2i\phi}, \qquad N_{3\pm 2} = \frac{\sqrt{105}}{\sqrt{32\pi}}$$

$$Y_{3\pm 3} = N_{3\pm 3} \sin^3\theta e^{\pm 3i\phi}, \qquad N_{3\pm 3} = \mp\frac{\sqrt{35}}{\sqrt{64\pi}}$$

Table 7.1: Spherical harmonics normalized according to (7.19).

Angular momentum

This inequality is a consequence of the fact that the left hand side of (7.24) may be written as

$$< \mathbf{L}^2 > = < L_x^2 > + < L_y^2 > + < L_z^2 > = \langle \hat{L}_x\psi|\hat{L}_x\psi\rangle + \langle \hat{L}_y\psi|\hat{L}_y\psi\rangle + \langle \hat{L}_z\psi|\hat{L}_z\psi\rangle, \tag{7.25}$$

where we have introduced the sum of the squares of the norm for the states $\hat{L}_i\psi$, $i = x, y, z$, using the notation of Chapter 3.

We shall write the eigenvalue equations for $\hat{\mathbf{L}}^2$ and $\hat{L}_z$ in the following form

$$\hat{\mathbf{L}}^2|\lambda,\mu\rangle = \lambda\hbar^2|\lambda,\mu\rangle \tag{7.26}$$

and

$$\hat{L}_z|\lambda,\mu\rangle = \mu\hbar|\lambda,\mu\rangle, \tag{7.27}$$

where λ and μ are real numbers. According to (7.24) the number λ must be non-negative, $\lambda \geq 0$. The square of the angular momentum is the sum of the squares of each component, and λ and μ must therefore satisfy the inequality

$$\lambda \geq \mu^2, \tag{7.28}$$

since

$$< L_x^2 > + < L_y^2 > + < L_z^2 > \geq < L_z^2 >. \tag{7.29}$$

In solving the eigenvalue problem given by (7.26) and (7.27) it is convenient to use operators that correspond to the creation and annihilation operators introduced in Chapter 2 in the context of the harmonic oscillator. In the present case we define

$$\hat{L}_+ = \hat{L}_x + i\hat{L}_y \tag{7.30}$$

and

$$\hat{L}_- = \hat{L}_x - i\hat{L}_y. \tag{7.31}$$

These operators are not Hermitian, since $\hat{L}_-$ is the Hermitian conjugate of $\hat{L}_+$. It follows from the commutation relations (7.21) that

$$\hat{L}_z\hat{L}_+ = \hat{L}_+(\hat{L}_z + \hbar) \tag{7.32}$$

and

$$\hat{L}_z\hat{L}_- = \hat{L}_-(\hat{L}_z - \hbar). \tag{7.33}$$

We shall also need the identities

$$\hat{L}_+\hat{L}_- = \hat{\mathbf{L}}^2 - \hat{L}_z(\hat{L}_z - \hbar) \tag{7.34}$$

and

$$\hat{L}_-\hat{L}_+ = \hat{\mathbf{L}}^2 - \hat{L}_z(\hat{L}_z + \hbar). \tag{7.35}$$

The operators $\hat{L}_+$ and $\hat{L}_-$ do not commute with each other, nor do they commute with $\hat{L}_z$. Both of them, however, commute with $\hat{\mathbf{L}}^2$. As a consequence, if the state $|\lambda, \mu\rangle$ is an eigenstate of $\hat{\mathbf{L}}^2$, then the states $\hat{L}_+|\lambda, \mu\rangle$ and $\hat{L}_-|\lambda, \mu\rangle$ are also eigenstates for this operator (unless they happen to be zero).

It follows from (7.32) that

$$\hat{L}_z \hat{L}_+ |\lambda, \mu\rangle = (\mu + 1)\hbar \hat{L}_+ |\lambda, \mu\rangle. \tag{7.36}$$

By repeating this n times we obtain

$$\hat{L}_z (\hat{L}_+)^n |\lambda, \mu\rangle = (\mu + n)\hbar (\hat{L}_+)^n |\lambda, \mu\rangle. \tag{7.37}$$

However, it is also true that

$$\hat{\mathbf{L}}^2 (\hat{L}_+)^n |\lambda, \mu\rangle = \lambda \hbar^2 (\hat{L}_+)^n |\lambda, \mu\rangle, \tag{7.38}$$

since $\hat{\mathbf{L}}^2$ commutes with $(\hat{L}_+)^n$. This shows that the use of the operators (7.30) and (7.31) does not change the length of the angular momentum. Sooner or later we shall therefore reach a value of n such that $(\mu+n)^2 > \lambda$ in contradiction to (7.28), unless the process stops at a certain value of n, which we shall denote by n_0. By definition the state

$$(\hat{L}_+)^{n_0} |\lambda, \mu\rangle \tag{7.39}$$

differs from zero, while

$$(\hat{L}_+)^{n_0+1} |\lambda, \mu\rangle \tag{7.40}$$

is equal to zero. Denoting $\mu + n_0$ by l we see that the state $|\lambda, l\rangle$ satisfies

$$\hat{L}_+ |\lambda, l\rangle = 0, \tag{7.41}$$

and

$$\hat{L}_z |\lambda, l\rangle = l\hbar |\lambda, l\rangle. \tag{7.42}$$

Now we let $\hat{L}_-$ operate on (7.41),

$$0 = \hat{L}_- \hat{L}_+ |\lambda, l\rangle = (\lambda - l(l+1))\hbar^2 |\lambda, l\rangle, \tag{7.43}$$

using the identity (7.35). Since $|\lambda, l\rangle$ differs from zero, cf. (7.39), we conclude from (7.43) that

$$\lambda - l(l+1) = 0. \tag{7.44}$$

In the following we shall label the states $|\lambda, \mu\rangle$ as $|l(l+1), \mu\rangle$ based on the result (7.44). Alternatively we could have used the notation $|l, \mu\rangle$, since l uniquely determines λ according to (7.44).

Angular momentum

Next we let the operator $\hat{L}_-$ act on the state $|l(l+1), l\rangle$. When the operator acts a sufficient number of times, the series must break off again, since the inequality (7.28) must always be satisfied. There exists therefore an integer n, such that

$$(\hat{L}_-)^n |l(l+1), l\rangle \tag{7.45}$$

is different from zero, while

$$\hat{L}_- (\hat{L}_-)^n |l(l+1), l\rangle = 0. \tag{7.46}$$

Letting the operator $\hat{L}_+$ act on (7.46) and using (7.34), we can deduce in analogy with (7.44) that

$$l(l+1) - (l-n)(l-n-1) = 0, \tag{7.47}$$

or

$$2l = n. \tag{7.48}$$

The other solution ($n = -1$) to the second order algebraic equation (7.47) must be discarded, since n by definition is greater than or equal to zero. As n is a non-negative integer, (7.48) shows that l may only assume integer or half-integer values. The eigenvalues for the operators $\hat{\mathbf{L}}^2$ and $\hat{L}_z$ are therefore given by

$$\lambda = l(l+1), \quad l = 0, \frac{1}{2}, 1, \frac{3}{2}, \cdots \tag{7.49}$$

while m for a given value of l may assume the $(2l+1)$ different values

$$m = l, l-1, l-2, \cdots, -l+1, -l. \tag{7.50}$$

When considering the orbital angular momentum we have already seen in (7.12) that m (and hence l) must be an integer, but the commutation relations (7.21), which were derived on the basis of the classical expression (7.1) for the angular momentum, evidently allow the possibility that l may assume half-integer values.

Before discussing the half-integer values further we shall close this subsection by deriving two relations that are particularly useful for constructing matrices of angular momentum operators. According to (7.36) the state $\hat{L}_+ |\lambda, m\rangle$ is an eigenstate for L_z corresponding to the eigenvalue $(m+1)\hbar$, that is

$$\hat{L}_+ |\lambda, m\rangle = c|\lambda, m+1\rangle, \tag{7.51}$$

where c is a constant. We shall choose the constant c to be real and determine it in analogy with (2.108-109) by using (7.35),

$$\langle \lambda, m | \hat{L}_- \hat{L}_+ | \lambda, m \rangle = c^2 = (\lambda - m(m+1))\hbar^2. \tag{7.52}$$

By inserting λ from (7.44) we obtain

$$\hat{L}_+|l(l+1),m\rangle = \hbar\sqrt{l(l+1) - m(m+1)}\,|l(l+1),m+1\rangle. \tag{7.53}$$

In precisely the same manner one finds by using (7.34) that

$$\hat{L}_-|l(l+1),m\rangle = \hbar\sqrt{l(l+1) - m(m-1)}\,|l(l+1),m-1\rangle. \tag{7.54}$$

Note that the right hand sides of (7.53) and (7.54) are zero when m is l and $-l$, respectively.

7.2 Spin

It is an empirical fact (and a consequence of the relativistic theory of the electron) that a magnetic field not only affects an electron because of its orbital motion, as expressed by the Hamiltonian in (7.2), but also because of the magnetic moment associated with its spin. In the last section of this chapter we shall see how the Hamiltonian is modified accordingly, starting from the relativistic description due to Dirac. For the moment we simply add the following term to the Hamiltonian (7.2)

$$\hat{H}_s = g_s \frac{e}{2m} B \hat{S}_z, \tag{7.55}$$

where $\hat{\mathbf{S}}$ is the operator for the electron spin, while the g-factor g_s is equal to 2 to a very good approximation. The result (7.55) will be derived in Section 7.5. Physically, it represents the energy of a magnetic dipole, the dipole moment operator being equal to $-g_s e\hat{\mathbf{S}}/2m$. According to the general theory of the previous section the eigenvalue of $\hat{\mathbf{S}}^2$ is given by $s(s+1)\hbar^2$, where $s = 1/2$ for an electron. The corresponding eigenvalues for $\hat{S}_z$ are $\hbar/2$ and $-\hbar/2$. The contribution from (7.55) to the Hamiltonian is thus comparable to the second term on the right hand side of (7.2) which arises from the orbital motion of the electron. In atoms these terms give rise to the so-called Zeeman effect.

In order to include the effect (7.55) of the electron spin in our description we may introduce eigenvectors of the form (2.79) and (2.80), corresponding to the eigenvalues $\hbar/2$ and $-\hbar/2$, respectively, for the operator $\hat{S}_z$. The matrix representation of this operator is $\hbar\sigma_z/2$, where the Pauli matrix σ_z is given in (2.76). For brevity one often denotes (2.79) by α and (2.80) by β. The eigenvalue equations for the z-component of the angular momentum then become

$$\hat{S}_z\alpha = \frac{\hbar}{2}\alpha \tag{7.56}$$

and

$$\hat{S}_z\beta = -\frac{\hbar}{2}\beta. \tag{7.57}$$

Angular momentum

The effect of operating with the two other components $\hat{S}_x$ and $\hat{S}_y$ may be obtained from the general theory of the previous section. We introduce the operators

$$\hat{S}_+ = \hat{S}_x + i\hat{S}_y; \quad \hat{S}_- = \hat{S}_x - i\hat{S}_y. \tag{7.58}$$

According to (7.53) one has

$$\hat{S}_+\alpha = 0, \quad \hat{S}_+\beta = \hbar\alpha, \tag{7.59}$$

while (7.54) shows that

$$\hat{S}_-\alpha = \hbar\beta, \quad \hat{S}_-\beta = 0. \tag{7.60}$$

By using the relation analogous to (7.34) together with (7.56), (7.57), (7.59) and (7.60) one sees that

$$\hat{\mathbf{S}}^2\alpha = \frac{3\hbar^2}{4}\alpha \tag{7.61}$$

and

$$\hat{\mathbf{S}}^2\beta = \frac{3\hbar^2}{4}\beta, \tag{7.62}$$

in agreement with the fact that α and β are eigenstates of the square of the angular momentum with the eigenvalue $\hbar^2 s(s+1)$, where $s = 1/2$.

The matrices for $\hat{S}_x$ and $\hat{S}_y$ are obtained by adding and subtracting, respectively, the matrices for $\hat{S}_+$ and $\hat{S}_-$, which may be deduced from (7.59) and (7.60), with the result $\hat{\mathbf{S}} = \hbar\boldsymbol{\sigma}/2$. The components of the electron spin have thus been represented by the Pauli matrices σ_x, σ_y and σ_z, which were introduced in (2.76). Note the similarity between the commutation relations (2.77) for the Pauli matrices and the classical Poisson brackets (2.68) for the two-dimensional harmonic oscillator.

7.3 Addition of angular momentum

In the applications of quantum theory one frequently needs to add angular momenta. An atomic electron not only possesses an angular momentum deriving from its spin, but from its orbital motion as well (unless the quantum number l characterizing the length of the orbital angular momentum is equal to zero). It may also be necessary to add the angular momenta of different particles in order to determine the total angular momentum of an atom or an atomic nucleus.

The general theory of the addition of angular momentum will not be given here. Instead we consider the simplest possible example, the addition of the spins of two electrons. The general result of the addition of angular momenta is stated below in (7.74). In mathematical terms the addition of angular momenta amounts to determining simultaneous eigenstates for the square of the total angular momentum and one of its components. The relevance of the corresponding eigenstates for the description of a physical system depends on the mutual interaction of the particles and the effect of external fields.

For a system consisting of two electrons the operator $\hat{\mathbf{S}}$ for the total spin is given by

$$\hat{\mathbf{S}} = \hat{\mathbf{S}}_1 + \hat{\mathbf{S}}_2, \qquad (7.63)$$

where the properties of the spin operators for each electron are given by the relations (7.56)-(7.60). In order to facilitate the notation we write the set of basis states for the total system in the form

$$\alpha\alpha, \alpha\beta, \beta\alpha, \beta\beta, \qquad (7.64)$$

using the convention that the first α in $\alpha\alpha$ refers to electron 1 and the second one to electron 2, etc. (alternatively we could have used the notation $\alpha_1\alpha_2, \alpha_1\beta_2, \beta_1\alpha_2, \beta_1\beta_2$). The state $\beta\alpha$ thus represents an eigenstate for $\hat{S}_{1z}\hat{S}_{2z}$ with eigenvalue $-\hbar^2/4$. The components of the spin operator $\hat{\mathbf{S}}_1$ commute with those of the spin operator $\hat{\mathbf{S}}_2$, and the square of the total angular momentum is therefore

$$\hat{\mathbf{S}}^2 = \hat{\mathbf{S}}_1^2 + \hat{\mathbf{S}}_2^2 + 2\hat{\mathbf{S}}_1 \cdot \hat{\mathbf{S}}_2, \qquad (7.65)$$

which may be rewritten as

$$\hat{\mathbf{S}}^2 = \hat{\mathbf{S}}_1^2 + \hat{\mathbf{S}}_2^2 + \hat{S}_{1+}\hat{S}_{2-} + \hat{S}_{1-}\hat{S}_{2+} + 2\hat{S}_{1z}\hat{S}_{2z}. \qquad (7.66)$$

We shall now use (7.66) to determine the eigenvalues and the corresponding eigenstates for the square of the total angular momentum and its component along the z-axis given by

$$\hat{S}_z = \hat{S}_{1z} + \hat{S}_{2z}. \qquad (7.67)$$

As our first step we let the operators (7.66) and (7.67) act on each of the states in (7.64) and observe that these are eigenstates for $\hat{S}_z$, but not for $\hat{\mathbf{S}}^2$ because of (7.59) and (7.60). The result is conveniently written in matrix form with $|i\rangle$ for $i = 1, 2, 3, 4$ denoting the four states in (7.64) in the order indicated. By using the orthonormality relation

$$\langle i|j\rangle = \delta_{ij}, \qquad (7.68)$$

which may be verified with the help of the representation (2.79-80), we shall now show that

$$\hat{S}_z/\hbar : \begin{pmatrix} 1 & 0 & 0 & 0 \\ 0 & 0 & 0 & 0 \\ 0 & 0 & 0 & 0 \\ 0 & 0 & 0 & -1 \end{pmatrix} \qquad (7.69)$$

and
$$\hat{S}^2/\hbar^2 : \begin{pmatrix} 2 & 0 & 0 & 0 \\ 0 & 1 & 1 & 0 \\ 0 & 1 & 1 & 0 \\ 0 & 0 & 0 & 2 \end{pmatrix}. \tag{7.70}$$

Let us verify (7.69) and (7.70) by working out the first two columns in each of the two matrices. The result of operating with $\hat{S}_z/\hbar = (\hat{S}_{1z} + \hat{S}_{2z})/\hbar$ on $\alpha\alpha$ is $\alpha\alpha$, while the corresponding result of acting with $\hat{S}_z/\hbar$ on $\alpha\beta$ is 0. This shows explicitly that $\hat{S}_z$ is diagonal in the basis (7.64). The result of operating with the square of the angular momentum (7.66) on $\alpha\alpha$ is $(3/4 + 3/4 + 2/4)\hbar^2\alpha\alpha$, since the third and the fourth terms in the sum (7.66) both yield zero on account of (7.59) and (7.60). This proves that the first column of the matrix is given by the set of numbers (2,0,0,0), since the states (7.64) are orthonormal. The second column is obtained by letting the operator (7.66) act on $\alpha\beta$ and using (7.56)-(7.60) with the result $\hbar^2(\alpha\beta + \beta\alpha)$. The second column is thus given by the set of numbers $(0, 1, 1, 0)$. The third and the fourth column may be obtained in precisely the same manner, or by using the evident symmetry.

The matrix (7.70) is not diagonal, but it is a simple matter to determine its eigenvalues which are 2 and 0. The eigenstates are

$$\alpha\alpha, \beta\beta, \frac{1}{\sqrt{2}}(\alpha\beta + \beta\alpha), \tag{7.71}$$

corresponding to the eigenvalue $2\hbar^2$, which shows that the quantum number for the total spin in this case is $1(= \frac{1}{2} + \frac{1}{2})$, while the eigenvalues associated with the z-component are $1, -1, 0$ in units of $\hbar$ for the three states in (7.71). The states (7.71) are said to constitute a triplet, since the eigenvalue for the square of the total spin is triply degenerate. The eigenvalue 0 for the square of the total spin is non-degenerate. The corresponding state is said to be a singlet, given by

$$\frac{1}{\sqrt{2}}(\alpha\beta - \beta\alpha), \tag{7.72}$$

which is also an eigenstate of $\hat{S}_z$ with the eigenvalue 0.

The process of diagonalizing the matrices (7.69) and (7.70) has evidently resulted in linear combinations (7.71) and (7.72) of a definite symmetry with respect to the interchange of the two electrons. The states (7.71) are seen to be symmetric, while (7.72) is antisymmetric. In the following chapter we shall see that identical spin-1/2 particles such as electrons must be described by states that are antisymmetric with respect to interchange of any two particles. This does not imply that the states (7.71) are unphysical, but only that the spatial part of the wave function must be antisymmetric to ensure the antisymmetry of the total wave function. Likewise, the antisymmetric singlet state (7.72) must be combined with a symmetric spatial wave function. This difference in the spatial wave functions provides a basis for understanding the difference in energy between singlet and triplet states, since a spatially antisymmetric wave function vanishes when the two electrons occupy the same point in space, while a spatially symmetric wave function in general remains non-zero. The contribution to the energy of the helium atom from the repulsion between its two electrons is therefore different for a symmetric and an antisymmetric spatial wave function. This may be expressed by means of an effective Hamiltonian which is

$$\hat{H}_{\text{eff}} = A\hat{S}_1 \cdot \hat{S}_2. \tag{7.73}$$

According to (7.65) the eigenvalues of $2\hat{S}_1 \cdot \hat{S}_2$ are equal to the eigenvalues of $(\hat{S}^2 - 3\hbar^2/2)$. This means that the eigenvalues of the Hamiltonian (7.73) are $A\hbar^2/4$ and $-3A\hbar^2/4$ corresponding to (7.71) and (7.72), respectively. One often uses a sum of terms of the form (7.73) as the Hamiltonian for magnetic materials (cf. Section 9.4).

The general method of adding angular momenta corresponds closely to the example given above. The expansion coefficients in (7.71) and (7.72) are called *Clebsch-Gordan coefficients*. In the example given they are 1 and $\pm 1/\sqrt{2}$. By adding two angular momenta $\hat{\mathbf{J}}_1$ and $\hat{\mathbf{J}}_2$ one finds in general that the eigenvalues of the square of the total angular momentum are given by $j(j+1)\hbar^2$, where j may assume the values

$$j_1 + j_2, j_1 + j_2 - 1, \cdots, |j_1 - j_2|, \tag{7.74}$$

with $j_1(j_1+1)\hbar^2$ and $j_2(j_2+1)\hbar^2$ being the eigenvalues for the squares of the two angular momentum vectors. The appropriate Clebsch-Gordan coefficients are found by diagonalizing a matrix with $(2j_1+1)(2j_2+1)$ rows and $(2j_1+1)(2j_2+1)$ columns corresponding to (7.70). Note that the two angular momenta may be an orbital and a spin angular momentum (in which case they are usually denoted by $\hat{\mathbf{L}}$ and $\hat{\mathbf{S}}$), but they may also both be spin angular momenta. The addition of an orbital angular momentum characterized by the quantum number $l = 1$ and a spin angular momentum with $s = 1/2$ is treated in Problem 7.1.

By adding the spin angular momenta of three spin-1/2 particles (Problem 7.5) one finds the following four states associated with $j = 3/2$,

$$\alpha\alpha\alpha, \frac{1}{\sqrt{3}}(\beta\alpha\alpha + \alpha\beta\alpha + \alpha\alpha\beta), \frac{1}{\sqrt{3}}(\alpha\beta\beta + \beta\alpha\beta + \beta\beta\alpha), \beta\beta\beta. \tag{7.75}$$

Addition of angular momenta appears in many different contexts within quantum physics. States of the form (7.75) appear for instance in the description of the quark structure of baryons.

7.4 Motion in a central field

When a particle moves in a central field it is possible, as shown in Section 7.1.1, to find simultaneous eigenfunctions for the Hamiltonian $\hat{H}$, the square $\hat{\mathbf{L}}^2$ of the orbital angular momentum and its component $\hat{L}_z$ along a given axis. The simultaneous eigenfunctions for $\hat{\mathbf{L}}^2$ and $\hat{L}_z$ are called spherical harmonics and denoted by $Y_{lm}(\theta, \phi)$. The eigenvalue equations are

$$\hat{\mathbf{L}}^2 Y_{lm}(\theta, \phi) = l(l+1)\hbar^2 Y_{lm}(\theta, \phi) \tag{7.76}$$

and

$$\hat{L}_z Y_{lm}(\theta, \phi) = m\hbar Y_{lm}(\theta, \phi). \tag{7.77}$$

According to (7.18) and (7.16) the energy operator is given by

$$\hat{H} = -\frac{\hbar^2}{2m}\left(\frac{1}{r^2}\frac{\partial}{\partial r}\left(r^2\frac{\partial}{\partial r}\right) - \frac{1}{\hbar^2 r^2}\hat{\mathbf{L}}^2\right) + V(r). \tag{7.78}$$

Since the potential energy V only depends on r, the solution ψ of the time-independent Schrödinger equation $\hat{H}\psi = E\psi$ may therefore be separated according to

$$\psi = R(r)Y_{lm}(\theta, \phi), \tag{7.79}$$

where the radial function R depends on the quantum number l as well as the energy eigenvalue E. The eigenvalue equation for the energy then becomes

$$-\frac{\hbar^2}{2m}\left(\frac{d^2}{dr^2} + \frac{2}{r}\frac{d}{dr} - \frac{l(l+1)}{r^2}\right)R(r) + V(r)R(r) = ER(r). \tag{7.80}$$

Instead of working with the radial equation in the form (7.80) we introduce as in Example 6 an auxiliary function, which will turn out to satisfy a one-dimensional Schrödinger equation. In the present case we introduce the auxiliary function $\chi(r) = rR(r)$, which must satisfy the boundary condition[1] $\chi \to 0$ for $r \to 0$. Since $d^2\chi/dr^2 = rd^2R/dr^2 + 2dR/dr$ the function χ is seen from (7.80) to satisfy the one-dimensional Schrödinger equation

$$\chi'' + \frac{2m}{\hbar^2}[E - V_{\text{eff}}(r)]\chi = 0, \tag{7.81}$$

where $\chi'' = d^2\chi/dr^2$ and V_{eff} denotes the effective potential given by

$$V_{\text{eff}} = V(r) + \frac{\hbar^2 l(l+1)}{2mr^2}. \tag{7.82}$$

---

[1] This boundary condition is satisfied when R tends towards a finite value or zero for $r \to 0$. For certain potentials the radial function $R(r)$ may diverge for $r \to 0$ (an example being $V(r) = -C/r^2$, where C is a positive constant), but the divergence in R for $r \to 0$ must be sufficiently weak that $\chi = rR \to 0$ for $r \to 0$.

7.4.1 The hydrogen atom

The solutions to (7.81) depend on the form of the potential. In the following we discuss hydrogen-like systems, for which $V(r) = -Ze_0^2/r$. Here Z is the atomic number, giving the charge of the nucleus, which is 1 for a hydrogen atom, while it is 2 for the He$^+$ ion and 3 for the Li^{++} ion. If we want to take into account the motion of the nucleus, we must replace the electron mass m by the reduced mass $\mu = mM/(m+M)$, where M is the mass of the nucleus, just as in the classical Kepler problem. The proof of this is obtained by expressing the operator for the total kinetic energy in relative coordinates and center-of-mass coordinates, cf. Problem 7.10.

The general solution of the eigenvalue problem (7.81) is discussed in Problem 7.11. Here we shall demonstrate how an approximate method valid for large quantum numbers allows a simple determination of the energies for the bound states. A similar procedure is used for other effective potentials, e. g. in the calculation of the vibrational energies of diatomic molecules. Subsequently we determine some simple exact solutions of the eigenvalue problem which will be used in the following chapter.

In the case $V(r) = -Ze_0^2/r$ the effective radial potential has the form

$$V_{\text{eff}}(r) = -\frac{A}{r} + \frac{B}{2r^2}, \tag{7.83}$$

where A and B are positive constants. The effective potential V_{eff} is seen to have a minimum for $r = r_{\min} = B/A$. By expanding V_{eff} to second order in $(r - r_{\min})$ we obtain

$$V_{\text{eff}} \simeq -\frac{A^2}{2B} + \frac{1}{2}K(r - \frac{B}{A})^2, \tag{7.84}$$

where $K = A^4/B^3$. To verify this result for the force constant K we may differentiate V_{eff} as given by (7.83) twice with respect to r and insert $r = r_{\min} = B/A$. This yields $K = -2A/r_{\min}^3 + 3B/r_{\min}^4 = A^4/B^3$ as stated. The energy eigenvalues are then given by the usual result for the harmonic oscillator,

$$E = -\frac{A^2}{2B} + (p + \frac{1}{2})\hbar\sqrt{\frac{A^4}{B^3 m}}, \tag{7.85}$$

where $p = 0, 1, 2, 3 \cdots$.

We now insert the expressions $A = Ze_0^2$ and $B = \hbar^2 l(l+1)/m$ and use the assumption $l \gg 1$, which implies that $1/l(l+1) \simeq l^{-2} - l^{-3}$. The energy eigenvalues may then be written as

$$E \simeq -\frac{mZ^2 e_0^4}{2\hbar^2}[\frac{1}{l^2} - \frac{1}{l^3} - 2(p+\frac{1}{2})\frac{1}{l^3}] \simeq -\frac{mZ^2 e_0^4}{2\hbar^2}\frac{1}{(l+p+1)^2}, \tag{7.86}$$

Angular momentum 169

provided that $p \ll l$ (we have used that $(1-\epsilon) \simeq (1+\epsilon)^{-1}$, if $\epsilon \ll 1$). With the definition $n = l + p + 1$ the result (7.86) then becomes

$$E_n = -\frac{mZ^2 e_0^4}{2\hbar^2} \frac{1}{n^2}, \tag{7.87}$$

the quantum number n being a positive integer. The energy spectrum (7.87) is the one derived by Bohr in 1913 (see Section 1.2.1). According to Bohr's model the number n may assume the values

$$n = 1, 2, 3 \cdots, \tag{7.88}$$

which is also the result obtained from the general solution to the differential equation (7.81), cf. Problem 7.11. Apart from the discrete negative eigenvalues (7.87) there is also a continuous spectrum of positive energies. Such positive-energy states describe an electron which is not bound to the hydrogen nucleus.

We now turn towards the determination of a few simple exact solutions to the eigenvalue equation (7.80) with the potential energy $V(r) = -Ze_0^2/r$. The differential equation (7.81) then becomes

$$\chi'' + \frac{2m}{\hbar^2} E\chi + \frac{2mZe_0^2}{\hbar^2 r}\chi - \frac{l(l+1)}{r^2}\chi = 0. \tag{7.89}$$

We are here only interested in the bound states, with energy $E < 0$, for which the solution goes exponentially towards zero for $r \to \infty$. In this region the third and the fourth term on the left hand side of (7.89) may be neglected compared to the second. At small r, however, the last term dominates the second and the third, and the equation is seen in this limit to be satisfied by functions of the type $\chi \propto r^{l+1}$ and $\chi \propto r^{-l}$, both of which satisfy $r^2\chi'' = l(l+1)\chi$. Since the boundary condition on χ at $r = 0$ is $\chi(0) = 0$ we may only use those solutions which are given by $\chi \propto r^{l+1}$ near $r = 0$.

Taking these considerations of the asymptotic behavior as our starting point we shall investigate whether functions of the form

$$\chi \propto r^{l+1} e^{-r/a} \tag{7.90}$$

satisfy (7.89) for all r. By inserting (7.90) into (7.89) one gets

$$\left(l(l+1) - \frac{2(l+1)r}{a} + \frac{r^2}{a^2}\right)r^{l-1}e^{-r/a} + \frac{2m}{\hbar^2}Er^{l+1}e^{-r/a}$$
$$+ \frac{2mZe_0^2}{\hbar^2 r}r^{l+1}e^{-r/a} - \frac{l(l+1)}{r^2}r^{l+1}e^{-r/a} = 0. \tag{7.91}$$

This equation is satisfied, if

$$a = \frac{(l+1)a_0}{Z}, \tag{7.92}$$

where a_0 is the Bohr radius given by

$$a_0 = \frac{\hbar^2}{me_0^2}, \qquad (7.93)$$

while

$$E = -\frac{e_0^2}{2a^2} = -\frac{mZ^2 e_0^4}{2\hbar^2 (l+1)^2} \qquad (7.94)$$

is the corresponding energy in agreement with the Bohr formula (7.87). The lowest energy is obtained for $l = 0$, namely the ground-state energy

$$E = -\frac{mZ^2 e_0^4}{2\hbar^2}. \qquad (7.95)$$

We have not proved that (7.95) is the lowest possible energy, but it appears plausible since the radial function has no zeros, while the corresponding spherical harmonic Y_{00} is a constant.

Even though the result (7.94) is identical to the Bohr formula (7.87), we have not found all possible solutions describing bound states. In Problem 7.11 we study the general solution to the eigenvalue equation (7.89), and we shall see how solutions of the form (7.90) are generalized by multiplication with a polynomial in r.

Let us illustrate this by showing that (7.89) with $l = 0$ has other solutions than the function $r\exp(-r/a)$, which is of the type (7.90). We may for instance insert a function of the form

$$\chi = (r + br^2)e^{-r/a} \qquad (7.96)$$

and attempt to determine the constants a, b and c, such that (7.89) with $l = 0$ is satisfied. The result is that the equation is satisfied if $b = -1/a$ and $a = 2a_0/Z$, while $E = -Z^2 e_0^2/8a_0$. The energy is thus given by the condition that the quantum number n in the Bohr formula (7.87) equals 2. The normalized wave function ψ_{200} corresponding to this energy is an eigenfunction of $\hat{L}^2$ and $\hat{L}_z$ with eigenvalues equal to zero, since the wave function only depends on r,

$$\psi_{200} = \frac{1}{2\sqrt{2\pi}} \frac{Z^{3/2}}{a_0^{3/2}} \left(1 - \frac{rZ}{2a_0}\right) e^{-rZ/2a_0}. \qquad (7.97)$$

The quantum numbers n, l and m are thus 2, 0 and 0, respectively. Note that the quantum number n equals $l + 1$ for states of the type (7.90). For a given value of l there are $2l + 1$ different values of m, cf. (7.50). In the case $n = 2$ we have thus determined four states with the same energy, namely ψ_{200}, ψ_{210}, ψ_{211} and ψ_{21-1}. Since there are no other linearly-independent solutions associated with this energy, the degree of degeneracy is $4 = 2^2$. In Table 7.2 we list all

Angular momentum

$$R_{10} = N_{10}e^{-\rho}, \qquad N_{10} = \frac{2}{a_0^{3/2}}$$

$$R_{20} = N_{20}(1 - \frac{\rho}{2})e^{-\rho/2}, \qquad N_{20} = \frac{1}{\sqrt{2}(a_0)^{3/2}}$$

$$R_{21} = N_{21}\rho e^{-\rho/2}, \qquad N_{21} = \frac{1}{2\sqrt{6}(a_0)^{3/2}}$$

$$R_{30} = N_{30}(1 - \frac{2\rho}{3} + \frac{2\rho^2}{27})e^{-\rho/3}, \qquad N_{30} = \frac{2}{3\sqrt{3}(a_0)^{3/2}}$$

$$R_{31} = N_{31}\rho(1 - \frac{\rho}{6})e^{-\rho/3}, \qquad N_{31} = \frac{4\sqrt{2}}{27\sqrt{3}(a_0)^{3/2}}$$

$$R_{32} = N_{32}\rho^2 e^{-\rho/3}, \qquad N_{32} = \frac{4}{81\sqrt{30}(a_0)^{3/2}}$$

$$R_{40} = N_{40}(1 - \frac{3\rho}{4} + \frac{\rho^2}{8} - \frac{\rho^3}{192})e^{-\rho/4}, \qquad N_{40} = \frac{1}{4(a_0)^{3/2}}$$

$$R_{41} = N_{41}\rho(1 - \frac{\rho}{4} + \frac{\rho^2}{80})e^{-\rho/4}, \qquad N_{41} = \frac{\sqrt{5}}{16\sqrt{3}(a_0)^{3/2}}$$

$$R_{42} = N_{42}\rho^2(1 - \frac{\rho}{12})e^{-\rho/4}, \qquad N_{42} = \frac{1}{64\sqrt{5}(a_0)^{3/2}}$$

$$R_{43} = N_{43}\rho^3 e^{-\rho/4}, \qquad N_{43} = \frac{1}{768\sqrt{35}(a_0)^{3/2}}$$

Table 7.2: Radial wave functions for the hydrogen atom with $\rho = r/a_0, a_0 = \hbar^2/me_0^2$, normalized according to (7.98).

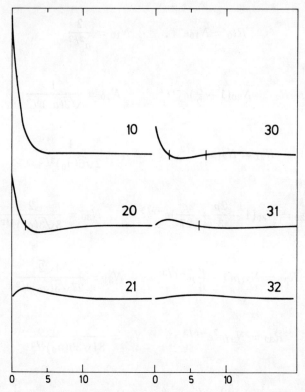

Figure 7.1: Normalized radial wave functions for hydrogen, plotted on the same vertical and horizontal scale as functions of r/a_0. The different functions are labelled by the quantum numbers n and l. The zeros are indicated by the vertical lines.

radial functions associated with the quantum numbers $n = 1, 2, 3$ and 4. The radial functions are normalized according to

$$\int_0^\infty dr\, r^2 |R_{nl}(r)|^2 = 1. \tag{7.98}$$

It is seen from the table that the different l-values belonging to a given n are $0, 1, 2, \cdots, n-1$. This rule is of general validity for an arbitrary n as shown in Problem 7.11. The degree of degeneracy g is therefore given by

$$g = \sum_{l=0}^{n-1}(2l+1) = n^2. \tag{7.99}$$

If in addition the spin of the electron is taken into account, the result (7.99) must be multiplied by 2, yielding $g = 2n^2$.

EXAMPLE 7. ZEEMAN EFFECT AND SPIN-ORBIT COUPLING.

The spin-orbit coupling, which is a relativistic effect, causes fine structure in the spectrum of hydrogen and other atoms. It is described by adding the term

$$H_1 = \frac{\hbar^2}{2m^2c^2} \frac{1}{r} \frac{dV(r)}{dr} \mathbf{L} \cdot \mathbf{S}, \qquad (7.100)$$

to the non-relativistic Hamilton operator. For simplicity we have here written the operators of angular momentum in dimensionless form, $\hbar\mathbf{L}$ being the operator for the orbital angular momentum of the electron, while $\hbar\mathbf{S}$ is the operator for the spin angular momentum. For convenience we leave out the 'hats' on the operators in this example.

For the hydrogen atom which we discuss in the present example the potential energy is given by $V(r) = -e_0^2/r$, where $e_0^2 = e^2/4\pi\epsilon_0$. According to (7.2) and (7.55) the magnetic field couples to both the orbital angular momentum of the electron and its spin angular momentum, corresponding to the term

$$H_2 = \mu_B B(L_z + 2S_z), \qquad (7.101)$$

where μ_B is the Bohr magneton, defined by $\mu_B = e\hbar/2m$. The term quadratic in B in the Hamiltonian (7.2) is of negligible importance in atoms: Its contribution may be estimated by replacing $\hat{x}^2 + \hat{y}^2$ with a_0^2, the square of the Bohr radius, yielding an energy of order $\mu_B B/(e_0^2/a_0)$ times $\mu_B B$.

In the following we shall determine the influence of H_1 and H_2 on the energy of the 2p-level, that is the energy of those states that are characterized by the quantum numbers $n = 2$ and $l = 1$ (in the commonly used spectroscopic notation s-states refer to $l = 0$, p-states to $l = 1$, d-states to $l = 2$, f-states to $l = 3$ etc.). We shall consider H_1 and H_2 to be small changes (so-called perturbations) of the Hamiltonian relative to the potential energy $-e_0^2/r$, which enters the unperturbed Hamiltonian H_0. We do not, however, make any assumptions about the relative significance of the two perturbations. This implies that we shall diagonalize the Hamiltonian $H = H_0 + H_1 + H_2$ within the subspace spanned by the eigenfunctions $|m_l, m_s\rangle$ in the case $l = 1, s = 1/2$. This procedure is equivalent to the use of first-order perturbation theory (see (8.17) in the following chapter) on the six ($= 3 \cdot 2$) times degenerate 2p-level, the unperturbed energy[2] being $E = -e_0^2/8a_0$.

The electron in the hydrogen atom moves in a central field, since the potential energy $V(r) = -e_0^2/r$ only depends on the length of the radius vector $\mathbf{r}$. The solution to the Schrödinger equation for the motion in the central field can be written in the form $R_{nl}(r)Y_{lm_l}(\theta, \phi)$, where Y_{lm_l} is a simultaneous eigenfunction for $\mathbf{L}^2$ and L_z with eigenvalue $l(l+1)$ and m_l, respectively.

The radial function of the 2p-state is R_{21}, which is proportional to $r\exp(-r/2a_0)$ cf. Table 7.2. This radial function belongs to all six linearly-independent basis states

---

[2] The remaining two states associated with the eightfold degenerate $n = 2$ level are s-states with quantum numbers $l = m_l = 0$ and $m_s = \pm 1/2$ and therefore unaffected by the spin-orbit interaction. The change in energy of these two states in a magnetic field is given by $\pm \mu_B B$.

$|m_l, m_s\rangle$ (for convenience we leave out the quantum numbers n and l), and the matrix associated with the operator H_1 is therefore proportional to the integral I given by

$$I = \int_0^\infty dr\, r e^{-r/a_0} = a_0^2, \qquad (7.102)$$

since $dV(r)/dr = e_0^2/r^2$. We conclude that it is sufficient to diagonalize the sum of the operators

$$H_a = a \mathbf{L} \cdot \mathbf{S} \qquad (7.103)$$

and

$$H_b = b(L_z + 2S_z). \qquad (7.104)$$

When the normalization constant in R_{21} is taken into account, the constant a is seen to be $\alpha^4 mc^2/48$, where α is the fine-structure constant (1.42), while b equals $\mu_B B$.

We have now arrived at the core of the problem: How does one diagonalize $H_a + H_b$ within the subspace under consideration? The operator H_b is trivial to handle, since it is 'born' diagonal with the given choice of basis. The eigenvalues are seen to be $2b, b, 0, -b, -2b$, with the eigenvalue 0 belonging to both $|1,-1/2\rangle$ and $|-1,+1/2\rangle$. If the spin-orbit coupling is ignored completely, we conclude that the 2p-level splits into five different energy levels. The distance between the neighboring levels is the same and seen to be proportional to the magnitude of the magnetic field. One of these levels, that belonging to the eigenvalue zero, is doubly degenerate.

The operator H_a is not diagonal. Its matrix elements may be determined by means of the identity

$$\mathbf{L} \cdot \mathbf{S} = L_z S_z + \frac{1}{2}(L_+ S_- + L_- S_+). \qquad (7.105)$$

The matrix elements are then found with the help of $L_- S_+ |0, -1/2\rangle = \sqrt{2}|-1, 1/2\rangle$ and $L_- S_+ |1, -1/2\rangle = \sqrt{2}|0, +1/2\rangle$ (as well as the corresponding relations for the Hermitian conjugate $L_+ S_-$).

When the basis states are ordered according to

$$|1, 1/2\rangle,\ |1, -1/2\rangle,\ |0, 1/2\rangle,\ |0, -1/2\rangle,\ |-1, 1/2\rangle,\ |-1, -1/2\rangle,$$

the matrix for the operator $H_a + H_b$ becomes

$$H_a + H_b : \begin{pmatrix} 2b + a/2 & 0 & 0 & 0 & 0 & 0 \\ 0 & -a/2 & a/\sqrt{2} & 0 & 0 & 0 \\ 0 & a/\sqrt{2} & b & 0 & 0 & 0 \\ 0 & 0 & 0 & -b & a/\sqrt{2} & 0 \\ 0 & 0 & 0 & a/\sqrt{2} & -a/2 & 0 \\ 0 & 0 & 0 & 0 & 0 & -2b + a/2 \end{pmatrix}.$$

This 6×6-matrix consists of four blocks, two 1×1-matrices and two 2×2-matrices, because of relations such as $L_- S_+ |-1, -1/2\rangle = 0$. The six eigenvalues λ are thus readily determined. We get

$$\lambda = \pm 2b + \frac{a}{2} \qquad (7.106)$$

Angular momentum 175

corresponding to the states $|1, 1/2\rangle$ and $|-1, -1/2\rangle$. The four remaining eigenvalues are

$$\lambda = \frac{1}{2}(\pm b - \frac{a}{2} + \sqrt{b^2 + (3a/2)^2 \pm ab}) \qquad (7.107)$$

together with

$$\lambda = \frac{1}{2}(\pm b - \frac{a}{2} - \sqrt{b^2 + (3a/2)^2 \pm ab}). \qquad (7.108)$$

In low magnetic fields ($b \ll a$) the result assumes a simple form; by performing a series expansion of λ/a in terms of b/a we see that four of the eigenvalues may be written as

$$\lambda = \frac{a}{2} + g_{3/2} m_j b, \quad \text{where} \quad m_j = \frac{3}{2}, \frac{1}{2}, -\frac{1}{2}, -\frac{3}{2}. \qquad (7.109)$$

Here we have introduced the g-factor $g_{3/2}$, which is seen from the series expansion to be $g_{3/2} = 4/3$. The two remaining levels become

$$\lambda = -a + g_{1/2} m_j b, \quad \text{where} \quad m_j = \frac{1}{2}, -\frac{1}{2}. \qquad (7.110)$$

In this case the g-factor is $g_{1/2} = 2/3$. For small fields ($b \ll a$) the splitting of the fine-structure levels may thus be described by associating different g-factors $g_{3/2} = 4/3$ and $g_{1/2} = 2/3$ with the levels belonging to $j = 3/2$ and $j = 1/2$, respectively. The quantum number j determines the eigenvalue $j(j+1)$ of the square $\mathbf{J}^2$ of the total angular momentum $\mathbf{J} = \mathbf{L} + \mathbf{S}$. The spin-orbit splitting in the absence of a magnetic field is determined by the value of j, since the eigenvalues of $\mathbf{L} \cdot \mathbf{S}$ may be obtained from

$$\mathbf{L} \cdot \mathbf{S} = (\mathbf{J}^2 - \mathbf{L}^2 - \mathbf{S}^2)/2. \qquad (7.111)$$

In the present case the eigenvalues of $\mathbf{L} \cdot \mathbf{S}$ are seen from (7.111) to be $j(j+1)/2 - 11/8$, which yields $1/2$ for $j = 3/2$ and -1 for $j = 1/2$, in agreement with (7.109) and (7.110) for $b = 0$.

The splitting described by (7.109) and (7.110) is called the anomalous Zeeman effect. The normal Zeeman effect is obtained when the fine-structure splitting is zero, $a = 0$, yielding the five different eigenvalues

$$\lambda = 2b, b, 0, -b, -2b, \qquad (7.112)$$

with the eigenvalue $\lambda = 0$ being doubly degenerate.

7.5 The spin of the electron

The Dirac equation for a freely-moving electron was given in Section 1.4.5 in the form

$$\hat{H}\psi = i\hbar \frac{\partial \psi}{\partial t}, \qquad (7.113)$$

where $\hat{H}$ is a linear operator in the spatial derivatives,

$$\hat{H} = c\boldsymbol{\alpha} \cdot \hat{\mathbf{p}} + \beta mc^2. \qquad (7.114)$$

The quantities α_i and β were represented by 4×4 matrices according to (1.135-139), and ψ is therefore a column vector with four components, cf. (1.140).

In the present section we shall start by discussing one of the consequences of the Dirac theory, namely the classical orbital angular momentum **L** is not a constant of the motion when the electron moves in a central field. This leads to a physical interpretation of the electron spin $\hbar\boldsymbol{\sigma}/2$ as an 'inner' angular momentum. The relativistic Hamiltonian in (7.113) for an electron moving in a central field described by the potential energy $V(r)$ is given by

$$\hat{H} = c\boldsymbol{\alpha}\cdot\hat{\mathbf{p}} + \beta mc^2 + V(r). \tag{7.115}$$

One might expect on the basis of classical mechanics that the operator for orbital angular momentum $\hat{\mathbf{L}} = \mathbf{r}\times\hat{\mathbf{p}}$ commutes with $\hat{H}$. This is however not the case since e. g. the commutator of $\hat{L}_z$ with $\hat{H}$ is nonvanishing,

$$[\hat{L}_z, \hat{H}] = i\hbar c(\alpha_x\hat{p}_y - \alpha_y\hat{p}_x). \tag{7.116}$$

The operator $\hat{L}_z$ should be interpreted as $(x\hat{p}_y - y\hat{p}_x)$ times a 4×4 unit matrix. Let us try to find an operator which does commute with $\hat{H}$. With this in mind we determine the commutator between $\hat{H}$ and the three 4×4 matrices

$$\sigma'_i = \begin{pmatrix} \sigma_i & 0 \\ 0 & \sigma_i \end{pmatrix}, \quad i = x, y, z. \tag{7.117}$$

Thus we find by using $[\sigma_z, \sigma_x] = 2i\sigma_y$ and $[\sigma_z, \sigma_y] = -2i\sigma_x$ that

$$[\sigma'_z, \hat{H}] = -2ic(\alpha_x\hat{p}_y - \alpha_y\hat{p}_x), \tag{7.118}$$

which together with (7.116) shows that $\hat{L}_z + \hbar\sigma'_z/2$ commutes with the Hamiltonian. Since the z-direction is equivalent to the x- or y-direction, it follows that all three components $\hat{L}_i + \hbar\sigma'_i/2$ commute with the Hamiltonian, thus being constants of the motion. We shall therefore interpret the quantity $\hat{L}_i + \hbar\sigma'_i/2$ as the total angular momentum of the particle.

When relativistic effects are treated as a small correction, it is only necessary to take into account the first two of the four components of the column vector ψ, since the last two are small (for a free particle with $E \simeq mc^2$ they are approximately p/mc times the first two, cf. (1.140)). This allows the total angular momentum **J** to be represented by 2×2-matrices, $\mathbf{J} = \mathbf{L} + \hbar\boldsymbol{\sigma}/2$, while the wave function is represented by a column vector with two elements ψ_1 and ψ_2.

7.5.1 The magnetic moment of the electron

The energy of an electron moving in a magnetic field **B** depends on the electron spin. This may be seen directly from the Hamiltonian (7.115), suitably

Angular momentum

modified to include the effect of the magnetic field through its vector potential **A**. According to Chapter 6 the Hamiltonian becomes

$$\hat{H} = c\boldsymbol{\alpha} \cdot (\hat{\mathbf{p}} + e\mathbf{A}) + \beta mc^2, \tag{7.119}$$

the charge of the electron being $-e$. For a constant magnetic field $(0, 0, B)$ the vector potential may be chosen as

$$A_x = 0, \quad A_y = Bx, \quad A_z = 0, \tag{7.120}$$

which is the Landau gauge discussed in Chapter 6. The wave equation corresponding to (7.113) is

$$(i\hbar \partial/\partial t - c\hat{p}_x \alpha_x - c\alpha_y(\hat{p}_y + eBx) - c\hat{p}_z \alpha_z - \beta mc^2)\psi = 0. \tag{7.121}$$

We multiply this equation with $(i\hbar\partial/\partial t + c\hat{p}_x\alpha_x + c\alpha_y(\hat{p}_y + eBx) + c\hat{p}_z\alpha_z + \beta mc^2)$,

$$(i\hbar\partial/\partial t + c\hat{p}_x\alpha_x + c\alpha_y(\hat{p}_y + eBx) + c\hat{p}_z\alpha_z + \beta mc^2)$$
$$(i\hbar\partial/\partial t - c\hat{p}_x\alpha_x - c\alpha_y(\hat{p}_y + eBx) - c\hat{p}_z\alpha_z - \beta mc^2)\psi = 0 \tag{7.122}$$

and obtain the result

$$(-\hbar^2 \partial^2/\partial t^2 - c^2(\hat{p}_x^2 + \hat{p}_z^2) - c^2(\hat{p}_y + eBx)^2 - \sigma'_z e\hbar c^2 B - (mc^2)^2)\psi = 0 \tag{7.123}$$

since

$$(\hat{p}_x \alpha_x + \alpha_y eBx)(-\hat{p}_x \alpha_x - \alpha_y eBx) = -\hat{p}_x^2 - e^2 B^2 x^2 - \hbar\sigma'_z eB. \tag{7.124}$$

We have used that $\sigma_x\sigma_y + \sigma_y\sigma_x = 0$ and $\sigma_x\sigma_y = i\sigma_z$ together with the commutation relation $[\hat{p}_x, x] = -i\hbar$. The Hamiltonian in the nonrelativistic limit is therefore obtained from the expansion

$$\hat{H} = \sqrt{(mc^2)^2 + c^2(\hat{p}_x^2 + (\hat{p}_y + eBx)^2 + \hat{p}_z^2) + c^2\hbar\sigma_z eB}$$
$$\simeq mc^2 + \frac{1}{2m}(\hat{p}_x^2 + (\hat{p}_y + eBx)^2 + \hat{p}_z^2) + \frac{\hbar}{2m}\sigma_z eB, \tag{7.125}$$

since the rest energy mc^2 in this limit is much greater than the energy due to the magnetic field. When this expression is compared with (7.2), we conclude that the electron spin s_i given by $s_i = \hbar\sigma_i/2$ corresponds to a magnetic moment $g_s(-e/2m)s_i$, where $g_s = 2$ is the g-factor introduced in (7.55).

In our discussion of the relativistic equation we have only considered the quantization of the motion of the electron, but treated the electromagnetic field as a classical quantity. The quantization of the electromagnetic field (resulting in photons) and the coupling between electrons and photons constitutes the topic of quantum electrodynamics (QED). As a result of the quantization of the electromagnetic field one finds that the g-factor of the electron differs slightly from $g_s = 2$. To first order in the fine-structure constant $\alpha = e_0^2/\hbar c$ the g-factor is given by

$$g_s = 2(1 + \frac{\alpha}{2\pi}) = 2.00232282. \tag{7.126}$$

It is possible to measure very accurately the g-factor of the electron, resulting in the value

$$g_s = 2.00231930438. \tag{7.127}$$

The difference between (7.126) and (7.127) is due to processes of higher order in α. When these are calculated according to quantum electrodynamics one finds perfect agreement with (7.127), to within the experimental error.

7.6 Problems

PROBLEM 7.1

a) Consider the set of operators

$$\hat{\mathbf{L}}, \ \hat{\mathbf{S}}, \ \hat{\mathbf{J}}, \ \hat{\mathbf{L}} \cdot \hat{\mathbf{S}}, \ \hat{\mathbf{L}}^2, \ \hat{\mathbf{S}}^2, \ \hat{\mathbf{J}}^2,$$

where $\hat{\mathbf{J}} = \hat{\mathbf{L}} + \hat{\mathbf{S}}$. Each component of $\hat{\mathbf{L}}$ commutes with each component of $\hat{\mathbf{S}}$, $[\hat{L}_i, \hat{S}_j] = 0$. Show that

$$\hat{L}_z, \ \hat{S}_z, \ \hat{\mathbf{L}}^2, \ \hat{\mathbf{S}}^2$$

is a commuting set of operators. Demonstrate that

$$\hat{\mathbf{J}}^2, \ \hat{J}_z, \ \hat{\mathbf{L}}^2, \ \hat{\mathbf{S}}^2$$

is also a commuting set of operators. Show finally that there are operators in the set

$$\hat{J}_z, \ \hat{L}_z, \ \hat{S}_z, \ \hat{\mathbf{J}}^2, \ \hat{\mathbf{L}}^2, \ \hat{\mathbf{S}}^2$$

which do not commute with each other.

b) Determine the 6 × 6 matrices corresponding to (7.69) and (7.70) for the total angular momentum obtained by adding orbital and spin angular momentum for a p-electron ($l = 1, s = 1/2$). Find the eigenvalues and the corresponding eigenstates.

Angular momentum

Problem 7.2
An electron moves in a harmonic oscillator potential $V(r) = Kr^2/2$. The Hamiltonian contains in addition to the kinetic and potential energy the spin-orbit coupling term

$$\hat{H}_{so} = \frac{1}{2m^2c^2}\frac{1}{r}\frac{dV}{dr}\hat{\mathbf{L}}\cdot\hat{\mathbf{S}} \qquad (7.128)$$

due to relativistic corrections.

a) Express $\hat{H}_{so}$ in terms of the total angular momentum $\hat{\mathbf{J}} = \hat{\mathbf{L}} + \hat{\mathbf{S}}$. Which of the operators $\hat{\mathbf{J}}, \hat{\mathbf{L}}, \hat{\mathbf{S}}, \hat{\mathbf{J}}^2, \hat{\mathbf{L}}^2$ and $\hat{\mathbf{S}}^2$ commute with the Hamiltonian?

b) Determine how the second-lowest energy level of the harmonic oscillator is split under the influence of the spin-orbit interaction and find the degree of degeneracy of the resulting levels. As in Problem 6.1 we assume that the spin-orbit coupling is sufficiently weak that the splitting may be determined by expressing the Hamiltonian as a matrix in the restricted basis of the eigenstates belonging to the particular energy level for the harmonic oscillator under consideration.

c) Examine the additional effect of a weak magnetic field (cf. Problem 6.1) on the energy levels determined in b). Investigate the cases where the effect of the magnetic field is either small or large compared to that of the spin-orbit interaction, as well as the general case.

Problem 7.3
It is assumed that the state of a particle is decribed by the normalized wave function,

$$\psi(r,\theta,\phi) = f(r)\sin^2\theta(\cos^2\phi - \sin^2\phi), \qquad (7.129)$$

in terms of the usual spherical coordinates, with $f(r)$ denoting a function of the radial variable r.

a) Examine whether ψ is an eigenfunction of the operators $\hat{\mathbf{L}}^2$, $\hat{L}_z$ and $\hat{L}_z^2$.

b) What is the probability of measuring the value $2\hbar$ for the z-component of the angular momentum?

c) What is the value of the mean-square deviation $\Delta^2(\mathbf{L}^2)$?

d) Show that the mean-square deviations satisfy

$$\Delta^2(L_x) = \Delta^2(L_y). \qquad (7.130)$$

Prove that $<L_\pm> = 0$ and therefore $<L_x> = <L_y> = 0$. Use this to demonstrate that

$$\Delta^2(L_x) = \Delta^2(L_y) = \frac{1}{4}\Delta^2(L_z). \qquad (7.131)$$

e) Sketch how the probability density associated with the wave function varies as a function of direction.

Problem 7.4
Show that the ground-state wave function for the isotropic three-dimensional harmonic oscillator is a simultaneous eigenfunction for $\hat{L}_x$, $\hat{L}_y$ and $\hat{L}_z$. Examine whether the three states $|1,0,0\rangle$, $|0,1,0\rangle$ and $|0,0,1\rangle$ associated with the second-lowest energy level (we use the same notation as in Problem 6.1) are eigenfunctions for these operators, and show how one may form linear combinations such that the resulting states are eigenfunctions for L_z. Determine the degree of degeneracy of the third-lowest energy level and show that one of the corresponding eigenstates may be chosen to have the form given in Problem 7.3.

Problem 7.5
Show that the states (7.75) are eigenstates for $\hat{S}^2$, where $\hat{S} = \hat{S}_1 + \hat{S}_2 + \hat{S}_3$ is the total spin angular momentum for three spin-1/2 particles, with eigenvalue $j(j+1)\hbar^2$, where $j = 3/2$.

Problem 7.6
In quantum mechanics one frequently employs the spherical coordinates defined by (7.14) when evaluating integrals. The volume element $dxdydz$ is given in spherical coordinates by
$$r^2 dr \sin\theta d\theta d\phi. \tag{7.132}$$
It is often convenient to use $\cos\theta$ as a variable in the integration (see below) by means of the identity $\sin\theta d\theta = -d(\cos\theta)$.

a) Use (7.15) and (7.16) to prove that the three functions
$$\cos\theta,\ \sin\theta e^{i\phi},\ \sin\theta e^{-i\phi} \tag{7.133}$$
are eigenfunctions for the square of the angular momentum and its component along the z-axis, and determine the corresponding eigenvalues.

b) Find the result of operating on each of these functions with the operator $\hat{O}$ given by
$$\hat{O} = e^{-i\phi}(i\frac{\partial}{\partial\theta} + \cot\theta\frac{\partial}{\partial\phi}). \tag{7.134}$$

c) We introduce an inner product $\langle\cdot|\cdot\rangle$ defined by
$$\langle\psi_i|\psi_j\rangle = \int_0^{2\pi} d\phi \int_{-1}^1 d(\cos\theta)\psi_i^* \psi_j. \tag{7.135}$$

By multiplying each of the functions given above with a normalization constant we are able to satisfy the orthonormality condition
$$\langle\psi_i|\psi_j\rangle = \delta_{ij}. \tag{7.136}$$

Angular momentum

Find the normalization constant for each of the three functions and determine the matrix for the operator $\hat{O}$ in this basis. Is the matrix Hermitian?

d) Determine the normalization constant for the function

$$\psi = \sin^2\theta(\cos^2\phi - \sin^2\phi). \tag{7.137}$$

Indicate the subspace of functions spanned by $\hat{O}^n\psi$, with n being an arbitrary positive integer.

Problem 7.7

We consider a particle in the state $|l,m\rangle$, which is a simultaneous eigenstate for $\hat{L}^2$ and $\hat{L}_z$. Show that the mean-square deviations $\Delta^2(L_x)$ and $\Delta^2(L_y)$ satisfy the relation

$$\Delta^2(L_x) + \Delta^2(L_y) = \hbar^2[l(l+1) - m^2]. \tag{7.138}$$

Find the smallest possible value of this quantity for a given value of l and discuss the classical limit, in which l tends towards infinity.

Problem 7.8

A particle of mass m is moving in a central field given by the potential energy $V(r)$ and a weak homogeneous magnetic field $\mathbf{B}$, described by the Hamiltonian (compare (7.2))

$$\hat{H} = \frac{\hat{\mathbf{p}}^2}{2m} + \frac{e}{2m}\mathbf{B}\cdot\hat{\mathbf{L}} + V(r). \tag{7.139}$$

Use (4.9) to show that the mean value of the angular momentum vector precesses with respect to the axis along $\mathbf{B}$ and determine the precession frequency. Compare the frequency of the motion with the cyclotron frequency. Show that $<L_+>$ satisfies a first-order differential equation in time and indicate its general solution.

Problem 7.9

In this problem we shall examine the motion of a particle of mass m in a three-dimensional well given by the potential

$$V(r) = -V_0 \text{ for } r \leq a; \quad V(r) = 0 \text{ otherwise,} \tag{7.140}$$

where V_0 is a positive constant. We shall only consider bound s-states in this problem, and the quantum number l is thus zero.

a) Write down the differential equation satisfied by the radial function $\chi = rR$ and determine the condition under which the well contains at least one s-state.

b) Determine an approximate expression for the energies E_n of the bound s-states in the case $|E_n| \ll V_0$.

Problem 7.10

The operator $\hat{T}$ for the total kinetic energy of two particles (mass m_1 and m_2) moving in one dimension is

$$\hat{T} = -\frac{\hbar^2}{2m_1}\frac{\partial^2}{\partial x_1^2} - \frac{\hbar^2}{2m_2}\frac{\partial^2}{\partial x_2^2}.$$

Introduce the center-of-mass coordinate

$$X = \frac{1}{m_1 + m_2}(m_1 x_1 + m_2 x_2)$$

and the relative coordinate

$$x = x_2 - x_1$$

and show that the kinetic energy can be written in the form

$$\hat{T} = -\frac{\hbar^2}{2M}\frac{\partial^2}{\partial X^2} - \frac{\hbar^2}{2\mu}\frac{\partial^2}{\partial x^2},$$

where $M = m_1 + m_2$, while μ is the reduced mass $\mu = m_1 m_2/(m_1 + m_2)$.

Problem 7.11

In this problem we shall discuss the solution of (7.89) in more detail and see how the general solution for the bound states with energy $E < 0$ may be determined. We introduce the following dimensionless variable

$$\epsilon = |E|\hbar^2/m_e Z^2 e_0^4, \qquad \rho = rmZe_0^2/\hbar^2 = rZ/a_0. \qquad (7.141)$$

With $'$ denoting the derivative of χ with respect to the dimensionless variable ρ the differential equation (7.89) becomes

$$\chi'' - 2\epsilon\chi + \frac{2}{\rho}\chi - \frac{l(l+1)}{\rho^2}\chi = 0, \qquad (7.142)$$

which is the starting point for the following. Note that ϵ is positive.

a) Show that the substitution

$$\chi = \rho^{l+1} L(\rho) \exp(-\sqrt{2\epsilon}\rho) \qquad (7.143)$$

results in the differential equation

$$\rho L'' + 2L'(l+1-\rho\sqrt{2\epsilon}) + 2L(1-(l+1)\sqrt{2\epsilon}) = 0. \qquad (7.144)$$

b) Discuss the case $L = $ const. and show how this recovers our earlier result for the energy (7.94).

Angular momentum

We shall now show that (7.142) may be satisfied by a polynomial solution of the form

$$L = \sum_{k=0}^{p} c_k \rho^k, \tag{7.145}$$

where p is a non-negative integer.

c) Show that ϵ is determined by setting the coefficient of ρ^p equal to zero according to

$$-p\sqrt{2\epsilon} + 1 - (l+1)\sqrt{2\epsilon} = 0 \tag{7.146}$$

and use this result to prove that the degree of degeneracy of an energy level with quantum number $n = l+p+1$ is equal to n^2 (it may be shown by the use of the power series method discussed in Problem 3.1 that there are no other normalizable bound-state solutions than those given by a polynomial of the form (7.145)).

d) Use (7.144) and (7.145) to demonstrate that the coefficients c_i may be expressed in terms of c_p. Determine the ratios c_0/c_2 and c_1/c_2 for the particular case $p = 2$ and compare with Table 7.2. The polynomials (7.145) are named Laguerre polynomials.

8 SYMMETRIES

Symmetry considerations are of fundamental importance in all areas of physics. When the Hamiltonian of a classical system is independent of one or more of the generalized coordinates, the corresponding generalized momentum is a constant of the motion. For a particle moving in an axially symmetric potential the angular momentum component along the symmetry axis is a constant of the motion. Spherical symmetry, which characterizes the motion in a central field, leads to conservation of all three components of the angular momentum.

In quantum mechanics there exist a number of important symmetries in addition to those known from classical mechanics. Some of these will be discussed in the present chapter. Symmetry operations in quantum mechanics are associated with linear operators. It is a characteristic of these operators that they are not necessarily Hermitian, implying that their eigenvalues are not necessarily real. We shall consider symmetry transformations described by unitary operators which preserve the norm of the states. The characteristic feature of a unitary operator U is that its inverse is equal to its Hermitian conjugate,

$$U^{-1} = U^\dagger, \tag{8.1}$$

implying that $UU^\dagger = 1$. Because of the property (8.1) the norm of a state is unchanged under a unitary operation,

$$||Uf||^2 = \langle Uf|Uf\rangle = \langle f|U^\dagger U f\rangle = \langle f|f\rangle = ||f||^2. \tag{8.2}$$

8.1 Parity

The potential energy $Kx^2/2$ of a one-dimensional harmonic oscillator is symmetric in x. This symmetry gives rise to the existence of a constant of the motion with no immediate classical analogue. The Hamiltonian for the harmonic oscillator is invariant under the symmetry transformation

$$x' = -x, \tag{8.3}$$

since

$$\hat{H} = -\frac{\hbar^2}{2m}\frac{d^2}{dx^2} + \frac{1}{2}Kx^2 = -\frac{\hbar^2}{2m}\frac{d^2}{dx'^2} + \frac{1}{2}Kx'^2. \tag{8.4}$$

The Hamiltonian evidently has the same appearance in terms of the primed and unprimed coordinates. This would not be true if the potential energy in addition to a quadratic term in x also contained a term of third order in x.

Quantum mechanics associates a linear operator with a symmetry transformation. The parity operator $\hat{P}$ is defined by

$$\hat{P}f(\mathbf{r}) = f(-\mathbf{r}). \tag{8.5}$$

Symmetries

The parity operator is both unitary and Hermitian. Its eigenvalues P' are ± 1. This may be seen as follows. For an arbitrary function $f(\mathbf{r})$ one has

$$\hat{P}^2 f = f. \tag{8.6}$$

If f is an eigenfunction for $\hat{P}$ belonging to the eigenvalue P' we have

$$\hat{P}^2 f = P'^2 f. \tag{8.7}$$

Taken together (8.6) and (8.7) result in $P'^2 = 1$ or $P' = \pm 1$.

The symmetry of the one-dimensional harmonic oscillator exhibited in (8.4) implies that the parity operator commutes with the Hamiltonian. The eigenstates for the Hamiltonian of the harmonic oscillator are eigenstates for the parity operator as well, with eigenvalues $(-1)^n$, where the quantum number n determines the energy eigenvalues according to (2.103). The ground-state wave function is thus an even function of x, the next one odd etc. Superpositions such as the minimal wave packets examined in Chapter 4 have no well-defined parity, unless the wave packet happens to be the ground state.

The fact that the parity operator may commute with the Hamiltonian does not, of course, imply that the eigenstates of the Hamiltonian necessarily have a definite parity. Let us consider as an example the electron in the hydrogen atom. The parity operator commutes with the Hamiltonian, since the potential energy is invariant under the replacement of $\mathbf{r}$ by $-\mathbf{r}$. The energy level associated with the quantum number $n = 2$, which determines the energy, is however fourfold degenerate, the degree of degeneracy being $4(= n^2)$, as long as spin is neglected. As we have seen in Section 7.4, the four linearly-independent eigenstates may be chosen to be

$$|0,0\rangle, |1,1\rangle, |1,0\rangle, |1,-1\rangle, \tag{8.8}$$

where $|l, m\rangle$ denotes[1] a simultaneous eigenstate for the square of the orbital angular momentum (with eigenvalue $l(l+1)\hbar^2$) and its z-component (with eigenvalue $m\hbar$). The parity of these states is $(-1)^l$, and it is therefore possible out of the four states to make linear combinations without any definite parity. The superpositions

$$\frac{1}{\sqrt{2}}(|0,0\rangle + |1,0\rangle), \quad \frac{1}{\sqrt{2}}(|0,0\rangle - |1,0\rangle) \tag{8.9}$$

have no definite parity. The mean value of the parity operator for each of the two superpositions (8.9) is zero. If we let the parity operator act on the first one, we get the last one and vice versa. In a basis consisting of the states

---
[1] Here we label the state with the value of the quantum number l instead of using $l(l+1)$ as in Chapter 7.

(8.9) the parity operator is therefore non-diagonal, equal to the Pauli matrix σ_x introduced in (2.76).

The fact that the states (8.9) are not eigenfunctions of the parity operator has important consequences for the effect of a small change in the Hamiltonian on their energies. Let us as an example consider a hydrogen atom in a constant electric field of field strength $\mathcal{E}$ in the direction of the z-axis. The potential energy V due to the electric field is given by

$$V = e\mathcal{E}z, \qquad (8.10)$$

where the coordinates of the electron relative to the position of the nucleus are (x, y, z). The contribution of (8.10) to the Hamiltonian has a definite parity (namely -1). The probability density associated with each of the four states in (8.8) has a definite parity (namely 1), and the mean value of the energy (8.10) is therefore zero in such a state. The probability density corresponding to the two states (8.9) does not, however, possess a definite parity. One therefore expects (as confirmed by the perturbation calculation given below), that the energy changes are proportional to $\mathcal{E}$, since the probability densities change under the replacement $z \to -z$. When the system is in a state given by one of the superpositions (8.9), it appears to possess a permanent electric dipole moment. Given that the energy change ΔE is proportional to the magnitude of the electric field strength $\mathcal{E}$, it follows from dimensional analysis that ΔE must have the form

$$\Delta E = \pm ce\mathcal{E}a_0, \qquad (8.11)$$

where c is a numerical constant. For reasons of symmetry one expects that the energy changes associated with each of the states (8.9) are equal in magnitude and opposite in sign[2]. The magnitude of the constant c turns out to be 3, when one calculates the expectation value of (8.10) using the known radial dependence of the wave functions associated with the states (8.8).

These considerations do not, however, take into account the possibility that linear combinations other than those given in (8.9) could be relevant. In order to show that this is not the case we shall make use of the fact that the contribution (8.10) to the Hamiltonian of the system commutes with $\hat{L}_z$, the z-component of the angular momentum of the electron. The mean value of z in a state based on an arbitrary linear combination of the states in (8.8) contains terms of the form

$$\langle l, m|z|l', m'\rangle, \qquad (8.12)$$

where $l \neq l'$ and $m \neq m'$. Now, the commutator $[z, \hat{L}_z]$ is zero, resulting in

$$\langle l, m|[z, \hat{L}_z]|l', m'\rangle = 0, \qquad (8.13)$$

---

[2] This is a consequence of the angular dependence of the corresponding spatial wave functions, resulting in probability densities of the form $(f(r) \pm g(r)\cos\theta)^2$, where f and g are functions of the length r of the position vector of the electron relative to the nucleus, while θ is the polar angle with respect to the direction of the electric field.

Symmetries

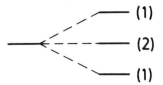

Figure 8.1: Splitting of the four times degenerate $n = 2$ level in the hydrogen atom.

which means that the matrix elements (8.12) must be zero, since the left hand side of (8.13) may be written as $(m' - m)\hbar$ times (8.12), and $m \neq m'$. Two of the four states in (8.8), corresponding to $m = \pm 1$, are therefore unaffected by the presence of a weak electric field.

The degeneracy without electric field is thus partially removed by (8.10). The level is split in three with a remaining doubly-degenerate level, which is unaffected by the electric field (Fig. 8.1).

We have seen that symmetry arguments may help us to understand how a hydrogen atom is affected by an electric field. A quantitative treatment of this problem requires a solution of the Schrödinger equation by the use of first-order perturbation theory. It is, however, of great importance in the applications of quantum theory to make use of the existence of symmetries, since these may lead to considerable simplifications in the actual calculation and a better understanding of the result.

8.1.1 Time-independent perturbation theory

Instead of relying on symmetry arguments as in the previous section we shall now demonstrate the use of perturbation theory in a direct calculation of the energy change due to a small disturbance. Let us consider a system with Hamiltonian $\hat{H}$ given by

$$\hat{H} = \hat{H}_0 + \lambda \hat{H}', \tag{8.14}$$

where $\hat{H}_0$ is the zero-order Hamiltonian, while $\lambda \hat{H}'$ represents the perturbation. The parameter λ is dimensionless. As in Section 5.4.2 it has been introduced to keep track of the order in the perturbation expansion. In the example mentioned in the previous section $\hat{H}_0$ is the Hamiltonian $\hat{p}^2/2m - e_0^2/r$ for the hydrogen atom, while $\hat{H}'$ is identified with $e\mathcal{E}z$. We wish to calculate the changes in energy to first order in $\hat{H}'$ (and hence to first order in the electric field strength $\mathcal{E}$ in the example considered). The wave function is separated into a zero-order part $\sum_{i=1}^{g} a_i \Phi_i$ and higher order terms. We have here allowed for a g-fold degeneracy of the unperturbed level by expanding the zero-order term in a basis consisting of orthonormal eigenfunctions Φ_i for $\hat{H}_0$ associated with the eigenvalue E_0. Thus

$$\psi = \sum_{i=1}^{g} a_i \Phi_i + \lambda \psi', \tag{8.15}$$

where $\lambda \psi'$ represents terms which are of first order in λ. For the moment the coefficients a_i are unknown. As we shall see, they are determined simultaneously with the energy changes $\lambda E'$, provided the original g-fold degenerate level splits into g different levels as a result of the perturbation.

The eigenvalue equation for $\hat{H}$ is now

$$(\hat{H}_0 + \lambda \hat{H}')(\sum_{i=1}^{g} a_i \Phi_i + \lambda \psi') = (E_0 + \lambda E')(\sum_{i=1}^{g} a_i \Phi_i + \lambda \psi').$$

In solving this to first order in λ we use that $\sum_{i=1}^{g} a_i \Phi_i$ is an eigenfunction for $\hat{H}_0$, thereby eliminating terms of zero order in λ. By discarding terms of higher than first order, multiplying the equation with Φ_j^* and integrating over the variables (which in the present example are the spatial coordinates of the electron) one arrives at the following matrix equation

$$\sum_{i=1}^{g} (\langle j|\hat{H}'|i\rangle - E'\delta_{ji}) a_i = 0, \tag{8.16}$$

where $\langle j|\hat{H}'|i\rangle = \langle \Phi_j|\hat{H}'\Phi_i\rangle$ is a $g \times g$ matrix. We have used that $\hat{H}_0$ is Hermitian, which implies that $\langle \Phi_j|\hat{H}_0 \psi'\rangle = E_0 \langle \Phi_j|\psi'\rangle$, and furthermore divided by λ. The system of equations (8.16) may be written in matrix form,

$$\begin{pmatrix} \langle 1|\hat{H}'|1\rangle - E' & \langle 1|\hat{H}'|2\rangle & \cdots & \langle 1|\hat{H}'|g\rangle \\ \langle 2|\hat{H}'|1\rangle & \langle 2|\hat{H}'|2\rangle - E' & \cdots & \langle 2|\hat{H}'|g\rangle \\ \vdots & \vdots & \cdots & \vdots \\ \langle g|\hat{H}'|1\rangle & \langle g|\hat{H}'|2\rangle & \cdots & \langle g|\hat{H}'|g\rangle - E' \end{pmatrix} \begin{pmatrix} a_1 \\ a_2 \\ \vdots \\ a_g \end{pmatrix} = \begin{pmatrix} 0 \\ 0 \\ \vdots \\ 0 \end{pmatrix} \tag{8.17}$$

The homogeneous system of equations (8.17) has non-trivial solutions (i. e. solutions differing from zero) provided the determinant of the matrix on the left hand side of (8.17) vanishes. The associated wave functions are determined for each root E' of the determinantal equation by inserting the root in question in (8.17) and solving for the coefficients $a_1, a_2, \cdots, a_g$. In the special case when the degree of degeneracy equals 1, the energy change E' is given by the mean value of $\hat{H}'$ in the state Φ, which is a solution of the eigenvalue equation $\hat{H}_0 \Phi = E_0 \Phi$,

$$E' = \langle \Phi|\hat{H}'\Phi\rangle. \tag{8.18}$$

Symmetries

> *Comment.* It is natural to enquire about the validity of first-order perturbation theory. The answer may be obtained by going to second or higher order in the perturbation (cf. Problem 8.6). As an alternative we may consider a system with only two non-degenerate eigenvalues E_1 and E_2, which are associated with a Hamiltonian $\hat{H}_0$. The corresponding wave functions are called Φ_1 and Φ_2. With the total Hamiltonian being $\hat{H}_0 + \hat{H}'$, the exact eigenvalues may be determined by diagonalizing the matrix
>
> $$\begin{pmatrix} E_1 + H'_{11} & H'_{12} \\ H'_{21} & E_2 + H'_{22} \end{pmatrix}.$$
>
> Here $H'_{ij} = \langle \Phi_i | \hat{H}' \Phi_j \rangle$ with $i = 1, 2$ and $j = 1, 2$ is the matrix for $\hat{H}'$ in the basis consisting of the eigenfunctions of $\hat{H}_0$. For simplicity we write the matrix in the form
>
> $$\begin{pmatrix} a & c \\ c^* & b \end{pmatrix},$$
>
> where $a = E_1 + H'_{11}$, $b = E_2 + H'_{22}$ and $c = H'_{12} = (H'_{21})^*$.
> The eigenvalues λ of this matrix are given by the solutions to the second order equation
>
> $$(a - \lambda)(b - \lambda) - |c|^2 = 0,$$
>
> resulting in
>
> $$\lambda = \frac{1}{2}(a + b \pm \sqrt{(a-b)^2 + 4|c|^2}).$$
>
> The result of using first-order perturbation theory is thus equivalent to neglecting the non-diagonal terms in the matrix for H', corresponding to the condition $|c| \ll |a - b|$. When this condition is satisfied, it is usually also true that $|H'_{11} - H'_{22}|$ is small compared to $|E_1 - E_2|$. The condition for the validity of first-order perturbation theory is thus
>
> $$|H'_{12}| \ll |E_1 - E_2|.$$
>
> Similar conditions are obtained in the case when the matrix for $\hat{H}$ has more than two rows and columns, cf. Problem 8.6, where the result of using second-order perturbation theory is derived for the non-degenerate case and applied to the ground-state energy of a hydrogen atom in an electric field.

The results (8.17-18) of first-order perturbation theory for the stationary case are among those used most frequently in quantum theory. We illustrate the use of (8.17) by determining the matrix $\langle j | \hat{H}' | i \rangle$ for the $n = 2$ level in the hydrogen atom with $\hat{H}' = e\mathcal{E}z$, cf. (8.10). As we saw on the basis of symmetry considerations, the only nonvanishing matrix elements are those involving $\langle 1, 0 | z | 0, 0 \rangle$ and $\langle 0, 0 | z | 1, 0 \rangle$. As shown in the previous chapter the radial function R_{20} associated with $n = 2$ and $l = m = 0$ is proportional to $(1 - r/2a_0) \exp(-r/2a_0)$. The normalized wave function corresponding to the state $|0,0\rangle$ is thus

$$|0,0\rangle : \quad \frac{1}{2\sqrt{2\pi}} \frac{1}{a_0^{3/2}} (1 - \frac{r}{2a_0}) e^{-r/2a_0}. \tag{8.19}$$

Similarly, we have seen that the radial function R_{nl} belonging to $n = 2$ and $l = 1$ is proportional to $r \exp(-r/2a_0)$. The normalized wave function associated with $|1, 0\rangle$ is therefore

$$|1,0\rangle : \quad \frac{1}{4\sqrt{2\pi}} \frac{1}{a_0^{3/2}} \frac{r}{a_0} e^{-r/2a_0} \cos\theta \tag{8.20}$$

since $\cos\theta$ is an eigenfunction for $\hat{\mathbf{L}}^2$ with eigenvalue $2\hbar^2$.

The matrix element $\langle 0, 0|z|1, 0\rangle$ may now be determined by integrating the product of the two wave functions (8.19) and (8.20) with $r\cos\theta$, resulting in $-3a_0$, as mentioned above. It follows that the distance between the energy levels in Fig. 8.1 is $3ea_0\mathcal{E}$. By inserting the roots of the determinantal equation in the system of equations it may be verified that the corresponding zero-order solutions are determined by the linear combinations in (8.9). Note that the double root $E' = 0$ does not determine any definite linear combination of $|1, 1\rangle$ and $|1, -1\rangle$. The remaining double degeneracy may be removed by applying a magnetic field.

8.2 Permutation

We now proceed to the consideration of the permutation symmetry of identical particles. Two electrons are identical in every respect (same mass, charge and spin), but an electron and a neutron are not, since the neutron may be distinguished from the electron by its mass and charge. This is true regardless of whether we describe the particles by classical or quantum mechanics. In classical mechanics, however, it is possible in principle to distinguish identical particles such as two electrons by following their orbits, no matter how close they might come to each other. In quantum mechanics this is not possible. The lack of commutativity between the position and momentum operators leads, as we have seen in Chapter 3, to the uncertainty relations and the description of particle motion in terms of wave packets, which have neither well-defined position nor momentum. Only in the classical limit does it make sense to introduce the concept of an orbit. When two electrons scatter against each other, it is impossible to keep track of the individual motion of the particles. It is therefore necessary in the description of such a scattering experiment to take into account the particular symmetry with respect to the interchange of identical particles.

The permutation operator $\hat{P}_{12}$ is a linear operator defined by

$$\hat{P}_{12}\psi(1,2) = \psi(2,1), \tag{8.21}$$

where 1 and 2 denote the spatial coordinates and spin for particle 1 and 2, respectively. Let us consider a state with two identical particles described by the wave function $\psi(1,2)$. As for the parity operator the eigenvalues are ± 1, since

$$\hat{P}_{12}^2 \psi(1,2) = \psi(1,2) \tag{8.22}$$

and

$$\hat{P}_{12}\psi(1,2) = P'_{12}\psi(1,2), \tag{8.23}$$

if $\psi(1,2)$ is an eigenfunction. It follows then from (8.22-23) that $P'_{12} = \pm 1$.

Particles with half-integer spin such as electrons, protons or neutrons are described by wave functions which change sign as a result of the interchange of two identical particles, corresponding to an eigenvalue $P' = -1$. Particles with integer spin, such as pions or ^{4}He atoms, are described by a wave function which is unchanged as a result of the interchange, corresponding to the eigenvalue $P' = 1$. The first kind of particles are called fermions, the second kind bosons. The connection between spin and permutation symmetry may be shown to be a consequence of Lorentz invariance, but within the framework of nonrelativistic quantum theory the connection is merely an empirical fact.

As an example of the significance of the symmetry with regard to the interchange of identical particles we shall now discuss the ground state of the He atom. The atom consists of a positively charged nucleus (nuclear charge $2e$) surrounded by two electrons. The Hamiltonian for the atom is

$$\hat{H} = \frac{\hat{\mathbf{p}}_1^2}{2m} + \frac{\hat{\mathbf{p}}_2^2}{2m} - \frac{2e_0^2}{r_1} - \frac{2e_0^2}{r_2} + \frac{e_0^2}{|\mathbf{r}_1 - \mathbf{r}_2|}. \tag{8.24}$$

Here $\hat{\mathbf{p}}_i$ denotes the momentum operator for each of the two electrons ($i = 1, 2$), while $\mathbf{r}_i$ is the position vector of the i'th electron with respect to the nucleus, and $e_0^2 = e^2/4\pi\epsilon_0$. Owing to the large ratio between the mass of the nucleus and the mass of the electron we shall disregard the motion of the nucleus. The Hamiltonian (8.24) is symmetric with respect to interchange of the two electrons, in accordance with the fact that electrons are identical particles.

We shall seek to determine the ground-state energy E_0 of the helium atom. By definition this is the smallest possible eigenvalue for the Hamiltonian, and it therefore follows that the operator $\hat{H} - E_0$ has eigenvalues, which are all either positive or equal to zero. The mean value of the operator $\hat{H} - E_0$ must therefore be positive or zero,

$$0 \leq < (H - E_0) > = < H > - E_0, \tag{8.25}$$

where we have used that the wave function is normalized. According to (8.25) the mean value $< H >$ is an upper limit to the ground-state energy. By choosing a properly normalized trial function and varying its parameters we obtain an approximation to E_0 which may be improved systematically by using other trial functions with a greater number of parameters than the original one.

This is the *variational method* for the determination of the ground-state energy of a quantum mechanical system. It is often convenient to write the inequality $E_0 \leq\, <H>$ in the form $E_0 \leq \langle\psi|\hat{H}\psi\rangle/\langle\psi|\psi\rangle$, since it may then be used with a wave function which is not normalized (cf. Example 8 below).

As a starting point for the selection of a reasonable trial function we shall now discuss the problem by neglecting completely the repulsive interaction between the electrons, represented by the last term in the Hamiltonian (8.24). This reduces the Hamiltonian to a sum of two mutually independent terms, each of which describes an electron moving around a positively charged nucleus with a charge twice as large as for the hydrogen atom.

Let us therefore first remind ourselves of the ground-state energy of the hydrogen-like atom with nuclear charge Ze. As discussed in the previous chapter it is possible to solve this eigenvalue problem exactly. Here we shall use a wave function that satisfies the time-independent Schrödinger equation with the lowest possible eigenvalue corresponding to $n = 1$,

$$\psi_{100}(\mathbf{r}) = \frac{1}{\sqrt{\pi a^3}} e^{-r/a}, \qquad (8.26)$$

where $a = a_0/Z$. Note that the state (8.26) is normalized.

It would seem advantageous from the point of view of minimizing the energy to choose a trial function with the property that both electrons occupy the lowest-energy state for the hydrogen-like system with $Z = 2$, but this appears to be in conflict with the requirement that the wave function should be antisymmetric. The product state

$$\psi_{100}(\mathbf{r}_1)\psi_{100}(\mathbf{r}_2) \qquad (8.27)$$

is symmetric with respect to the interchange of 1 and 2. However, we have not taken the spin of the electron into account in these considerations. As we saw in the previous chapter it is possible to construct spin states which are either symmetric (triplet, $s = 1$) or antisymmetric (singlet, $s = 0$) with respect to the interchange of the particles. By multiplying the spatially symmetric wave function (8.27) with the antisymmetric spin state (7.72) we obtain a state which has the right permutation symmetry, since the total wave function changes sign, when the two particles are interchanged.

The resulting wave function Ψ may be written in determinantal form

$$\Psi = \frac{1}{\sqrt{2}} \begin{vmatrix} \psi_{100}(\mathbf{r}_1)\alpha_1 & \psi_{100}(\mathbf{r}_2)\alpha_2 \\ \psi_{100}(\mathbf{r}_1)\beta_1 & \psi_{100}(\mathbf{r}_2)\beta_2 \end{vmatrix}, \qquad (8.28)$$

where we have used the notation for the spin states introduced below equation (7.64). This state is not an eigenstate for the operator (8.24), but it does possess the correct symmetry. We therefore expect it to give reasonable results, when it is used as a trial function in (8.25). One finds (cf. (8.44) in Example 8 below), that the energy of the He atom must be less than

$$-\frac{11}{4}\frac{e_0^2}{a_0}, \tag{8.29}$$

which is 0.95 times the experimentally determined value. Even better results are obtained if Z in (8.27) is considered to be a parameter, which may be varied to find the smallest upper bound (Example 8). The minimum is obtained for $Z = 27/16$, which may be interpreted as a *screening* of the nuclear charge to a value which is less than 2. The corresponding energy is

$$-(\frac{27}{16})^2\frac{e_0^2}{a_0} = -2.85\frac{e_0^2}{a_0}. \tag{8.30}$$

This result is 0.98 times the experimental value.

In the following chapter we shall make use of a wave function of the type (8.28) by generalizing it to the case of N particles, where N may denote the number of conduction electrons in a metal. In practice N may be of the order of 10^{23}. To generalize (8.28) we only need to replace $1/\sqrt{2}$ in the normalization factor in front of the determinant by $1/\sqrt{N!}$ and construct the determinant itself as in (8.28) by letting the rows denote N possible single-particle states in an arbitrarily chosen order, while the columns specify the coordinates of the N identical particles corresponding to an arbitrarily chosen numbering. Such a state is an eigenfunction for the Hamiltonian of a noninteracting gas of fermions. Instead of specifying the state in the form (8.28), which is called a *Slater determinant*, one may use the notation

$$|1_{i_1}, 1_{i_2}, \cdots, 1_{i_N}, 0, 0, \cdots\rangle. \tag{8.31}$$

In this N-particle state the single-particle states labelled $i_1, i_2, \cdots i_N$ are occupied, while the remaining ones are unoccupied.

EXAMPLE 8. THE HE ATOM AND THE H$^-$ ION.

By means of the variational method it is possible to find an upper limit to the ground-state energy E_0, since the mean value of the Hamiltonian in an arbitrary state is greater than or equal to E_0.

In the following we shall determine an upper limit to the ground-state energy for two-electron atoms such as the He atom or the H$^-$ ion. We use a trial function of the form

$$\psi(r_1, r_2) = \exp(-\lambda_1 r_1 - \lambda_2 r_2) + \exp(-\lambda_1 r_2 - \lambda_2 r_1), \tag{8.32}$$

where λ_1 and λ_2 are variational parameters. It is not necessary to normalize ψ, since the ground-state energy E_0 satisfies the inequality

$$E_0 \leq \frac{\langle\psi|H\psi\rangle}{\langle\psi|\psi\rangle}, \tag{8.33}$$

where H is the Hamiltonian[3]. The right hand side of (8.33) is unchanged if the trial function ψ is multiplied by an arbitrary constant. The inner product in (8.33) is defined in terms of integrals (extending from $-\infty$ to $+\infty$) over the cartesian coordinates of the two electrons,

$$\langle \Psi | \Phi \rangle = \int dx_1 \int dy_1 \int dz_1 \int dx_2 \int dy_2 \int dz_2$$
$$\Psi^*(x_1, y_1, z_1, x_2, y_2, z_2)\Phi(x_1, y_1, z_1, x_2, y_2, z_2).$$

The Hamiltonian for a two-electron system (with nuclear charge Ze) is given in atomic units ($\hbar = m_e = e_0 = 1$) as

$$H = -\frac{\nabla_1^2}{2} - \frac{\nabla_2^2}{2} - \frac{Z}{r_1} - \frac{Z}{r_2} + \frac{1}{r_{12}}. \tag{8.34}$$

Here $r_{12} = |\mathbf{r}_1 - \mathbf{r}_2|$, with $\mathbf{r}_i$ ($i = 1, 2$) denoting the position vectors for the two electrons. Note that neither relativistic effects nor the motion of the nucleus have been taken into account.

The Hamiltonian (8.34) is symmetric with respect to the interchange of the two electrons. Since the electrons are identical fermions, the trial function must be multiplied by an antisymmetric spin state, in order that the requirement of antisymmetry of the total wave function is fulfilled. The Hamiltonian (8.34) does not depend on the spin operators of the electrons, and we may therefore suppress the spin state (which is an eigenstate for the square of the total spin with eigenvalue zero) when determining the mean value.

In calculating the mean value (8.33) we separate the Hamiltonian into two parts, one involving the kinetic energy of the two electrons and their interaction with the nucleus, while the other represents their repulsive interaction. Thus we define

$$H_i^0 = -\frac{\nabla_i^2}{2} - \frac{Z}{r_i}, \tag{8.35}$$

where $i = 1, 2$, and determine first the mean value of $H_1^0 + H_2^0$. It is advantageous to use polar rather than cartesian coordinates, since the trial function (8.32) only involves r_1 and r_2. The angular part of the Laplace operators gives therefore no contribution to the mean value of the kinetic energy, which is determined solely by the radial part. In calculating the mean value of the kinetic energy we use that

$$-\frac{1}{2r_i^2}\frac{d}{dr_i}\left(r_i^2 \frac{d}{dr_i} e^{-\lambda r_i}\right) = -\left(\frac{\lambda^2}{2} - \frac{\lambda}{r_i}\right)e^{-\lambda r_i}. \tag{8.36}$$

The subsequent integrations over the radial variables r_1 and r_2 are elementary and result in

$$\frac{\langle \psi | (H_1^0 + H_2^0) \psi \rangle}{\langle \psi | \psi \rangle} = -\frac{\lambda_1^2 + \lambda_2^2}{2} + \left(\frac{1}{\lambda_1^3 \lambda_2^3} + \frac{64}{(\lambda_1 + \lambda_2)^6}\right)^{-1}$$
$$\cdot \left[(\lambda_1 - Z)\left(\frac{1}{\lambda_1^2 \lambda_2^3} + \frac{32}{(\lambda_1 + \lambda_2)^5}\right) + (\lambda_2 - Z)\left(\frac{1}{\lambda_1^3 \lambda_2^2} + \frac{32}{(\lambda_1 + \lambda_2)^5}\right)\right]. \tag{8.37}$$

---

[3] For convenience we leave out the 'hats' on the operators everywhere in this and the following Example 9.

Symmetries

Note that this expression is symmetric with respect to interchange of λ_1 and λ_2. The smallest value of the function (8.37) is obtained when $\lambda_1 = \lambda_2 = Z$, corresponding to the energy $-Z^2$. This is what we should expect; in the absence of any interaction between the electrons the total energy equals twice the ground-state energy $-Z^2/2$ for a hydrogen-like system with nuclear charge Z.

Now we proceed to the calculation of the mean value of the interaction $H' = 1/r_{12}$ between the electrons. This interaction depends not only on the length r_1 and r_2 of the position vectors of the electrons but also on the angle between them, since

$$H' = \frac{1}{\sqrt{r_1^2 + r_2^2 - 2r_1 r_2 \cos\theta}}, \tag{8.38}$$

where θ is the angle between $\mathbf{r}_1$ and $\mathbf{r}_2$. First we carry out the integration over θ by choosing a polar axis along $\mathbf{r}_1$. By introducing the variable $x = \cos\theta$ we obtain

$$\frac{1}{2}\int_{-1}^{+1} dx \frac{1}{\sqrt{a - bx}} = (\sqrt{a+b} - \sqrt{a-b})/b, \tag{8.39}$$

where $a = r_1^2 + r_2^2$ and $b = 2r_1 r_2$. The value of (8.39) is seen to be $1/r_1$ if $r_1 > r_2$, and $1/r_2$ if $r_1 < r_2$. It is therefore necessary to divide the subsequent integration over the radial variables in two regions corresponding to $r_1 > r_2$ and $r_1 < r_2$.

The radial integrations therefore involve integrals of the form

$$I(\alpha,\beta) = \int_0^\infty dy\, y\, e^{-\alpha y} \int_0^y dz\, z^2 e^{-\beta z} \tag{8.40}$$

and

$$J(\alpha,\beta) = \int_0^\infty dy\, y^2 e^{-\alpha y} \int_y^\infty dz\, z\, e^{-\beta z}. \tag{8.41}$$

By carrying out these integrations it is seen that the sum $I + J$ becomes

$$I(\alpha,\beta) + J(\alpha,\beta) = \frac{2}{\alpha^2 \beta^3} - \frac{2}{(\alpha+\beta)^2 \beta^3} - \frac{2}{\beta^2(\alpha+\beta)^3}. \tag{8.42}$$

The resulting mean value of the interaction is therefore

$$\frac{\langle\psi|H'\psi\rangle}{\langle\psi|\psi\rangle} = \left(\frac{1}{\lambda_1^3 \lambda_2^3} + \frac{64}{(\lambda_1+\lambda_2)^6}\right)^{-1}\left[\frac{\lambda_1+\lambda_2}{2\lambda_1^3 \lambda_2^3} - \frac{\lambda_1^3 + \lambda_2^3}{2(\lambda_1+\lambda_2)^2 \lambda_1^3 \lambda_2^3}\right.$$
$$\left. - \frac{\lambda_1^2 + \lambda_2^2}{2(\lambda_1+\lambda_2)^3 \lambda_1^2 \lambda_2^2} + \frac{20}{(\lambda_1+\lambda_2)^5}\right]. \tag{8.43}$$

As in (8.37) the quantity raised to the exponent -1 originates in the denominator $\langle\psi|\psi\rangle$ of (8.33).

This concludes the determination of the mean value of the Hamiltonian (8.34) in the state given by (8.32), since the right hand side of (8.33) is identified with the sum of (8.37) and (8.43). In the following we shall determine different upper limits on E_0 corresponding to different choices of the parameters λ_1 and λ_2.

A. $\lambda_1 = \lambda_2 = Z$.

This choice of parameters represents the exact solution for the case where we neglect H' entirely. The value of (8.43) then becomes $5Z/8$. For the He atom ($Z=2$) the corresponding ground-state energy is determined by the inequality

$$E_0 \leq -\frac{11}{4}, \quad \text{(He)} \tag{8.44}$$

while the ground-state energy of the H^- ion has the upper limit given by

$$E_0 \leq -\frac{3}{8}. \quad (H^-) \tag{8.45}$$

These results are in agreement with the result of using first-order perturbation theory, since the trial function has been chosen as a solution to the eigenvalue equation for the Hamiltonian $H_1^0 + H_2^0$. The upper limit given by the variational expression is thus an exact result to first order in the perturbation, when the trial function has been chosen as the ground-state wave function for the unperturbed Hamiltonian. We further note, that the H^- ion is unstable in this approximation, since (8.45) is larger than the ground-state energy $-1/2$ for a hydrogen atom and an isolated electron. This result does not, of course, imply that the H^- ion is unstable, but only that our choice of trial function is inadequate.

B. $\lambda_1 = \lambda_2 = \lambda$.

In this case the expression (8.37) equals $\lambda^2 - 2\lambda Z$, while (8.43) becomes $5\lambda/8$. The minimum value of the sum of these two terms is achieved for $\lambda = Z - 5/16$, resulting in $-(Z - 5/16)^2$. It follows from this that

$$E_0 \leq -(Z - \frac{5}{16})^2. \tag{8.46}$$

For the helium atom the right hand side of (8.46) equals -2.848, which is an improvement on the perturbation result (8.44). For the H^- ion the right hand side equals -0.4727, which is still greater than -0.5. To obtain a stable system with energy less than -0.5 it is therefore necessary to let the two parameters λ_1 and λ_2 vary independently of each other.

C. Variation of both λ_1 and λ_2.

By calculating the sum of (8.37) and (8.43) as functions of λ_1 and λ_2 one finds for the H^- ion ($Z = 1$) that the smallest value is obtained for $\lambda_1 = 1.039$ and $\lambda_2 = 0.283$. Below we give a table of different values of the sum obtained for parameter choices λ_1, λ_2 in the vicinity of the set which gives the lowest energy. It is evident from these numbers that

$$E_0 \leq -0.51330. \quad (H^-) \tag{8.47}$$

We have thus shown that the H^- ion is stable with a binding energy which is greater than or equal to 0.01330 in atomic units ($\hbar = m_e = e_0 = 1$) or 0.3619 eV.

λ_1	λ_2	energy (a.u.)
1.038	0.282	-0.51330163
1.039	0.282	-0.51330217

1.040	0.282	-0.51330176
1.038	0.283	-0.51330218
1.039	0.283	-0.51330284
1.040	0.283	-0.51330256
1.038	0.284	-0.51330176
1.039	0.284	-0.51330254
1.040	0.284	-0.51330238

The members of the best pair of λ-values are evidently very different, the characteristic length $1/\lambda_2$ being almost four times as long as $1/\lambda_1$. From a classical point of view we might be tempted to interpret the wave function associated with the optimum pair of parameters $(\lambda_1, \lambda_2) = (1.039, 0.283)$ as indicating that one electron orbits the nucleus at a distance of the order of the Bohr radius, while the orbit of the other electron has a radius almost four times as large. Such a picture should not be taken seriously; the wave function (8.32) is a sum of products of single-particle wave functions. The absolute square of the wave function, which represents a probability density in the six-dimensional space $(x_1, y_1, z_1; x_2, y_2, z_2)$, contains in addition to the squares of each single-particle wave function also a double-product term. Therefore it cannot be identified with a simple product of probability densities associated with each electron.

The following example involves both the parity and permutation symmetries discussed above. We shall study the motion of a particle in a symmetric potential with two equivalent minima, using the ammonia molecule as an example, and discuss the related problem involving the H_2^+ ion. The results for the H_2^+ ion will be used to illustrate the nature of the covalent chemical bond for the hydrogen molecule.

EXAMPLE 9. THE AMMONIA MOLECULE, THE MASER AND THE CHEMICAL BOND.

In the present example we shall first examine a simple model for the ammonia molecule and use it to investigate how the inversion frequency depends on the height of the barrier separating the two potential minima for the nitrogen atom. Subsequently we discuss the maser as an example of a system which may be described in a basis consisting of two states. Our results will be used to obtain a qualititative understanding of the nature of the chemical bond, using the hydrogen molecule as an example.

The ammonia molecule is illustrated in Fig. 8.2. The inversion involves the motion of the nitrogen atom from the position shown in the figure to the symmetric position below the plane of the three hydrogen atoms. In the following we shall assume that the potential energy of the nitrogen atom in its motion perpendicular to the plane of the hydrogen atoms is given by

$$V(z) = \frac{V_0}{z_0^4}(z^2 - z_0^2)^2, \qquad (8.48)$$

where V_0 is a positive constant, while z denotes the perpendicular distance of the nitrogen atom from the plane in which the three hydrogen atoms are situated.

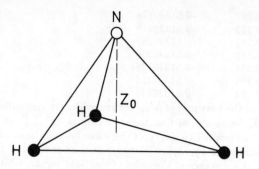

Figure 8.2: The ammonia molecule.

The Hamiltonian is given by

$$H = T + V(z), \tag{8.49}$$

where T is the kinetic energy,

$$T = -\frac{\hbar^2}{2\mu}\frac{d^2}{dz^2}. \tag{8.50}$$

Here μ is an effective mass, which is considerably less than the mass of the nitrogen atom (see below).

In the region around $z = \pm z_0$ the potential V may be approximated by the harmonic oscillator potential $4V_0(z \mp z_0)^2/z_0^2$. The force constant is $K = 8V_0/z_0^2$, which means that the associated energy quantum $\hbar\omega$ becomes

$$\hbar\omega = \sqrt{\frac{8V_0\hbar^2}{\mu z_0^2}}. \tag{8.51}$$

The characteristic length a of the oscillator is

$$a = \sqrt{\frac{\hbar}{\mu\omega}}. \tag{8.52}$$

The quantity

$$\epsilon = \frac{a^2}{z_0^2} = \sqrt{\frac{\hbar^2}{8\mu V_0 z_0^2}} \tag{8.53}$$

is a measure of the size of the oscillator wave function relative to the distance between the potential minima. From comparing (8.51) with (8.53) it is seen that

$$\epsilon = \frac{\hbar\omega}{8V_0}. \tag{8.54}$$

When ϵ is small, the inversion frequency is much less than ω.

Symmetries

The model potential (8.48) cannot be used for a quantitative description of the ammonia molecule, but its simple form makes it suitable for illustrating the difference between the characteristic inversion and vibration frequencies. The potential energy (8.48) is an even function of z and therefore commutes with the parity operator. The eigenfunctions of the Hamiltonian are either even or odd, since the energy levels are nondegenerate. The ground-state wave function is an even function of z. We shall determine the lowest and the second-lowest energy eigenvalues approximately by using the variational principle with trial functions of the form

$$\psi_\pm = N_\pm (e^{-(z-z_0)^2/2a^2} \pm e^{-(z+z_0)^2/2a^2}), \tag{8.55}$$

where $N_\pm$ is a real normalization constant. According to the variational principle $\langle \psi_+|H\psi_+\rangle$ is an upper limit to the ground-state energy. Correspondingly, $\langle \psi_-|H\psi_-\rangle$ is an upper limit on the lowest eigenvalue associated with the odd eigenstates. The difference ΔE_1 between these two energies is therefore approximately given by

$$\Delta E_1 = \langle \psi_-|H\psi_-\rangle - \langle \psi_+|H\psi_+\rangle. \tag{8.56}$$

The normalization condition on $\psi_\pm$ is

$$\int_{-\infty}^{\infty} dz\, \psi_\pm^2 = 1, \tag{8.57}$$

which determines the constants $N_\pm$ to be given by

$$N_\pm^{-2} = 2a\sqrt{\pi}(1 \pm e^{-z_0^2/a^2}). \tag{8.58}$$

We may now evaluate $\langle \psi_\pm|T\psi_\pm\rangle$ and $\langle \psi_\pm|V\psi_\pm\rangle$ by using that

$$\frac{d^2\psi_\pm}{dz^2} = N_\pm [((z-z_0)^2 a^{-4} - a^{-2})e^{-(z-z_0)^2/2a^2} \pm ((z+z_0)^2 a^{-4} - a^{-2})e^{-(z+z_0)^2/2a^2}]. \tag{8.59}$$

In the calculation of $\langle \psi_\pm|T\psi_\pm\rangle$ one needs the following integrals

$$\int_{-\infty}^{\infty} dz\, e^{-z^2} = \sqrt{\pi}, \quad \int_{-\infty}^{\infty} dz\, z^2 e^{-z^2} = \frac{\sqrt{\pi}}{2}, \quad \int_{-\infty}^{\infty} dz\, z^4 e^{-z^2} = \frac{3\sqrt{\pi}}{4}. \tag{8.60}$$

The mean value of the kinetic energy is therefore

$$\langle \psi_\pm|T\psi_\pm\rangle = \frac{1}{4}\hbar\omega \frac{1 \mp (\frac{2}{\epsilon} - 1)e^{-1/\epsilon}}{1 \pm e^{-1/\epsilon}}. \tag{8.61}$$

In a similar manner one obtains for the potential energy

$$\langle \psi_\pm|V\psi_\pm\rangle = \frac{1}{4}\hbar\omega \frac{(1 + \frac{3\epsilon}{8}) \pm (\frac{1}{2\epsilon} - \frac{1}{2} + \frac{3\epsilon}{8})e^{-1/\epsilon}}{1 \pm e^{-1/\epsilon}}. \tag{8.62}$$

By treating both $\exp(-1/\epsilon)$ and ϵ as small quantities one obtains

$$\langle \psi_\pm|H\psi_\pm\rangle \simeq \frac{1}{2}\hbar\omega(1 \mp \frac{3}{4\epsilon}e^{-1/\epsilon}) \tag{8.63}$$

and hence
$$\Delta E_1 = \hbar\omega \frac{3}{4\epsilon} e^{-1/\epsilon}. \tag{8.64}$$

The quantity ΔE_1 is thus the splitting in energy of the degenerate level $\hbar\omega/2$ appropriate to the situation when the two potential minima are infinitely separated. The degeneracy reflects the fact that the energy is the same whether the particle is at one or the other minimum, when the distance between them tends towards infinity.

For the excited oscillator level $3\hbar\omega/2$ one finds in a similar way the energy difference ΔE_2 given by
$$\Delta E_2 = \langle \phi_-|H\phi_-\rangle - \langle \phi_+|H\phi_+\rangle \tag{8.65}$$
by use of the functions
$$\phi_\pm = C_\pm((z-z_0)e^{-(z-z_0)^2/2a^2} \mp (z+z_0)e^{-(z+z_0)^2/2a^2}). \tag{8.66}$$

The result corresponding to (8.64) then becomes
$$\Delta E_2 = \hbar\omega \frac{3}{2\epsilon^2} e^{-1/\epsilon}. \tag{8.67}$$

It should be emphasized that the expressions (8.64) and (8.67) are not exact, since they have been determined from the mean value of the Hamiltonian H in states which are not eigenstates of the Hamiltonian. As long as ϵ is small, it is reasonable to assume that the expressions (8.64) and (8.67) are good approximations to the splitting of the two oscillator levels $\hbar\omega/2$ and $3\hbar\omega/2$.

The calculated energy differences will now be compared with experiment. The vibrational wave number $\bar{\nu} = \omega/2\pi c$ is determined experimentally to be 950 cm^{-1}, while the measured values of the inversion wave numbers $\bar{\nu}_1 = \Delta E_1/2\pi\hbar c$ and $\bar{\nu}_2 = \Delta E_2/2\pi\hbar c$ are 0.794 cm^{-1} and 36 cm^{-1}, respectively. By inserting the experimentally determined values of $\hbar\omega$ and ΔE_1 in (8.64) we obtain $\epsilon = 0.111$, which may be inserted into the expression
$$\frac{\Delta E_1}{\Delta E_2} = \frac{\epsilon}{2} \tag{8.68}$$

with the result 0.056. The measured value of this ratio is 0.794/36 or 0.022. We may also use the value $\epsilon = 0.111$ to determine V_0 from $\hbar\omega$ by inserting ϵ into (8.54). The result is $V_0 = 0.1326$ eV. Now we can use (8.53) to determine z_0 from the parameters given together with the reduced mass μ, which we shall assume to be equal to $3m_H m_N/(m_N + 3m_H) = 2.49$u, where $m_H(m_N)$ denotes the mass of a hydrogen atom (nitrogen atom). As a result one finds that z_0 is equal to 0.36 Å, while the experimentally determined value is 0.38 Å.

We have used the measured inversion frequency of 23.8 GHz (corresponding to the wave number 0.794 cm^{-1}) and the vibrational frequency $\hbar\omega$ to determine ϵ and V_0. When these parameters are used to calculate other measurable quantities such as the inversion frequency associated with the second-lowest level, we do not get quantitative agreement, although the calculated value of z_0 is in reasonable agreement with the known structure of the molecule. In order to improve the agreement one may choose a potential with more adjustable parameters to be determined by comparison with experimental results.

To elucidate the importance of the reduced mass we may consider the molecule NT$_3$, where T denotes tritium. One would expect the inversion frequency to be considerably lowered because of the increase in reduced mass, which in the present case is $\mu_T = 5.50$ u. If the parameters V_0 and z_0 are unchanged, $\epsilon_T = 0.0747$ and the calculated value of ΔE_1 for NT_3 is therefore $\exp(1/0.0747)/\exp(1/0.111) = 81$ times as small as for NH_3. Experimentally one finds that the inversion frequency for NT_3 is 306 MHz or 78 times as small as for NH_3.

Instead of using a definite Hamiltonian for the ammonia molecule with a potential energy such as (8.48), which in any case does not describe the molecule quantitatively, we shall now describe the inversion by approximating the Hamiltonian by a 2×2 matrix. As a basis we use the states $|\uparrow\rangle$ and $|\downarrow\rangle$, which correspond to the nitrogen atom being situated above and below the plane of the hydrogen atoms, respectively. With this assumption the state vector $|\psi\rangle$ may be expressed by the superposition

$$|\psi\rangle = c_\uparrow |\uparrow\rangle + c_\downarrow |\downarrow\rangle. \tag{8.69}$$

We now insert $|\psi\rangle$ in the time-dependent Schrödinger equation

$$i\hbar \frac{\partial}{\partial t} |\psi\rangle = H |\psi\rangle. \tag{8.70}$$

By forming the inner product of this equation with the basis vectors $|\uparrow\rangle$ and $|\downarrow\rangle$ we obtain the two coupled equations

$$i\hbar \dot{c}_\uparrow = \langle \uparrow |H| \uparrow \rangle c_\uparrow + \langle \uparrow |H| \downarrow \rangle c_\downarrow \tag{8.71}$$

and

$$i\hbar \dot{c}_\downarrow = \langle \downarrow |H| \uparrow \rangle c_\uparrow + \langle \downarrow |H| \downarrow \rangle c_\downarrow. \tag{8.72}$$

We shall try to find solutions that correspond to stationary states, with the time dependence given by

$$c_\uparrow = c_\uparrow^0 e^{-iEt/\hbar}, \quad c_\downarrow = c_\downarrow^0 e^{-iEt/\hbar}. \tag{8.73}$$

Let us consider an ammonia molecule situated in an external electric field $\mathcal{E}$, which is directed along the z-axis. We assume that the field is sufficiently weak that we may neglect its influence on the nondiagonal matrix elements. The diagonal matrix elements contain a term, which is linear in $\mathcal{E}$, arising from the electric dipole moment in the states $|\uparrow\rangle$ and $|\downarrow\rangle$. The matrix elements of H are thus given by

$$\langle \uparrow |H| \uparrow \rangle = a + p\mathcal{E}, \quad \langle \downarrow |H| \downarrow \rangle = a - p\mathcal{E}, \quad \langle \uparrow |H| \downarrow \rangle = -b, \quad \langle \downarrow |H| \uparrow \rangle = -b. \tag{8.74}$$

Here a, b and p are positive constants.

By inserting (8.73) in the equations (8.71-8.72) and requiring that the resulting homogeneous system of equations has non-trivial solutions one finds that the determinant of the matrix

$$\begin{pmatrix} a + p\mathcal{E} - E & -b \\ -b & a - p\mathcal{E} - E \end{pmatrix} \tag{8.75}$$

must vanish. The solutions of the resulting quadratic equation are

$$E = a \pm \sqrt{p^2 \mathcal{E}^2 + b^2}. \tag{8.76}$$

Let us examine the significance of (8.76). If $\mathcal{E}$ is zero, then the energy difference $2b$ is identical to ΔE_1 introduced above. Note that the matrix element b is a constant which may be adjusted to yield the calculated size of ΔE_1. When $\mathcal{E} \neq 0$ the equation (8.76) shows that the electric field increases the splitting due to tunnelling. If $\mathcal{E}$ is large compared to b/p, the energy level splitting is given by $\pm p\mathcal{E}$, corresponding to the energy of a dipole in two opposite directions.

According to (8.76) a stationary electric field yields the possibility of separating molecules in the upper energy level from molecules in the lower one. When the molecules move in a region of space, where $\nabla(\mathcal{E}^2)$ differs from zero, the forces on molecules in the two states will be oppositely directed. This principle is utilized in the maser to create an inverted population of the two states. Let us denote the energy of the highest (lowest) energy eigenstate by $E_2(E_1)$. If we neglect entirely the static electric field these energies are given by

$$E_2 = a + b, \quad E_1 = a - b. \tag{8.77}$$

Now let us assume that the molecules experience an oscillating electric field in the direction of the z-axis,

$$\mathcal{E} = \mathcal{E}_0(e^{i\omega t} + e^{-i\omega t}). \tag{8.78}$$

Next we expand the state in a basis consisting of the state vectors $|i\rangle$, which are eigenstates for the energy operator without an oscillating electric field,

$$|\psi\rangle = \sum_{i=1,2} a_i(t)|i\rangle e^{-iE_i t/\hbar}. \tag{8.79}$$

By inserting this state in the Schrödinger equation we obtain the following equations for the expansion coefficients

$$i\hbar \frac{da_k(t)}{dt} = \sum_{l=1,2} \langle k|H'|l\rangle e^{i\omega_{kl}t} a_l, \tag{8.80}$$

where $\omega_{kl} = (E_k - E_l)/\hbar$ (cf. equation (5.57) in Section 5.4.2). We assume that the matrix element $\langle k|H'|l\rangle$ only differs from zero when $(l,k) = (1,2)$ or $(2,1)$, in which case it is $p\mathcal{E}$, while the diagonal elements are zero. The equations (8.80) then become

$$i\hbar \frac{da_1(t)}{dt} = p\mathcal{E}_0(e^{i(\omega-\omega_0)t} + e^{-i(\omega+\omega_0)t})a_2 \tag{8.81}$$

and

$$i\hbar \frac{da_2(t)}{dt} = p\mathcal{E}_0(e^{i(\omega+\omega_0)t} + e^{-i(\omega-\omega_0)t})a_1, \tag{8.82}$$

where we have introduced the abbreviation $\omega_0 = \omega_2 - \omega_1$.

In the vicinity of resonance, for $\omega \simeq \omega_0$, we may neglect the rapidly varying terms which are proportional to $\exp(\pm i(\omega + \omega_0)t)$. This yields the simpler equations

$$i\hbar \frac{da_1(t)}{dt} = p\mathcal{E}_0 e^{i(\omega-\omega_0)t} a_2 \tag{8.83}$$

and

$$i\hbar \frac{da_2(t)}{dt} = p\mathcal{E}_0 e^{-i(\omega-\omega_0)t} a_1. \tag{8.84}$$

Symmetries

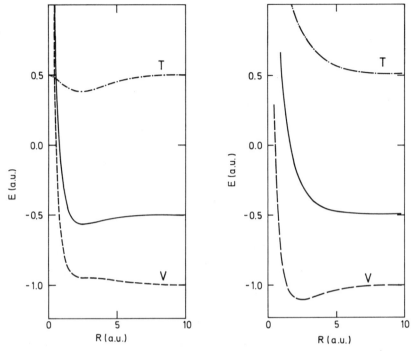

Figure 8.3: Kinetic (T), potential (V) and total energy for the H_2^+ ion, as functions of the separation between the nuclei. The figure on the left is obtained for the bonding state ψ_+, the one on the right for the anti-bonding state ψ_-.

By solving (8.83) for a_2, differentiating a_2 with respect to time and inserting the result in (8.84) we obtain

$$a_1 + \frac{\hbar^2}{p^2 \mathcal{E}_0^2}(-i\dot{a}_1(\omega - \omega_0) + \ddot{a}_1) = 0. \tag{8.85}$$

At resonance, $\omega = \omega_0$, the differential equation (8.85) assumes the simple form

$$\ddot{a}_1 + \omega_p^2 a_1 = 0 \tag{8.86}$$

with

$$\omega_p = \frac{p\mathcal{E}_0}{\hbar}. \tag{8.87}$$

The general solution of (8.86) is

$$a_1 = \alpha \cos \omega_p t + \beta \sin \omega_p t, \tag{8.88}$$

where α and β are constants. The corresponding solution for a_2 is

$$a_2 = i\beta \cos \omega_p t - i\alpha \sin \omega_p t. \tag{8.89}$$

With the initial conditions $a_1 = 0$ and $a_2 = 1$ at time $t = 0$, corresponding to $\alpha = 0$ and $\beta = -i$, the state 2 will be depopulated in one quarter of a period, $\pi/2\omega_p$, and the energy transferred to the electromagnetic field. This is the basic mechanism for the amplification of microwaves in the ammonia maser.

We shall complete this example with a discussion of the chemical bond, taking the H_2 molecule as an example. The H_2^+ ion is analogous to the ammonia molecule in the sense that the single electron moves under the attractive force of two protons. In the ground state the wave function of the electron is symmetric with respect to the midpoint between the two protons, just as for ψ_+ given in (8.55). A state like ψ_+ is called a bonding state, while ψ_- is called an anti-bonding state. The significance of the two types of states may be illustrated for the H_2^+ ion by determining the mean value of the energy in the states

$$\psi_\pm(\mathbf{r}) = C_\pm(u_0(r) \pm u_0(|\mathbf{r} - \mathbf{R}|)). \tag{8.90}$$

Here $\mathbf{R}$ is the position vector for one of the protons in a system of coordinates where the other is placed at the origin, while $u_0(r) = \exp(-r/a_0)/\sqrt{\pi a_0^3}$ is the ground-state wave function for the hydrogen atom. The integrals are calculated as in Example 8 by using polar coordinates with $\mathbf{R}$ chosen as the polar axis. We encounter integrals of the type

$$\int_{-1}^1 dx e^{-\sqrt{\alpha - \beta x}},$$

where x is the cosine of the angle between $\mathbf{r}$ and $\mathbf{R}$, while $\alpha = (R^2 + r^2)/a_0^2$ and $\beta = 2Rr/a_0^2$. These are determined by introducing $y = \sqrt{\alpha - \beta x}$ as a new variable. Since $\sqrt{\alpha - \beta} = |R - r|/a_0$ it is necessary to divide the integration over r into two regions: $r < R$ and $r > R$ (cf. Example 8). By carrying out these integrals one finds that the mean value of the kinetic energy of the electron is given, in atomic units (a.u.) defined by $e_0 = m_e = \hbar = 1$, by

$$\langle \psi_\pm | T \psi_\pm \rangle = \frac{1}{2} \mp \frac{R^2 e^{-R}}{3(1 \pm S)}, \tag{8.91}$$

where

$$S = e^{-R}(1 + R + \frac{R^2}{3}). \tag{8.92}$$

The mean value of the potential energy is

$$\langle \psi_\pm | V \psi_\pm \rangle = \frac{1}{R} - 1 + (1 \pm S)^{-1}[\pm(\frac{R^2}{3} - R - 1)e^{-R} - \frac{1}{R}(1 - e^{-2R}(1 + R))], \tag{8.93}$$

where we have added the contribution $1/R$ from the mutual repulsion of the protons to the mean value of the potential energy of the electron.

In Fig. 8.3 we plot the mean value of the kinetic energy and the mean value of the potential energy, together with their sum, the total energy, for the bonding and the anti-bonding state. Note that the minimum in the total energy of the bonding state is due the existence of a minimum for the kinetic energy. It is evident from (8.91), that the kinetic energy is reduced in the bonding state ψ_+ compared to the value $1/2$ appropriate to $R \to \infty$. The total energy in the bonding state ψ_+ has a minimum for $R = 1.32$ Å with a corresponding binding energy of 1.77 eV (the exact value is 2.79 eV).

The binding energy of the hydrogen molecule may now be determined approximately by letting the two electrons occupy the bonding state ψ_+ with the spin state being a singlet, thereby making the total wave function antisymmetric with respect to interchange of the two electrons. When this trial function is used for calculating approximately the binding energy of the hydrogen molecule, the answer comes out to be 2.68 eV with a corresponding equilibrium distance 0.8 Å. Experimentally, the binding energy is found to be 4.75 eV, while the equilibrium distance is 0.74 Å. The simple description in terms of bonding and anti-bonding states given above thus provides a qualitative understanding of the nature of the covalent chemical bond.

8.3 Translation

As our next examples of symmetry operations in quantum mechanics we now discuss translations in both space and time.

From a mathematical point of view symmetry operations constitute a group. A group is a set of elements $a, b, c, \cdots$, for which there exists a rule of combination (symbolized by ab, etc.). The elements constitute a group, if the following four conditions are satisfied: 1) Every result of combining two elements is an element in the group, 2) the associative law is valid in the sense that $(ab)c = a(bc)$ for all a, b and c, 3) there exists a unit element e such that $ea = a$ for every a, and 4) corresponding to every element a there exists an inverse a^{-1} such that $a^{-1}a = e$. The group of translations in space is a continuous group in the sense that the elements may be labelled by continuously varying parameters. The group of translations is also commutative or, as one also says, abelian. By this is meant that $ab = ba$ for every pair of elements a, b. Another example of an abelian group is the group of gauge transformations which we discuss in the next section. Not every group is abelian. In Section 8.5 we discuss rotations, which form a non-abelian group.

The unitary operator associated with translation may be introduced by considering a translation along one of the axes, for instance the x-axis. The operator $\hat{T}_{x_0}$ is then defined by

$$\hat{T}_{x_0} f(x) = f(x - x_0), \tag{8.94}$$

where f is an arbitrary function. By carrying out a Taylor expansion of the right hand side of (8.94) we obtain

$$\hat{T}_{x_0} f(x) = f(x) - x_0 \frac{df}{dx} + \frac{1}{2} x_0^2 \frac{d^2 f}{dx^2} + \cdots. \tag{8.95}$$

The right hand side of (8.95) may be written in the form

$$e^{-x_0 d/dx} f(x), \tag{8.96}$$

which shows that $\hat{T}$ may be expressed in terms of the momentum operator as follows

$$\hat{T}_{x_0} = e^{-ix_0 \hat{p}_x/\hbar}. \tag{8.97}$$

By exploiting the fact that $\hat{p}_x$ is a Hermitian operator, the unitarity of the operator (8.97) may be verified.

For an infinitesimal translation of ϵ the corresponding operator is thus

$$\hat{T}_\epsilon = 1 - \frac{i}{\hbar} \epsilon \hat{p}_x. \tag{8.98}$$

As a consequence of (8.98) the momentum operator is called a generator of infinitesimal spatial translations. When a system moves freely, in the absence of external forces, the Hamiltonian commutes with the translation operator as well as the generators of infinitesimal spatial translations. This corresponds to the fact that within classical mechanics the momentum is a constant of the motion for such a system.

Translations in time may be treated in an similar fashion. In analogy with (8.94) we define an operator $\hat{U}_{t_0}$ by

$$\hat{U}_{t_0} f(t) = f(t - t_0). \tag{8.99}$$

For an infinitesimal translation this equation becomes

$$\hat{U}_{t_0} f(t) = f(t) - t_0 \frac{\partial f}{\partial t}. \tag{8.100}$$

If f represents a physical state, which develops in time according to the Schrödinger equation, the operator in (8.100) may evidently be expressed in terms of the Hamiltonian according to

$$\hat{U}_{t_0} = 1 + \frac{it_0 \hat{H}}{\hbar}. \tag{8.101}$$

Let us assume that $\hat{H}$ does not depend explicitly on time. The operator associated with a finite translation t_0 in time is then seen to be

$$\hat{U}_{t_0} = e^{it_0 \hat{H}/\hbar}. \tag{8.102}$$

In this case the operators $\hat{H}$ and $\hat{U}$ commute with each other. When the Hamiltonian of the system is independent of time, a state obtained by a translation of t_0 also represents a possible state of the system, and it is readily shown that the displaced state satisfies the Schrödinger equation because $\hat{U}$ and $\hat{H}$ commute.

8.4 Gauge transformation

In quantum mechanics a magnetic field is introduced via its vector potential, which enters the classical Hamiltonian as we have seen in Chapter 6. The vector potential **A** determines the magnetic induction **B** according to

$$\mathbf{B} = \mathbf{\nabla} \times \mathbf{A}. \tag{8.103}$$

As we saw in Chapter 6 this is not sufficient to determine the vector potential uniquely from a given magnetic field. The transformation

$$\mathbf{A}' = \mathbf{A} + \mathbf{\nabla}\chi, \tag{8.104}$$

where χ is an arbitrary function, leaves **B** unchanged. We have already discussed two different choices of vector potential or gauge, namely (6.9) and (6.10). The difference in vector potential is given by setting the function χ in (8.104) equal to $\chi = -Bxy/2$. The two choices of vector potential yield the same energy spectrum, but the wave functions appear quite different. The choice of the Landau gauge (6.9) and the solution of the corresponding Schrödinger equation leads one to consider wave functions with an absolute value which is constant along a straight line perpendicular to the magnetic field. Likewise, the symmetric gauge leads one to consider wave functions with an absolute value which is constant on a circle in the plane perpendicular to the magnetic field. The degree of degeneracy of a given energy level as well as the value of the energy itself are, however, independent of the choice of gauge (cf. Example 6 in Chapter 6).

When the gauge transformation (8.104) is accompanied by a corresponding transformation of the wave function, the Schrödinger equation (3.22) preserves its form under such a transformation. To see this we consider an electron of charge $-e$. As shown in Chapter 6 the operator $\hat{\mathbf{p}}$ occurs together with the vector potential in the combination

$$\hat{\mathbf{p}} + e\mathbf{A}, \tag{8.105}$$

cf. (6.19). The effect of a change in the vector potential corresponding to (8.104) is therefore cancelled by the effect of the transformation

$$\psi' = \psi e^{-ie\chi/\hbar}, \tag{8.106}$$

which means that the eigenvalue equation for the transformed Hamiltonian

$$\hat{H}' = \frac{1}{2m}(\hat{\mathbf{p}} + e\mathbf{A}')^2 \tag{8.107}$$

is

$$\hat{H}'\psi' = E\psi'. \tag{8.108}$$

This may be verified by using that $\hat{\mathbf{p}}$ is given by $\hbar\nabla/i$, resulting in

$$(\hat{\mathbf{p}} + e\mathbf{A}')\psi' = e^{-ie\chi/\hbar}(\hat{\mathbf{p}} + e\mathbf{A})\psi. \tag{8.109}$$

The group of gauge transformations is clearly commutative[4].

8.5 Rotation

Rotations differ in an essential way from our previous examples of symmetry transformations, since the group is not commutative. We shall only consider the continuous case, although rotations through a finite angle have many important applications in solid state physics due to the existence of the symmetries of a crystalline lattice[5].

A series of rotations about different axes through a finite angle yield different results, depending upon the order in which the rotations are carried out. This lack of commutativity is closely related to the lack of commutativity for the generators of infinitesimal rotations. These are, as we shall see now, given by the different components of the angular momentum operator $\hat{\mathbf{L}}$.

A rotation in space is performed by the transformation

$$\mathbf{r}' = \mathsf{R}\mathbf{r}. \tag{8.110}$$

Here R denotes a 3×3 matrix. For a rotation of the angle ϕ about the z-axis the rotation matrix is given by

$$\mathsf{R} : \begin{pmatrix} \cos\phi & -\sin\phi & 0 \\ \sin\phi & \cos\phi & 0 \\ 0 & 0 & 1 \end{pmatrix} \tag{8.111}$$

The rotation matrices satisfy all the requirements for the elements forming a group: The rule of combination is matrix multiplication, which fulfills the associative law, the unit element is the unit matrix, while the inverse element is the matrix R^{-1}, which in the example given above represents a rotation of $-\phi$ about the z-axis. An arbitrary rotation may be specified by three continuously varying parameters, two describing the direction about which the rotation is carried out, and one the value of the angle of rotation.

As in (8.94) we define the unitary operator associated with the symmetry transformation (8.110) by

$$\hat{U}_{\mathsf{R}} f(\mathbf{r}) = f(\mathsf{R}^{-1}\mathbf{r}). \tag{8.112}$$

---

[4] Generalized non-abelian gauge transformations play an important role in the theories of weak and strong interactions.

[5] By means of the theory of irreducible representations it is possible to determine the nature of the splitting of the energy levels of an atom in a crystalline environment, without having to solve the Schrödinger equation.

Symmetries

For an infinitesimal rotation about the z-axis through the angle ϕ we have

$$\mathsf{R}: \begin{pmatrix} 1 & -\phi & 0 \\ \phi & 1 & 0 \\ 0 & 0 & 1 \end{pmatrix} \tag{8.113}$$

The vector $(0, 0, \phi)$ is denoted by $\boldsymbol{\phi}$, allowing us to write

$$\mathsf{R}^{-1}\mathbf{r} = \mathbf{r} - \boldsymbol{\phi} \times \mathbf{r} \tag{8.114}$$

to lowest order in ϕ, thereby casting the result of the infinitesimal rotation in a form which applies to an arbitrary direction of rotation. By expanding the right hand side of (8.112) to first order we obtain

$$\hat{U}_\mathsf{R} f(\mathbf{r}) = f(\mathbf{r}) - \frac{i}{\hbar}(\boldsymbol{\phi} \times \mathbf{r}) \cdot \hat{\mathbf{p}} f(\mathbf{r}). \tag{8.115}$$

This shows that the generators of infinitesimal rotations are the components of the angular momentum operator $\hat{\mathbf{L}}$, since

$$\hat{U}_\mathsf{R} = 1 - \frac{i}{\hbar}\boldsymbol{\phi} \cdot \hat{\mathbf{L}}. \tag{8.116}$$

It should be stressed that (8.116) only applies to an infinitesimal rotation. A rotation associated with an arbitrary rotation matrix R cannot be written in this form.

In conclusion, we have seen that just as $\hat{\mathbf{p}}$ is the generator for infinitesimal translations, $\hat{\mathbf{L}}$ is the generator for infinitesimal rotations. The difference between the two cases is that translations along different axes commute, while rotations about different directions do not. For a rotationally invariant system the unitary operator $\hat{U}$ in (8.116) commutes with the Hamiltonian, since each of the generators $\hat{\mathbf{L}}$ commutes with the Hamiltonian.

8.6 Problems

PROBLEM 8.1
In this problem we use the variational method for determining an upper limit on the ground-state energy E_0 for the motion in the potential given in Problem 1.11. It may be convenient to use trial functions that are not normalized. In this case the upper limit on E_0 is given by

$$E_0 \leq \frac{\int_0^\infty dz\, \psi^* \hat{H} \psi}{\int_0^\infty dz\, \psi^* \psi}, \tag{8.117}$$

since the integration is extended over the region $0 < z < \infty$ in which the wave function $\psi = \psi(z)$ differs from zero.

a) Use the trial function
$$\psi = ze^{-bz}, \tag{8.118}$$
where b is a constant, to determine an upper limit on E_0 as a function of b. Find the value of b which gives the best result, and determine the corresponding value of the ground-state energy. Compare with the result of using dimensional analysis (Problem 1.11).

b) We now change the trial function to
$$\psi = ze^{-cz^2}. \tag{8.119}$$

Find the best value of c and compare the corresponding value of the ground-state energy with that determined under a).

PROBLEM 8.2
Two identical particles, each of mass m, move in one dimension. The potential energy is
$$V(x_1, x_2) = \frac{\alpha}{2}(x_1^2 + x_2^2) + \frac{\beta}{2}(x_1 - x_2)^2. \tag{8.120}$$

Here α and β are positive constants, while x_1 and x_2 denote the positions of the particles.

a) Express the Hamiltonian in terms of the variables $(x_1 + x_2)$ and $(x_1 - x_2)$. Show that the Schrödinger equation separates in these variables and find the eigenvalues and the corresponding eigenfunctions for the Hamiltonian.

b) Indicate the symmetry of the eigenfunctions with respect to the interchange of the two particles. Which states are possible if the particles are identical spin-zero particles?

PROBLEM 8.3
A particle of mass m moves in two dimensions in a potential given by
$$V(x, y) = \frac{1}{2}K(x^2 + y^2) \quad \text{for} \quad x > 0, \ y > 0; \quad V(x, y) = \infty \text{ otherwise}, \tag{8.121}$$

where K is a positive constant. Find the ground-state energy of the system.

PROBLEM 8.4
A particle of mass m moves in a potential given by
$$V(z) = \frac{1}{2}Kz^2 \quad \text{for} \quad z > 0, \quad V(z) = \infty \quad \text{for} \quad z < 0, \tag{8.122}$$

where K is a positive constant. Use the trial functions from Problem 8.1 to find approximate values for the ground-state energy and compare with the known exact result.

Symmetries

PROBLEM 8.5
Use (8.17) to determine how the magnetic field of Problem 6.1 affects the lowest and the second-lowest level, when the magnetic field is considered to be a perturbation. Compare the result with the answer to question c) in Problem 6.1.

PROBLEM 8.6
Use the Schrödinger equation

$$(H_0 + \lambda H')(\psi^{(0)} + \lambda \psi^{(1)} + \lambda^2 \psi^{(2)} + \cdots) = $$
$$(E^{(0)} + \lambda E^{(1)} + \lambda^2 E^{(2)} + \cdots)(\psi^{(0)} + \lambda \psi^{(1)} + \lambda^2 \psi^{(2)} + \cdots) \qquad (8.123)$$

to show that the energy correction $E_n^{(2)}$ to a non-degenerate level to second order in the perturbation is given by

$$E_n^{(2)} = \sum_{n' \neq n} \frac{|\langle n|H'|n'\rangle|^2}{E_n^{(0)} - E_{n'}^{(0)}}, \qquad (8.124)$$

where E_n^0 is the unperturbed energy associated with the unperturbed state $\psi^{(0)}$. Use the result (8.124) to show that an electric field with field strength $\mathcal{E}$ lowers the ground-state energy of the hydrogen atom and estimate the order of magnitude of the effect for $\mathcal{E} = 1$ kV/cm.

9 FERMIONS AND BOSONS

The atoms in a crystalline solid are ordered in a regular lattice. The distance between neighboring atoms is of the order of 10^{-10} m, which is comparable to the characteristic length $\hbar^2/me_0^2$ of the hydrogen atom. The atoms oscillate around their equilibrium positions in the lattice, and these oscillations may to a first approximation be represented by a set of coupled harmonic oscillators. In this chapter we shall see how the quantum theory of a single harmonic oscillator is used to explain the magnitude and temperature dependence of the specific heat of solids.

Metals and semiconductors contain freely-moving electrons, in addition to the atoms which remain near their position in the ideal lattice during their oscillation. A metal such as aluminum consists of a lattice of positively charged ions and a gas of freely-moving conduction electrons. The conduction electrons are responsible for the transport of electricity through the metal. When the oscillations of the lattice atoms around their equilibrium position are neglected, the positive ions in the ideal lattice give rise to a periodic potential. At first sight one might believe that the electrons were scattered in their motion by the positive ions of the periodic lattice. As we shall see in the present chapter, this is not the case. The reason is that it is possible to find stationary states which are extended throughout the crystal, in close resemblance to the states of an electron in a constant potential (Section 4.3). The electrical resistivity is due to deviations from the periodic potential arising from the oscillations of the atoms in the lattice or the presence of impurities, dislocations and other crystal imperfections.

In the present chapter we shall first examine the motion of electrons in solids when the lattice atoms are at rest and consequently give rise to a periodic potential. Due to the small value of the ratio between the mass of an electron and the mass of an atom it is possible in a first approximation to regard the lattice as being static, and then determine the energy eigenstates for the motion of an electron. This is the *Born-Oppenheimer approximation*. We shall determine the form of these energy eigenstates by using the particular symmetry characterizing the motion in a periodic potential. Subsequently we treat the oscillations of the atoms around their equilibrium positions in the lattice and show how these oscillations may be resolved into normal modes. The quantized lattice vibrations may be described as particles which are named phonons, in analogy with light quanta or photons. Like photons the phonons are bosons, and in equilibrium they are distributed according to the Planck law. Because of the oscillations of the lattice atoms around their equilibrium position there exists the possibility of transitions between the stationary states for an electron moving in the periodic potential. Such processes are responsible for the temperature dependence of the electrical resistivity of metals.

Electrons and phonons are not independent particles. An electron may emit or absorb a phonon, because the motion of the lattice gives rise to deviations from the strictly periodic potential. It is also possible for two electrons to interact by the exchange of a phonon, one electron emitting a phonon, which is absorbed by the other electron. This interaction is attractive under certain conditions and is responsible for the superconductivity of many metals.

In the following subsection we shall disregard the motion of the lattice atoms and exploit the particular symmetry which characterizes the periodic potential; invariance with respect to translations through a lattice vector. In addition to this symmetry it is crucial to take into account that electrons are identical fermions. It is therefore necessary to describe electrons in terms of wave functions which are antisymmetric with respect to the interchange of any two electrons. This is often formulated as a Pauli exclusion principle: In a given quantum state it is possible to put at most one electron. Note however, that this formulation implicitly assumes that it is possible to neglect the interaction between the electrons themselves (cf. Section 8.2).

9.1 Free electron gas

The conduction electrons in a metal constitute a gas of freely-moving particles. In the simplest possible free-electron model one neglects entirely the presence of the ions in the lattice as well as the interaction between the electrons themselves, except in so far as the existence of the positively charged ionic lattice makes the metal electrically neutral as a whole. In the free-electron model one therefore represents this lattice by a uniform background of positive charge, resulting in a constant potential energy for a conduction electron. The potential energy is thus a constant in the free-electron model. By choosing this constant as the zero of energy we may immediately take over the results from Section 4.3, where the density of states for a free particle in a box was determined as a function of energy. The eigenstates were given in terms of plane waves, cf. (4.54), with the wave vector $\mathbf{k}$ being determined by periodic boundary conditions.

The ground state of the electron system as a whole is determined by using the Pauli principle for the occupation of the single-particle quantum states. In the state of lowest energy the electrons occupy all single-particle states inside a sphere in $\mathbf{k}$-space. The radius k_F of the sphere is given by

$$N = 2\frac{4\pi}{3}k_F^3 \frac{V}{(2\pi)^3}, \tag{9.1}$$

where N is the number of conduction electrons and V is the volume of the metal. The factor of 2 in (9.1) is due to the electron spin. The remaining factors were discussed in Section 4.3. In terms of the density $n = N/V$ of the conduction electrons the radius k_F may according to (9.1) be written as

$$k_F = (3\pi^2 n)^{1/3}, \tag{9.2}$$

which shows that k_F depends solely on the density of fermions. The quantity k_F is called the Fermi wave number. The corresponding Fermi momentum p_F is given by

$$p_F = \hbar k_F, \qquad (9.3)$$

while the Fermi energy is

$$E_F = \frac{\hbar^2 k_F^2}{2m}, \qquad (9.4)$$

with m denoting the electron mass. The Fermi energy is thus the kinetic energy of the conduction electrons occupying the single-particle states of highest energy in the ground state of the system. The equation $\epsilon(\mathbf{k}) = E_F$ defines a surface in $\mathbf{k}$-space separating the occupied states from the empty ones. This surface is called the Fermi surface. In the simple free-electron model under consideration the Fermi surface is spherical, since $\epsilon(\mathbf{k}) = \hbar^2 k^2/2m$, but other shapes are possible when the periodic potential is taken into account, as we shall see below. Finally one introduces the Fermi velocity v_F by the definition

$$v_F = \frac{\hbar k_F}{m}. \qquad (9.5)$$

The ground-state energy E_0 for the electron system as a whole may be found by summing the kinetic energy $\hbar^2 k^2/2m$ over all occupied states. By changing the sum to an integral in the usual manner, cf. (4.55-56), we obtain

$$E_0 = \frac{V}{\pi^2} \int_0^{k_F} dk\, k^2 \frac{\hbar^2 k^2}{2m} = \frac{3}{5} N E_F. \qquad (9.6)$$

It is to be expected that the ground-state energy per particle, $3E_F/5$, is less than the Fermi energy, since all states inside the Fermi surface are occupied, while all states outside are empty. If the electrons move in two rather than three dimensions - which may happen on a surface of liquid helium - one obtains a similar result but with a different numerical constant (Problem 9.1).

Apart from the numerical constants the expressions for the Fermi energy, the Fermi momentum and the Fermi velocity all follow from dimensional analysis (Section 1.3), once it is assumed that they involve only $\hbar$, the electron density n and the electron mass m.

The conduction electron density n thus determines the magnitude of the Fermi wave number k_F and the associated energy, momentum and velocity. For a metal like aluminum the conduction electron density n is determined once it is assumed that each aluminum atom in the metal gives off three electrons to the gas of conduction electrons, the valency of an aluminum atom being three. From the known mass density and molar mass of aluminum one then finds n to be $1.8 \cdot 10^{29}$ m^{-3}. The Fermi wave number is thus

$$k_F = 1.8 \cdot 10^{10}\,\text{m}^{-1}, \qquad (9.7)$$

while the Fermi velocity is

$$v_F = 2.0 \cdot 10^6 \, \text{m/s}. \tag{9.8}$$

The Fermi energy is the kinetic energy of an electron with Fermi wave number k_F and is seen to be

$$E_F = 1.9 \cdot 10^{-18} \, \text{J}. \tag{9.9}$$

When translated into a Fermi temperature according to the definition $T_F = E_F/k$ the Fermi energy (9.9) corresponds to $T_F = 1.4 \cdot 10^5$ K. Here k is the Boltzmann constant, $k = 1.38 \cdot 10^{-23}$ joule/K. The size of the Fermi temperature compared to ordinary room temperature shows that the electron gas cannot be described as a classical gas, since this would require a temperature large compared to T_F. This has important consequences for the contribution of the electrons to the specific heat, which is much reduced - by a factor of order T/T_F - compared to the value one would expect according to the classical equipartition law. It is not difficult to understand that the specific heat of the electron gas should be much less than the classical value of $3k/2$ per particle. When the temperature differs from zero, the energy of the electron system is increased as a result of the thermal motion. However, due to the Pauli principle only the electrons with energy near E_F may be raised to higher-lying states due to the thermal motion. The electrons occupying states with energies far below the Fermi energy have no empty states at their disposal, since all the neighboring states within an energy interval of the order of kT are fully occupied. The number of thermally excited electrons is therefore given approximately by NT/T_F when $T \ll T_F$, yielding a total energy increase of order NkT^2/T_F. It follows from this that the heat capacity per electron is of the order of magnitude kT/T_F, which is much less than the constant-volume result $3k/2$ for the particles in an ideal classical gas.

So far we have used the conduction electrons of a metal as an example of a Fermi gas, but the preceding results may be used for other Fermi gases as well. By similar considerations one may explain some of the essential features in models of stellar evolution. For the so-called white dwarfs the balance between the pressure due to the freely-moving electrons and the gravitational energy of the star plays a decisive role, since the density in these stars is so large that the electrons become separated from the nuclei and move about independently of them. Because of the small mass of the electrons the kinetic energy and hence the pressure in the interior of the star is dominated by the electrons. The nuclei contribute in this limit only to the gravitational energy.

To obtain a quantitative estimate let us first assume that the density of the electrons in the star is sufficiently low that the non-relativistic result (9.6) may be used for evaluating the kinetic energy. Accordingly the dependence of the kinetic energy on the total mass of the star is given by $E_\text{kin} \propto N^{5/3}/V^{2/3} \propto M^{5/3}/R^2$. Here R is the radius of the star ($\propto V^{-1/3}$), while M is the mass of the star which depends on the number N of freely-moving electrons according to $M = Nm'$, where m' is the mass per electron (typically about two nucleon masses). The gravitational energy is given by $E_{\text{grav}} \propto -M^2/R$, in analogy with the electrostatic energy of a uniformly charged sphere. An estimate of the dependence of the mass M on the radius R is then obtained by requiring that the sum of E_{kin} and E_{grav} is as small as possible, leading to $M \propto R^{-3}$.

When the electron density is so high that it becomes necessary to use the ultrarelativistic expression $c\hbar k$, instead of $\hbar^2 k^2/2m$ for the kinetic energy of an electron with momentum $\hbar k$ it is readily seen that (9.6) is replaced by $E_0 = 3N E_F/4$. The Fermi energy is then $E_F = c\hbar k_F$, while k_F is given by (9.2). This makes the total kinetic energy inversely proportional to R, just as is the gravitational energy. Under these conditions we obtain $E_{\text{kin}} \simeq \hbar c (M/m')^{4/3}/R$, while $E_{\text{grav}} \propto -M^2/R$. Since the two terms in the energy have the same dependence on the radius R, equilibrium may only be obtained for a definite value M_c of the mass M. Except for a constant the critical mass M_c is determined by balancing the kinetic and the gravitational energy according to $E_{\text{kin}} = |E_{\text{grav}}|$, implying that $M = M_c = (\hbar c/G)^{3/2}/m'^2$, as mentioned in Section 1.3.5.

9.1.1 Sound in metals

The quantum free-electron model for metals was introduced by Sommerfeld in 1928. It was successful in accounting for the specific heat of simple metals and their electrical and thermal conductivity. Due to its simplicity the free-electron model remains a useful starting point in many different contexts, and we shall see in the present section that it may be used to calculate approximately the sound velocity in simple metals.

In the following we shall use the result (9.6) for the ground-state energy of the gas of free electrons as our starting point for the determination of the sound velocity. In a sense we thus abandon the assumption that the positive ions form an inert background of uniform positive charge, since the propagation of sound involves oscillations in the mass density of the metal. The free-electron model allows us to determine the pressure gradient which causes the motion of the ions, the pressure being related to the volume dependence of the energy (9.6).

The sound velocity in ordinary gases is found from the continuity equation

$$\frac{\partial \rho}{\partial t} + \nabla \cdot \mathbf{j} = 0, \tag{9.10}$$

where ρ is the mass density and $\mathbf{j} = \rho \mathbf{v}$ is the mass current density. We have introduced a velocity field $\mathbf{v} = \mathbf{v}(\mathbf{r}, t)$, which specifies the local drift velocity. According to Newton's second law the time derivative of $\mathbf{j}$ is determined by the pressure gradient ∇p,

$$\frac{\partial \mathbf{j}}{\partial t} = -\nabla p. \tag{9.11}$$

We have neglected friction, which leads to damping of the sound, as well as nonlinear terms in the equation of motion. By taking the derivative of (9.10) with respect to time we obtain

$$\frac{\partial^2 \rho}{\partial t^2} = \nabla^2 p. \tag{9.12}$$

Since the equation (9.12) involves both pressure and density, we must now relate small changes in the pressure p to small changes in the mass density ρ. Let us first assume that the propagation of sound occurs isothermally, thereby allowing us to use the equation of state

$$p = \frac{\rho}{M} kT, \tag{9.13}$$

where M is the (average) mass of an air molecule. It follows from (9.12-13) that the sound velocity s is given by

$$s^2 = (\frac{\partial p}{\partial \rho})_T = \frac{kT}{M}. \tag{9.14}$$

Under usual circumstances the propagation of sound does not take place isothermally (at constant temperature), but adiabatically (at constant entropy). As a result (9.14) should be multiplied by the ratio C_p/C_V between the heat capacity at constant pressure and at constant volume. The resulting expression for the sound velocity, $s = \sqrt{C_p kT/C_V M}$, is in very good agreement with measured sound velocities in gases.

Let us now return to the determination of sound velocities in metals. The equation of state of a metal differs from (9.13). This is hardly surprising in view of the difference in specific heat discussed above[1]. At the relatively low frequencies and long wavelengths characterizing sound it is possible to regard the metal as a continuum and use the equations (9.10-11) leading to (9.12). The connection between pressure and volume is not, however, given by (9.13), which holds for an ideal classical gas. In order to determine the pressure as a function of volume we identify the ground-state energy E_0 in our simple model with the internal energy U at $T = 0$ K. Thus

$$U = \frac{3}{5} N E_F \propto V^{-2/3}. \tag{9.15}$$

As a consequence the pressure becomes proportional to $V^{-5/3}$, since it is given by

$$p = -\frac{\partial U}{\partial V} = \frac{2U}{3V} = \frac{2}{5} nk T_F. \tag{9.16}$$

---

[1] One of the consequences of this is that the difference between C_p and C_V is negligible for the conduction electrons. It is therefore permissible to consider the propagation of sound to be isothermal.

This equation of state for the electron gas differs fundamentally from the equation of state (9.13) for an ideal classical gas. Apart from numerical factors the difference is that the absolute temperature in (9.13) is replaced by the Fermi temperature T_F which is a measure of the density of the electron gas.

The mass density ρ is given to a good approximation by

$$\rho = n_{\text{ion}} M, \tag{9.17}$$

where M is the mass of an ion in the crystal lattice, while n_{ion} is the ion density. We asssume that there are z conduction electrons per ion, which implies that n_{ion} is equal to n/z. Consequently

$$s^2 = \frac{\partial p}{\partial \rho} = \frac{z}{3}\frac{m}{M} v_F^2. \tag{9.18}$$

This result is in reasonable agreement with measurements of sound velocities in simple metals such as Al, Na and Cu. One finds for these three metals that the value calculated according to (9.18) is 9.1, 3.0 and 2.7 km/s, respectively, while the measured longitudinal sound velocities are 6.8, 3.1 and 4.7 km/s. Sound propagation in metals is thus a quantum phenomenon, the compressibility $-(\partial V/\partial P)/V$ involving the Planck constant.

In our simple model it is the compressibility of the electron gas which enters the sound velocity. The physical reason for this is that the motion of the ions and the electrons is not independent. The electrons follow the motion of the ions in order to neutralize their charges locally. As a consequence the electrons are compressed and expanded, resulting in pressure oscillations which are determined locally by changes in the energy of the electron gas due to changes in its density.

9.2 Periodic potential

So far in this chapter we have only taken into account the kinetic energy of the conduction electrons. Due to the presence of the positive ions in the crystalline lattice the potential energy of a conduction electron varies periodically in space. The potential energy $V(\mathbf{r})$ has the symmetry property

$$V(\mathbf{r} + \mathbf{R}) = V(\mathbf{r}), \tag{9.19}$$

where $\mathbf{R}$ is an arbitrary lattice vector. The lattice vectors $\mathbf{R}$ specify the spatial location of the lattice points. They are constructed from linear combinations of three linearly-independent basis vectors $\mathbf{a}_i$, $i = 1, 2, 3$ according to $\mathbf{R} = n_1 \mathbf{a}_1 + n_2 \mathbf{a}_2 + n_3 \mathbf{a}_3$, where n_i are integers. We shall now use this symmetry to show that the eigenstates for the Hamiltonian of an electron in a periodic potential may be chosen in the form

$$\psi_{\mathbf{k}}(\mathbf{r}) = \exp(i\mathbf{k} \cdot \mathbf{r}) u_{\mathbf{k}}(\mathbf{r}). \tag{9.20}$$

Fermions and bosons

Here u is a function with the same periodicity as that of the potential energy, i. e. $u(\mathbf{r} + \mathbf{R}) = u(\mathbf{r})$. The eigenstates (9.20) are thus plane waves that have been modulated by a periodic function with the same periodicity as that of the lattice. This is *Bloch's theorem*.

To simplify the discussion we prove Bloch's theorem in one dimension. The potential energy of the electron is assumed to satisfy the condition

$$V(x + na) = V(x). \tag{9.21}$$

Here a denotes the distance between two neighboring atoms in the one-dimensional lattice, while n numbers the lattice points from 1 to N, the total number[2] of lattice atoms being N. In analogy with (8.94) we define the translation operator $\hat{T}_n$ by

$$\hat{T}_n f(x) = f(x - na) \tag{9.22}$$

corresponding to a displacement of na. The Hamiltonian for a single electron is

$$\hat{H} = -\frac{\hbar^2}{2m}\frac{d^2}{dx^2} + V(x). \tag{9.23}$$

Evidently $\hat{T}$ and $\hat{H}$ commute with each other, since

$$\hat{T}\hat{H}f(x) = -\frac{\hbar^2}{2m}\frac{d^2}{d(x-na)^2}f(x-na) + V(x-na)f(x-na), \tag{9.24}$$

while

$$\hat{H}\hat{T}f(x) = -\frac{\hbar^2}{2m}\frac{d^2}{dx^2}f(x-na) + V(x)f(x-na), \tag{9.25}$$

which are seen to be identical on account of (9.21). It is therefore possible to find simultaneous eigenfunctions for $\hat{T}$ and $\hat{H}$. Before doing this we note that the translations through the set of lattice vectors na constitute a group. According to the periodic boundary conditions $\hat{T}_N = \hat{T}_0$ and $\hat{T}_{N+1} = \hat{T}_1$. The group is abelian in one, two and three dimensions.

Let us assume that ψ is a simultaneous eigenfunction for $\hat{T}_n$ and $\hat{H}$. We denote the eigenvalue associated with $\hat{T}_n$ by $\lambda(n)$,

$$\hat{T}_n \psi = \lambda(n)\psi. \tag{9.26}$$

Since

$$\hat{T}_{n+n'} = \hat{T}_n \hat{T}_{n'} \tag{9.27}$$

we have

$$\lambda(n + n') = \lambda(n)\lambda(n'). \tag{9.28}$$

---

[2] We use periodic boundary conditions, thereby effectively turning the finite linear chain into an infinite one, for which the symmetry (9.21) is obeyed for any n.

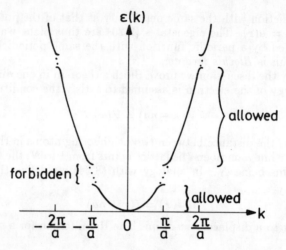

Figure 9.1: Energy eigenvalue $\epsilon(k)$ in one dimension.

According to the periodic boundary conditions

$$\lambda(N) = 1, \; \lambda(N+1) = \lambda(1). \tag{9.29}$$

In order to satisfy the condition (9.28) we write $\lambda(n)$ in the form

$$\lambda(n) = e^{-ikna}, \tag{9.30}$$

and conclude from (9.29) that k must be real.

Let us consider the function

$$u(x) = e^{-ikx}\psi(x), \tag{9.31}$$

which is clearly periodic, $u(x - na) = u(x)$, since

$$\psi(x - na) = e^{-ikna}\psi(x). \tag{9.32}$$

Thus we have proved Bloch's theorem in one dimension,

$$\psi(x) = e^{ikx}u(x). \tag{9.33}$$

Since the translation operators associated with three different directions in a crystal commute with each other, the proof may be readily generalized to three dimensions. We also remark that the use of periodic boundary conditions

Fermions and bosons

proceeds in the same way as in Section 4.3. In particular, the equation (4.55) remains valid for states such as (9.20).

The energy associated with the state (9.20) depends on the particular form of the periodic potential. In one dimension the connection between the energy eigenvalue $\epsilon(k)$ and k might have the form given in Fig. 9.1. Note the existence of energy gaps, in which there are no allowed eigenstates, separating intervals or *bands* of allowed energies.

To conclude this introduction to the motion in a periodic potential we shall now solve the Schrödinger equation for a very simple model, in which the potential energy is represented by a series of delta functions. Although the model is physically unrealistic, it provides an insight into some of the important aspects of the motion in the periodic potential.

EXAMPLE 10. A ONE-DIMENSIONAL LATTICE.

We shall examine the motion of a particle in a potential of the form

$$V(x) = \sum_n I\delta(x - (n + \frac{1}{2})a), \qquad (9.34)$$

where I is a positive constant, n an integer, while a represents the distance between neighboring atoms. When the potential becomes infinitely large, we cannot demand that the derivative of the wave function be continuous. It follows from the equation

$$-\frac{d^2\psi}{dx^2} + C_1\delta(x)\psi = C_2\psi, \qquad (9.35)$$

where C_1 and C_2 are constants, that the difference between $d\psi/dx$ at $x = \epsilon_1$ and at $x = -\epsilon_1$ must equal $C_1\psi$ at $x = 0$ (ϵ_1 is a positive infinitesimal). This may be shown by integrating the equation (9.35) from $-\epsilon_1$ to ϵ_1.

In order to solve the Schrödinger equation with the given potential it is convenient to introduce dimensionless variables by

$$x' = x/a, \quad \epsilon' = 2ma^2\epsilon/\hbar^2, \quad I' = 2maI/\hbar^2. \qquad (9.36)$$

The eigenvalue equation for the Hamiltonian then becomes

$$[-\frac{d^2}{dx^2} + I\sum_n \delta(x - (n + \frac{1}{2}))]\psi = \epsilon\psi, \qquad (9.37)$$

after one has removed for convenience the primes on the dimensionless variables.

In the region

$$-\frac{1}{2} < x < \frac{1}{2} \qquad (9.38)$$

the solution is

$$\psi = A\cos\sqrt{\epsilon}x + B\sin\sqrt{\epsilon}x. \qquad (9.39)$$

According to Bloch's theorem

$$\psi(\frac{1}{2}) = e^{ik}\psi(-\frac{1}{2}). \qquad (9.40)$$

For the derivative we obtain similarly

$$e^{ik}\frac{d\psi}{dx}\Big|_{x=-\frac{1}{2}+\epsilon_1} = \frac{d\psi}{dx}\Big|_{x=\frac{1}{2}+\epsilon_1}. \tag{9.41}$$

The value of the derivative at $x = \frac{1}{2} + \epsilon_1$ is, as demonstrated below (9.35), connected to the value in $x = \frac{1}{2} - \epsilon_1$ by

$$\frac{d\psi}{dx}\Big|_{x=\frac{1}{2}+\epsilon_1} = \frac{d\psi}{dx}\Big|_{x=\frac{1}{2}-\epsilon_1} + I\psi\Big|_{x=\frac{1}{2}}. \tag{9.42}$$

From (9.39) and (9.40) we then obtain

$$A\cos(\sqrt{\epsilon}/2) + B\sin(\sqrt{\epsilon}/2) = e^{ik}A\cos(\sqrt{\epsilon}/2) - e^{ik}B\sin(\sqrt{\epsilon}/2), \tag{9.43}$$

while (9.39) and (9.41-42) result in

$$e^{ik}\sqrt{\epsilon}A\sin(\sqrt{\epsilon}/2) + e^{ik}\sqrt{\epsilon}B\cos(\sqrt{\epsilon}/2) = -\sqrt{\epsilon}A\sin(\sqrt{\epsilon}/2) + \sqrt{\epsilon}B\cos(\sqrt{\epsilon}/2) + $$
$$I[A\cos(\sqrt{\epsilon}/2) + B\sin(\sqrt{\epsilon}/2)]. \tag{9.44}$$

The two equations (9.43-44) form a homogeneous system of equations for the two unknowns A and B. The condition for the existence of non-trivial solutions to these equations is that the determinant should vanish, i. e.

$$(1-e^{ik})^2\sqrt{\epsilon}\cos^2(\sqrt{\epsilon}/2) + (1+e^{ik})^2\sqrt{\epsilon}\sin^2(\sqrt{\epsilon}/2) - 2e^{ik}I\sin(\sqrt{\epsilon}/2)\cos(\sqrt{\epsilon}/2) = 0, \tag{9.45}$$

which may be written as

$$4\sin^2(k/2)\sqrt{\epsilon}\cos^2(\sqrt{\epsilon}/2) + 4\cos^2(k/2)\sqrt{\epsilon}\sin^2(\sqrt{\epsilon}/2) = I\sin\sqrt{\epsilon}. \tag{9.46}$$

When $\sin^2(k/2)$ and $\cos^2(k/2)$ are expressed in terms of $\cos k$ (9.46) becomes

$$\cos k = \cos\sqrt{\epsilon} + \frac{I}{2}\frac{\sin\sqrt{\epsilon}}{\sqrt{\epsilon}}. \tag{9.47}$$

We may now choose an ϵ and use (9.47) to find the corresponding k (if it exists), cf. Problem 9.3. This determines ϵ as a function of k,

$$\epsilon = F(k). \tag{9.48}$$

The result may be plotted and compared to the sketch in Fig. 9.1.

In two dimensions the problem separates, when the potential energy is chosen to be

$$V = V(x) + V(y). \tag{9.49}$$

The energy ϵ is then determined by

$$\epsilon = F(k_x) + F(k_y), \tag{9.50}$$

where F is the function given in (9.48). By choosing an ϵ it is a simple matter to determine k_y as a function of k_x with a pocket calculator (Problem 9.4) and hence

Figure 9.2: Linear chain.

the curves of constant energy in the $k_x k_y$-plane. When $I \to 0$ the curves become circles corresponding to $\epsilon = k^2 = k_x^2 + k_y^2$.

In three dimensions the connection between ϵ and $\mathbf{k}$ may be quite complicated for a realistic periodic potential, but the principles involved in the solution of the eigenvalue problem remain the same. Bloch's theorem is employed as a boundary condition on the solution of the Schrödinger equation within a unit cell of the crystal. The occupation of the different energy levels in the ground state determines whether the solid is metallic or insulating. If all states within and below a given band are occupied, while the higher bands are empty, the solid is an insulator or, possibly, a semiconductor, depending upon the size of the energy gap separating the highest occupied from the lowest unoccupied band. If a band is only partially filled the solid is metallic, since the conduction electrons are then free to move under the influence of external fields.

9.3 Lattice vibrations

Each of the atoms in a crystalline solid carries out vibrations around its equilibrium position, but these vibrations are not independent, since the vibrations of a given atom will influence those of its neighbors. A crystalline solid is therefore a large system of coupled oscillators, in fact as many as the number of atoms times three. Though this appears to complicate the description enormously compared to the single harmonic oscillator it will turn out that the result (2.103) for the energy eigenvalues of a single harmonic oscillator forms the basis for the description of the vibrations of the crystal as a whole.

To see this we proceed in three steps, the first one involving only classical mechanics, the second quantum mechanics, and the third one statistical mechanics. Our aim is to turn the problem of coupled oscillators into a simpler one, for which the oscillations are independent of each other. Such a transformation is achieved by a transition to normal coordinates, and the resulting oscillations are called normal modes (cf. Chapter 2). As a simple example we shall see how the normal modes are determined for the simplest possible 'crystal' consisting of a linear chain of atoms.

9.3.1 Normal modes

In this subsection we treat the classical equations of motion for a linear chain of identical atoms which are connected by springs with the same force constant K (Fig. 9.2). The mass of an atom is called M. By u_n we denote the deviation of the n'th atom from its equilibrium position. The atoms are only allowed to move along the direction of the chain. Later on we shall see how the model is generalized to the case when the atoms occupy positions in a three-dimensional lattice and are allowed to move in different directions.

According to Newton's second law the acceleration $\ddot{u}_n$ of the n'th atom is given by

$$M\ddot{u}_n = -K(u_n - u_{n-1}) - K(u_n - u_{n+1}). \quad (9.51)$$

For a chain of N atoms we shall use the periodic boundary conditions

$$u_N = u_0; \quad u_{N+1} = u_1 \quad (9.52)$$

on the solutions to (9.51). The distance between two neighboring atoms in equilibrium is a, which means that the total length L of the chain is $L = Na$. The boundary conditions (9.52) evidently correspond to joining together the two ends of the linear chain.

In solving the equations of motion it is convenient to regard u_n as the real part of a complex quantity. We note that (9.51) is satisfied by

$$u_n = u_0 e^{iqna - i\omega t} \quad (9.53)$$

provided ω and q are related by the condition

$$M\omega^2 = 4K \sin^2 \frac{qa}{2}. \quad (9.54)$$

The solution (9.53) has the form of a plane wave

$$e^{iqx - i\omega t}, \quad (9.55)$$

where the position coordinate x assumes the discrete values $x = na$ with $n = 1, \cdots, N$. The wave number is evidently $q/2\pi$, while the angular frequency is ω. The actual deviation u_n is of course real, and (9.53) should strictly speaking be written as

$$u_n = \Re u_0 e^{iqna - i\omega t} = u_0 \cos(qna - \omega t), \quad (9.56)$$

where u_0 is chosen to be real. The symbol $\Re$ denotes that the real part is taken, but it is usually suppressed with the understanding that the real part is taken at the end. By inserting (9.56) in (9.51) we obtain again the condition (9.54), which may be written as

$$\omega = 2\sqrt{\frac{K}{M}} |\sin \frac{qa}{2}|, \quad (9.57)$$

the sign of ω being chosen to be positive. An equation such as (9.57), which connects the wave vector with the frequency, is called a dispersion relation.

Fermions and bosons

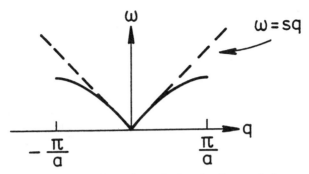

Figure 9.3: Dispersion relation for linear chain.

The particular pattern of oscillations u_n associated with a definite value of q (and hence ω), is called a normal mode. The dispersion relation (9.57) is shown in Fig. 9.3. Note that the connection between ω and q is linear for small values of q. In this limit the mode corresponds to a sound wave, with dispersion relation

$$\omega = sq, \tag{9.58}$$

where the sound velocity s is given by

$$s = a\sqrt{\frac{K}{M}}. \tag{9.59}$$

The reason why the frequency ω becomes small for small values of q is that neighboring atoms oscillate very nearly in phase and therefore only move slightly with respect to each other. However, if q assumes the value π/a, neighboring atoms oscillate in opposite phase ($u_n = -u_{n-1}$), giving rise to the maximum frequency equal to $2\sqrt{K/M}$.

The boundary conditions (9.52) determine the allowed values of q. This is of course a purely classical consideration. The condition (9.52) requires that

$$e^{iqNa} = 1, \tag{9.60}$$

which implies that q may assume the values

$$q = \frac{m}{N}\frac{2\pi}{a}, \tag{9.61}$$

where m is an integer (positive, negative or zero).

Besides (9.61) the allowed wave vectors must satisfy another condition, which is a consequence of the fact that the only physically significant values of the plane wave (9.55) are those associated with the lattice points $x = na$. If we replace in (9.53) the wave vector q by $q + 2\pi/a$, the mode u_n remains the same. The two different values of q thus correspond to exactly the same mode pattern. We therefore limit q to the interval

$$-\frac{\pi}{a} < q < \frac{\pi}{a}, \qquad (9.62)$$

noting that q may then assume precisely N different values.

In summary, we have shown that the oscillations of the atoms in the crystal may be described in terms of normal modes such as (9.53), provided ω and q are related by the dispersion relation (9.57). There exist N different modes, namely as many as the allowed values of q in the interval $-\pi/a < q < \pi/a$.

Before moving on to the quantum mechanical treatment of the coupled vibrations we emphasize that the discussion given above is purely classical. We have found that q may only assume discrete values according to (9.61), but this is simply a consequence of the periodic boundary conditions. We shall now see how the Hamiltonian of the system is obtained from the classical Hamiltonian.

9.3.2 Phonons

We have now solved the classical linear-chain problem of the motion of N atoms, each with mass M, interacting via harmonic forces which are represented by springs with force constant K. A general solution may be represented as a superposition of normal modes such as (9.56), with ω and q being related by (9.57). The complicated motion of the lattice atoms has thus been reduced to something far simpler, a superposition of independent normal modes given by (9.56-57). On the basis of this classical result it is reasonable to assume that the Hamiltonian $\hat{H}$ for the linear chain is a sum over q of terms of the form (2.91), corresponding to

$$\hat{H} = \sum_q \hbar\omega(q)(\hat{a}_q^\dagger \hat{a}_q + \frac{1}{2}). \qquad (9.63)$$

This result may be proved without difficulty by starting from the classical Hamiltonian $\sum_n (p_n^2/2M + K(u_n - n_{n-1})^2/2)$ and the commutation relations $[u_n, p_m] = i\hbar\delta_{nm}$, but we omit the detailed proof here.

The Hamiltonian for a linear chain has thus a very simple form, since it is a sum of operators associated with each independent harmonic oscillator that belongs to one of the N possible values of q. The connection between $\omega(q)$ and q is the classical one given by (9.57).

The different possible energy eigenvalues are obtained from (9.63) in a way analogous to the single harmonic oscillator by specifying the number of quanta n_q associated with each normal mode. The resulting contribution to the energy is thus $\hbar\omega(q)(n_q + 1/2)$, and the total energy in the state under consideration is therefore the sum over q of all these energies. The state is labelled as a ket with the values of the quantum numbers n_{q_i}, where the index i runs through N different values 1 to N corresponding to (9.78),

$$|n_{q_1}, n_{q_2}, \cdots, n_{q_N}\rangle. \tag{9.64}$$

The energy quanta in a normal mode are called phonons. In the long wavelength limit ($qa \ll 1$) the normal modes have linear dispersion just like sound, cf. (9.58). Instead of considering the different energy eigenvalues of the total energy as a sum of energies associated with each normal mode specified by the quantum numbers n_q, one may just as well denote the eigenstate by giving the number of phonons, n_q, associated with each normal mode. The lattice has thus been transformed from a system of N atoms that are mutually interacting to a non-interacting gas of phonons. The phonon gas is non-interacting, because the Hamiltonian only contains terms that are quadratic in the deviations u_n of the atoms from their equilibrium position. If higher-order terms are retained in the Hamiltonian, it is possible to describe these to a first approximation as perturbations. This gives the phonons a finite lifetime. A given phonon may for instance decay into two other phonons. Such processes play an important role in the thermal conduction in insulators. In the absence of interaction between the phonons and neglecting any effects of boundaries such an insulator would have infinite thermal conductivity.

The number of phonons is not a conserved quantity, unlike the number of molecules in ordinary gases such as air in a closed container. The number of phonons depends on the temperature. In the ground state of the lattice the number of phonons is zero, since $n_{q_i} = 0$ for all i. In the following section we shall see that the population of the energy states at a given temperature is determined by the Planck distribution. This will allow us to determine the magnitude and the temperature dependence of the specific heat.

9.3.3 Lattice specific heat

A linear chain of atoms is not a satisfactory model for a solid. In a real three-dimensional crystal each atom has typically six, eight or twelve nearest neighbors, depending on the crystal structure. The determination of the classical lattice vibrations must take the crystal structure into account and consider the coupling between an atom and its nearest neighbors - and possibly its more distant neighbors as well. This complicates the calculation of the frequencies of the normal modes, but involves no conceptual changes. In the three-dimensional case a normal mode is characterized by a wave vector $\mathbf{q}$ with an associated angular frequency $\omega(\mathbf{q})$, corresponding to the replacement of (9.53) by

$$\mathbf{u} = \mathbf{u}_0 \exp(i\mathbf{q}\cdot\mathbf{R} - i\omega t), \tag{9.65}$$

where $\mathbf{R}$ is a lattice vector. Since the atoms may vibrate in three directions, there are three normal modes associated with each value of $\mathbf{q}$. The lattice vibrations are therefore characterized not only by the wave vector $\mathbf{q}$, but also by polarization vectors, which may be chosen to be orthogonal to each other. If the direction of $\mathbf{q}$ is along a symmetry axis in the crystal, it becomes possible to choose the polarization vectors to be parallel with or perpendicular to $\mathbf{q}$, and the phonons are called longitudinal and transverse, respectively. Many crystals have a unit cell containing more than one atom. In this case one has not only branches that are acoustic as in the one-dimensional case (meaning that the frequency goes to zero when q goes to zero), but there are also optical branches, in which the frequency tends towards a non-zero value as q goes to zero.

The quantization of the lattice vibrations proceeds as in the one-dimensional case. The Hamiltonian is given by (9.63), provided q is interpreted as $(\mathbf{q}, \lambda)$, where λ is a polarization index. According to the discussion given above λ may assume three different values ($\lambda = 1, 2, 3$), which label the three unit vectors $\mathbf{e}_1$, $\mathbf{e}_2$ and $\mathbf{e}_3$ giving the polarization directions for the normal modes associated with a given $\mathbf{q}$. This enables us to write down the different possible energy eigenvalues E_i for the crystal as a whole,

$$E_i = \sum_{\mathbf{q},\lambda} \hbar \omega_\lambda(\mathbf{q})(n_\lambda(\mathbf{q}) + \frac{1}{2}), \qquad (9.66)$$

where i is an index labelling the set of quantum numbers $n_\lambda(\mathbf{q})$.

Having determined the possible energy eigenvalues (9.66) we must use a fundamental result of statistical mechanics, the so-called canonical distribution, in order to be able to determine the specific heat. The canonical distribution applies to a system in contact with a heat reservoir with temperature T. For such a system the probability p_i for finding the system in a state with energy E_i may be shown to be

$$p_i = \frac{e^{-E_i/kT}}{\sum_i e^{-E_i/kT}}. \qquad (9.67)$$

Note that the sum of p_i over all i equals 1. The constant k appearing in (9.67) is the Boltzmann constant.

We shall not derive the canonical distribution (9.67) here, but rather seek to make it plausible. It is clearly necessary to describe the thermal properties of large systems by statistical methods. It is a hopeless task to solve Hamilton's equations or the Schrödinger equation for 10^{23} particles interacting among themselves and with their surroundings. Instead of specifying the state of the system in as much detail as quantum mechanics allows, we resort to a statistical description based on an ensemble of identical systems. This allows us to introduce the probability p_i that the system is in a state with energy E_i, where E_i denotes the set of energy eigenvalues obtained from the quantum mechanical description of an isolated system.

It is reasonable to expect for a system in contact with a heat reservoir that there is a greater probability of finding the system in a state with lower

Fermions and bosons

energy than in a state with higher energy. This may be illustrated by a gas of diatomic molecules. As mentioned in Chapter 1 these molecules have rotational and vibrational degrees of freedom. If the energy difference $\hbar\omega$ between the second-lowest and the lowest vibrational state is large compared to kT, it is highly probable that the molecules occupy the lowest vibrational state, for which the quantum number n in the energy spectrum $E_n = (n + 1/2)\hbar\omega$ is zero. On the other hand, if $\hbar\omega$ is comparable with kT (which is roughly the mean value of the kinetic energy per molecule), we may expect a significant population of the higher-energy states.

A well-known example of a canonical distribution is the Maxwell-Boltzmann distribution, according to which the number $f(\mathbf{p})d\mathbf{p}$ of molecules with momentum components in the intervals $(p_x, p_x + dp_x)$, $(p_y, p_y + dp_y)$ and $(p_z, p_z + dp_z)$ is given by

$$f(\mathbf{p})d\mathbf{p} = C_0 e^{-p^2/2MkT} d\mathbf{p}, \tag{9.68}$$

where C_0 is a constant and M is the mass of a molecule. In this case the energy $p^2/2M$ entering the exponential function is purely kinetic. Another example is the so-called barometric formula, which relates the density n to the potential energy of the molecules in a gravitational field. The barometric formula is

$$n(z) = n_0 e^{-mgz/kT}, \tag{9.69}$$

where n_0 is the density at $z = 0$, while z is the height above this level. In this case the distribution is determined by the potential energy mgz which enters the exponential function. These examples of a canonical distribution are taken from classical statistical mechanics, but the distribution (9.67) is a general result valid for any system which is in contact with a heat reservoir.

We shall now see how the use of the canonical distribution for a harmonic oscillator leads to the Planck distribution, which forms the basis for the description of the specific heat of solids, radiation from black bodies, and many other phenomena. In the following we consider a single oscillator in contact with a heat reservoir and seek to determine the mean energy as a function of temperature. The spectrum of energy eigenvalues is given by (2.103). The probability p_n of finding the oscillator in the state labelled by n is thus

$$p_n = \frac{e^{-(n+\frac{1}{2})\hbar\omega/kT}}{\sum_{n=0}^{\infty} e^{-(n+\frac{1}{2})\hbar\omega/kT}}, \tag{9.70}$$

and the mean energy $\bar{E}$ is therefore

$$\bar{E} = \sum_{n=0}^{\infty} E_n p_n. \tag{9.71}$$

The sum (9.71) is carried out most easily by introducing the partition function Z, which is defined generally by

$$Z = \sum_i e^{-E_i/kT}, \qquad (9.72)$$

where E_i are the energy eigenvalues of the system. When these energy eigenvalues are given by (2.103), the sum becomes very simple, since the terms in (9.72) form a geometric series. By summing the series we obtain

$$Z = \frac{e^{-\hbar\omega/2kT}}{1 - e^{-\hbar\omega/kT}}. \qquad (9.73)$$

On comparing (9.71) and (9.72) it is seen that the mean energy may be written as

$$\bar{E} = -\frac{\partial \ln Z(\beta)}{\partial \beta}, \qquad (9.74)$$

where the parameter β is defined by $\beta = 1/kT$. When (9.73) is inserted into (9.74), the mean energy of the oscillator becomes

$$\bar{E} = \hbar\omega\left(\frac{1}{e^{\hbar\omega/kT} - 1} + \frac{1}{2}\right) = \hbar\omega\left(\bar{n} + \frac{1}{2}\right). \qquad (9.75)$$

The function $\bar{n}$ given by

$$\bar{n} = \frac{1}{e^{\hbar\omega/kT} - 1} \qquad (9.76)$$

is the Planck distribution, cf. (1.1). It is equal to the mean number of energy quanta in a harmonic oscillator with classical frequency ω. At temperatures high compared to $\hbar\omega/k$, the distribution $\bar{n}$ becomes approximately $kT/\hbar\omega$. In this case the mean energy is seen to be independent of the Planck constant,

$$\bar{E} \simeq kT. \qquad (9.77)$$

In the opposite limit, $kT \ll \hbar\omega$, the mean energy $\bar{E}$ is approximately equal to the zero-point energy of the oscillator, $\hbar\omega/2$. In this limit the population of the higher-energy states of the oscillator does not affect the mean value, since the probability of finding the oscillator in a state of higher energy is very small, given by $1 - p_0 \simeq \exp(-\hbar\omega/kT)$.

Having determined the mean energy of a single oscillator in contact with a heat reservoir we are now able to find the mean energy of a crystal containing N atoms. In the classical limit, when $kT/\hbar$ is large compared to any phonon frequency $\omega(\mathbf{q})$, the mean energy is simply obtained by multiplying the result (9.77) for a single oscillator by the number of oscillators $3N$. Note that each atom corresponds to three oscillators, one for each direction in space. This yields

$$\bar{E} = 3NkT. \qquad (9.78)$$

Fermions and bosons 231

The heat capacity C is given in general by the thermodynamic relation

$$C = \frac{\partial \bar{E}}{\partial T}. \tag{9.79}$$

It is understood that the volume of the crystal is held fixed when the differentiation with respect to temperature is carried out. As long as the lattice vibrations are treated in the harmonic approximation corresponding to the Hamiltonian (9.63) there is no difference between the heat capacity at constant volume and at constant pressure. In the classical limit, where $\bar{E}$ is given by (9.78), the heat capacity is therefore

$$C = 3Nk. \tag{9.80}$$

According to (9.80) the molar heat capacity is a universal quantity equal to $3N_A k = 3R$, where N_A is the Avogadro number and R the gas constant, or 25 joule/(mol K). The result (9.80) is generally in good agreement with measurements of the heat capacity at room temperature. The law was established empirically by Dulong and Petit in 1819. To illustrate the validity of Dulong and Petit's law we mention that the heat capacity C at room temperature has been measured to be $28, 26, 24$ and 8.5 joule/(mol K) for Na, Ca, Al and C (diamond), respectively. The large difference from the value 25 joule/(mol K) found for diamond indicates that quantum effects are important in this substance even at room temperature, as shown by Einstein in 1907. In the following we shall see that the heat capacity of these different elements may be described to a good approximation by a universal function of kT divided by a characteristic energy for the element in question.

For a crystal with an energy spectrum given by (9.66) the mean energy of the entire crystal equals the sum of the mean energies of the individual oscillators associated with the normal modes,

$$\bar{E} = \sum_{\mathbf{q},\lambda} \hbar \omega_\lambda(\mathbf{q})(\bar{n}_\lambda(\mathbf{q}) + \frac{1}{2}). \tag{9.81}$$

The function $\bar{n}_\lambda(\mathbf{q})$ is the Planck distribution (9.76) with $\omega = \omega_\lambda(\mathbf{q})$. The validity of (9.81) may be shown directly from (9.74) by forming the partition function Z, which is a product of $3N$ factors $Z_\lambda(\mathbf{q})$, where each factor $Z_\lambda(\mathbf{q})$ is given by (9.73) with $\omega = \omega_\lambda(\mathbf{q})$. The mean energy is obtained from (9.74) by using that $\ln Z$ equals the sum of $\ln Z_\lambda(\mathbf{q})$ over $\mathbf{q}$ and λ, which results in (9.81).

The specific heat may thus be determined from (9.81) and (9.79) on the basis of a definite phonon spectrum $\omega_\lambda(\mathbf{q})$. Since the allowed values of $\mathbf{q}$ are very closely spaced in a crystal of macroscopic size, the sum over $\mathbf{q}$ is turned into an integral according to

$$\sum_{\mathbf{q}} \cdots = \frac{V}{(2\pi)^3} \int d\mathbf{q} \cdots \tag{9.82}$$

where V is the volume of the crystal. The prescription (9.82) may be verified by generalizing (9.61) to three dimensions as follows. We consider a cubic crystal with one atom per unit cell of side-lengths a_1, a_2 and a_3. The number of unit cells in the volume V is $N = N_1 N_2 N_3$ with $N_i a_i$ being the side-length of the crystal in the i-direction ($i = x, y, z$). The components of the vector $\mathbf{q}$ may then assume the values

$$q_x = \frac{m_1}{N_1}\frac{2\pi}{a_1}, \; q_y = \frac{m_2}{N_2}\frac{2\pi}{a_2}, \; q_z = \frac{m_3}{N_3}\frac{2\pi}{a_3}, \tag{9.83}$$

where m_1, m_2 and m_3 are integers in order that the periodic boundary conditions may be satisfied.

It is in general an arduous task to calculate the lattice specific heat from (9.81) on the basis of a realistic phonon spectrum, since $\omega_\lambda(\mathbf{q})$ not only depends on the length of the wave vector $\mathbf{q}$, but also on its direction. However, at temperatures that are sufficiently low that $kT/\hbar$ is much less than the maximum phonon frequency, the calculation simplifies considerably. To see this we shall first consider the case where the three acoustic branches are characterized by the same (isotropic) sound velocity s. The temperature-dependent part of the mean energy is then at low temperatures given by

$$\bar{E} = 3\frac{V}{(2\pi)^3}\int_0^\infty dq \, 4\pi q^2 \frac{\hbar s q}{e^{\hbar s q/kT} - 1}. \tag{9.84}$$

Here V is the volume of the crystal. The factor of three arises from the summation over the three polarization directions, while $4\pi q^2$ is the surface area of a sphere with radius q. In going from (9.81) to (9.84) we have left out the temperature-independent zero-point energy, which is the energy of the crystal in the ground state, for which all $n_\lambda(\mathbf{q})$ are equal to zero. The integration over q in (9.84) has been extended to infinity. The validity of this approximation depends on the temperature being sufficiently low that the contribution to the mean energy from the part of $\mathbf{q}$-space in which the phonon frequencies do not depend linearly on q, is vanishingly small. As seen from (9.81), it is the exponential dependence of the Planck distribution on $\omega_\lambda(\mathbf{q})$, for $\hbar\omega_\lambda(\mathbf{q}) \gg kT$, which enables one to carry out this approximation at low temperatures.

The temperature dependence of the mean energy $\bar{E}$ may now be conveniently determined by introducing the dimensionless variable

$$x = \frac{\hbar s q}{kT}, \tag{9.85}$$

which turns (9.84) into

$$\bar{E} = \frac{3}{2\pi^2}\frac{V(kT)^4}{(\hbar s)^3}\int_0^\infty dx \, \frac{x^3}{e^x - 1}. \tag{9.86}$$

Fermions and bosons 233

The value of the integral in (9.86) is $\pi^4/15$, and the heat capacity may then be determined from (9.79) with the result

$$C = \frac{2\pi^2}{5}\frac{Vk^4T^3}{(\hbar s)^3}. \qquad (9.87)$$

We note that this simple expression for C is derived on the assumption that the dispersion relations of the three acoustic branches are identical, given by $\omega = sq$. The same temperature dependence ($\propto T^3$) is obtained if the velocities s_λ for the three acoustic branches depend on direction in q-space according to $\omega_\lambda = s_\lambda(\theta,\phi)q$, with θ and ϕ denoting polar angles in q-space.

The expression (9.87) for the heat capacity is only valid at low temperatures. It is possible to obtain a formula valid at all temperatures, provided one makes a highly simplifying (and somewhat unrealistic) assumption about the dispersion relation by taking it to be linear, $\omega(q) = sq$, for all q up to a maximum value q_D. The value of q_D depends on the number density N/V of the atoms in the lattice according to

$$q_D^3 = \frac{6\pi^2 N}{V}. \qquad (9.88)$$

This value of q_D is determined by the requirement that the number of different normal modes must equal $3N$ or N per polarization direction, which according to (9.82) implies that

$$N = \frac{V}{(2\pi)^3}\frac{4\pi q_D^3}{3}, \qquad (9.89)$$

resulting in (9.88). The mean energy is then obtained from (9.84) by replacing the upper limit in the integral with q_D. The resulting heat capacity becomes

$$C = 3NkF(T/\Theta_D), \qquad (9.90)$$

where the function $F(x)$ is given by

$$F(x) = \frac{3}{4}x^3\int_0^{1/x} dy\frac{y^4}{\sinh^2(y/2)} \qquad (9.91)$$

and the parameter Θ_D is defined by

$$\Theta_D = \hbar s q_D/k. \qquad (9.92)$$

The function F is shown in Fig. 9.4.

This simplified model due to Debye yields a good description of the temperature dependence of the measured heat capacity, if one adjusts the parameter Θ_D, which is named the Debye temperature. This temperature is typically

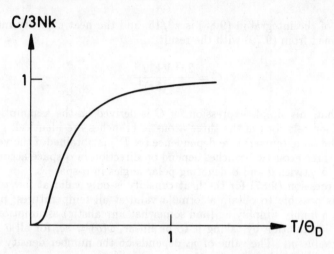

Figure 9.4: The heat capacity in the Debye model, (9.90-91).

a few hundred degrees in most materials. At temperatures somewhat below the Debye temperature quantum effects become significant. According to the Debye expression (9.90) the specific heat is equal to 0.95 times its classical value when the temperature is equal to the Debye temperature Θ_D, while it has dropped to about one half of its classical value when the temperature is equal to $\Theta_D/4$, cf. Fig. 9.4. The values of the Debye temperature obtained from such a fitting procedure for the elements Na, Ca, Al and C, are 150, 230, 390 and 1860 K. These values are determined by fitting the calculated Debye curve (9.90) to the measured heat capacity as a function of temperature. A crude estimate of the magnitude of the Debye temperature is obtained from (1.56), since the force constant for the motion of the lattice may be expected to be of the same order of magnitude as for a molecule. By inserting the atomic mass for the elements in question into (1.56) one finds energies that correspond to temperatures between 1000 and 2000 K.

9.4 Spin waves

In the previous section we characterized the possible states of a crystalline lattice by the number of phonons associated with each normal mode. We shall now see how ferromagnets at low temperatures may be described in terms of spin waves, the so-called magnons. Like phonons the magnons represent excited states of the system, i. e. states with higher energy than the ground state. We shall first determine the energy of the magnons as a function of their wave vector and then proceed to use the formula (9.81) to find the mean

Figure 9.5: Ground state of a linear magnetic chain.

energy and the specific heat of a ferromagnetic material at low temperatures.

To simplify matters we consider a linear chain of atoms interacting with their nearest neighbors. The Hamiltonian of the magnetic chain is

$$\hat{H} = -J\sum_n \hat{\mathbf{S}}_n \cdot \hat{\mathbf{S}}_{n+1} = -J\sum_n (\hat{S}_{n,z}\hat{S}_{n+1,z} + \frac{1}{2}(\hat{S}_{n,+}\hat{S}_{n+1,-} + \hat{S}_{n,-}\hat{S}_{n+1,+})),$$
(9.93)

where J is a positive constant. The Hamiltonian is thus a sum of terms of the form (7.73) which describe the exchange interaction between neighboring atoms. We shall take this model Hamiltonian as our starting point without considering the question of how well it applies to a given magnetic material. As usual $\hat{S}_+ = \hat{S}_x + i\hat{S}_y$ and $\hat{S}_- = \hat{S}_x - i\hat{S}_y$. For convenience we have made the eigenvalues of the spin operators dimensionless by dividing them by $\hbar$. This gives J the dimension of energy. The positions of the atoms in the chain have been numbered by $n = 1, 2, 3, \cdots, N$, where N is the number of atoms. As in the phonon case we use periodic boundary conditions. By using the properties of the operators of angular momentum one finds that the state $|0\rangle$ given by

$$|0\rangle = |s_1, s_2, \cdots s_N\rangle \tag{9.94}$$

is an eigenstate of $\hat{H}$,

$$\hat{H}|0\rangle = -JNs^2|0\rangle, \tag{9.95}$$

with $s_i = s$ denoting the maximum value of the z-component of the spin of the i'th atom ($s = 1/2$ for a single electron, but in general s may be larger than this value, since the atomic spin is the result of the addition of the angular momenta of the atomic electrons). The state $|0\rangle$ is thus characterized by the eigenvalue of each $\hat{S}_{n,z}$ being s. This state is illustrated in Fig. 9.5. It is not difficult to show that the eigenvalue in (9.95) is the lowest one possible, corresponding to the ground state, but we leave out the proof. From a classical point of view it is plausible that the lowest energy is obtained when the spins all line up in the same direction, since J is taken to be positive. In an antiferromagnetic material J is negative which makes the description far more complicated. In this case it is not possible to find a simple expression for the ground state of the system corresponding to (9.94-95).

Let us now try to find excited states of the system, i. e. eigenstates for the Hamiltonian with energies larger than the ground-state energy $-JNs^2$. We

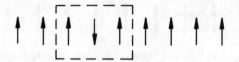

Figure 9.6: Illustration of (9.96) for $s = \frac{1}{2}$.

consider a (normalized) state $|n\rangle$ given by

$$|n\rangle = \frac{1}{\sqrt{2s}} \hat{S}_{n,-} |0\rangle. \tag{9.96}$$

This state is illustrated in Fig. 9.6 for the case $s = 1/2$. It is readily seen that the state is not an eigenstate for $\hat{H}$, since one obtains

$$\hat{H}|n\rangle = -Js(|n+1\rangle + |n-1\rangle) - J(N-2)s^2|n\rangle - s(s-1)2J|n\rangle \tag{9.97}$$

(cf. Fig. 9.6) by using the properties of the angular momentum operators. Let us, however, attempt to build an eigenstate by forming the linear combination

$$|q\rangle = \frac{1}{\sqrt{N}} \sum_n e^{iqna} |n\rangle, \tag{9.98}$$

where a is the distance between neighboring atoms. We find that (9.98) is indeed an eigenstate, since

$$\hat{H}|q\rangle = (-JNs^2 + 2Js(1 - \cos qa))|q\rangle. \tag{9.99}$$

Evidently $|q\rangle$ is an energy eigenstate with energy

$$\hbar\omega(q) = 2Js(1 - \cos qa) = 4Js \sin^2 qa/2 \tag{9.100}$$

relative to the ground state. Note that this state is also an eigenstate of the z-component of the total spin with eigenvalue $Ns - 1$ (in units of $\hbar$), since it is a superposition (9.98) of states such as (9.96) which are eigenstates of the z-component of the total spin with eigenvalue $Ns - 1$, the operator $\hat{S}_{n,-}$ reducing the eigenvalue by 1 relative to the ground state.

The determination of the magnon energies in the three-dimensional case proceeds just as for one dimension. For simplicity we assume that the magnetic material is ordered in a simple cubic lattice. In three dimensions the magnon energy corresponding to (9.100) becomes

$$\hbar\omega(\mathbf{q}) = 4Js(\sin^2(q_x a/2) + \sin^2(q_y a/2) + \sin^2(q_z a/2)), \tag{9.101}$$

since the magnon state is characterized by a three-dimensional wave vector **q** as in the case of phonons. The dispersion relation (9.101) has cubic symmetry, since the magnon energy does not change by interchange of the axes in **q**-space. It is anisotropic in the sense that the energy depends not only on the length of **q**, but also on its direction. At low temperatures ($T \ll J/k$) the thermodynamic quantities are primarily determined by the long-wavelength magnons with small energy, since the mean number of magnons is given by the Planck distribution

$$\bar{n}(\mathbf{q}) = \frac{1}{\exp(\hbar\omega(\mathbf{q})/kT) - 1}, \qquad (9.102)$$

just as for phonons, cf. (9.76). At low temperatures it is therefore sufficient to use the magnon energy (9.101) in the limit of small wave vectors satisfying $qa \ll 1$,

$$\hbar\omega(q) = Jsq^2a^2. \qquad (9.103)$$

The temperature dependent part of the mean energy $\bar{E}$ is

$$\bar{E} = \sum_{\mathbf{q}} \hbar\omega(\mathbf{q})\bar{n}(\mathbf{q}). \qquad (9.104)$$

By inserting (9.103) in this expression for the mean energy and using (9.82), where $V = Na^3$ is the volume of the magnetic material containing N atoms, we obtain

$$\bar{E} = \frac{Na^3}{(2\pi)^3} \int_0^\infty dq\, 4\pi q^2 \frac{Jsq^2a^2}{e^{Jsq^2a^2/kT} - 1}, \qquad (9.105)$$

cf. (9.84). Upon introduction of the dimensionless variable $x = Jsq^2a^2/kT$ the mean energy becomes

$$\bar{E} = \frac{1}{4\pi^2} N \frac{(kT)^{\frac{5}{2}}}{(Js)^{\frac{3}{2}}} \int_0^\infty dx \frac{x^{\frac{3}{2}}}{e^x - 1}. \qquad (9.106)$$

Since the value of the integral in (9.106) is $\Gamma(5/2)\zeta(5/2)$, where $\Gamma(z)$ and $\zeta(z)$ are the gamma and the zeta function, respectively[3], the specific heat (9.79) becomes

$$C = \frac{5\Gamma(\frac{5}{2})\zeta(\frac{5}{2})}{8\pi^2} Nk\left(\frac{kT}{Js}\right)^{\frac{3}{2}}. \qquad (9.107)$$

This law is in good agreement with measurements of the specific heat of iron at temperatures much less than J/k.

The magnetization is found in a similar manner. As argued above, the state (9.98) is an eigenstate of the z-component of the total spin with eigenvalue $(Ns-1)\hbar$. The mean number of spin waves at a given temperature is therefore a

---
[3] Numerically $\Gamma(5/2) = 3\sqrt{\pi}/4$ and $\zeta(5/2) = 1.341$.

measure of the relative reduction in magnetization $M(T)$ at low temperatures. The magnetization at low temperatures is consequently given by

$$M(T) = M(0)(1 - \frac{1}{Ns}\sum_{\mathbf{q}} \frac{1}{\exp(\hbar\omega(\mathbf{q})/kT) - 1}) \simeq M(0)(1 - \frac{\Gamma(\frac{3}{2})\zeta(\frac{3}{2})}{4\pi^2 s^{\frac{5}{2}}}(\frac{kT}{J})^{\frac{3}{2}}).$$
(9.108)

This characteristic $T^{3/2}$-law is also in agreement with experiments.

9.5 The blackbody radiation

Having determined the heat capacity of solids at low temperatures it is now a simple matter to find the mean energy and heat capacity of the blackbody radiation. We only need to replace the velocity s in (9.87) by the velocity of light c and multiply the right hand side by 2/3 in order to take into account that the electric field in a plane electromagnetic wave is perpendicular to the direction of propagation. This will be shown in the following.

The electric field $\mathbf{E}$ in a plane electromagnetic wave may be written in our complex notation as

$$\mathbf{E} = \mathbf{E}_0 \exp(i\mathbf{k} \cdot \mathbf{r} - i\omega t).$$
(9.109)

The expressions (9.109) and (9.65) differ in several respects: While the lattice vector $\mathbf{R}$ only assumes discrete values, the electric field $\mathbf{E}$ is defined everywhere in space. As a consequence there is no upper limit to the magnitude of the wave vector $\mathbf{k}$, the dispersion relation being given by

$$\omega = ck$$
(9.110)

for all $\mathbf{k}$. The thermodynamic properties of blackbody radiation are therefore closely related to those of a solid at low temperatures, cf. (9.84), where the q-integration was extended to infinity. For solids this procedure is an approximation valid only at low temperatures, but it is exact for the blackbody radiation since the dispersion relation is given by (9.110) for all k. There exist, however, only two linearly independent polarization vectors for the electric field. The reason is that the divergence of the electric field in vacuum is zero, $\nabla \cdot \mathbf{E} = 0$, resulting in $\mathbf{k} \cdot \mathbf{E} = 0$ when $\mathbf{E}$ is given by (9.109). The electric field is thus perpendicular to the direction of propagation of the electromagnetic wave.

We are now able to determine the frequency distribution for the electromagnetic field by enclosing the radiation in a volume V and using periodic boundary conditions. This means that the number $g(\omega)d\omega$ of oscillations with angular frequency between ω and $\omega + d\omega$ is given by (cf. (9.82))

$$g(\omega)d\omega = 2\frac{V}{(2\pi)^3}4\pi k^2 dk = \frac{V\omega^2}{\pi^2 c^3}d\omega.$$
(9.111)

The mean energy $\bar{E}$ is then obtained from

$$\bar{E} = \int_0^\infty d\omega g(\omega)\hbar\omega \frac{1}{e^{\hbar\omega/kT} - 1}. \quad (9.112)$$

The result (9.112) is seen to be identical with (9.84) except for the factor of 2/3 and the replacement of the sound velocity s by the velocity of light c.

9.5.1 The background radiation of the universe

The remarkable feature of the Planck formula (9.76) is that the frequency distribution is determined solely by the temperature. In the first chapter we have discussed the significance of measurements of blackbody radiation for the development of quantum mechanics. In concluding the present chapter we mention a related phenomenon, the observation of the background radiation of the universe with its far-reaching implications for cosmology.

In 1964 the two American radio astronomers, Arno Penzias and Robert Wilson, decided to measure radio signals from the Milky Way using a radio telescope, which had been employed earlier for receiving radio signals from satellites. The antenna had an unusually high sensitivity and was especially suited for receiving radio waves which were uniformly distributed in all directions, while at the same time avoiding signals from the many different sources on earth. When they examined the radiation coming from outside the plane of the Milky Way, Penzias and Wilson noticed an excess source of noise which they could neither identify nor eliminate. The radiation appeared to be independent of direction and season. Shortly after the publication of their discovery it became generally accepted that the observed radiation had to be interpreted as a signal from the 'Big Bang', which is presumed to have started the expansion of the universe. Later measurements of the background radiation at different wavelengths made it clear that the energy of the radiation was distributed in frequency in exactly the same manner as the blackbody radiation. The original measurement by Penzias and Wilson was carried out at the wavelength 7.35 cm, in the classical region where $\hbar\omega$ is much less than kT. Measurements obtained in 1989 from the satellite COBE, the Cosmic Background Explorer, show that the background radiation to a very good approximation corresponds to a blackbody radiation curve with a temperature of 2.736 ± 0.06 K.

9.6 Quantum liquids and superconductors

Ordinary liquids solidify when they are cooled to low temperatures. Water becomes ice at 273 K, and liquid nitrogen solidifies at 63 K. The only known fluid which does not solidify upon cooling is liquid helium. This is a consequence of the quantum mechanical zero-point motion and the weakness of the attractive van der Waals forces between the helium atoms.

The element helium exists in two stable isotopic forms, ^{3}He and ^{4}He, the latter being by far the most abundant. The nucleus of a ^{4}He atom contains two protons and two neutrons, while that of the ^{3}He atom contains two protons and only one neutron. Because of the difference between the atomic masses liquid ^{3}He has a lower boiling point, $T = 3.2$ K, than that of liquid ^{4}He which boils at 4.2 K. The low boiling points make the liquids useful as cooling agents. At the start of this century it was learned how to liquify helium gas. By cooling the liquid to lower temperatures the surprising discovery was made that it remains in the liquid state down to the lowest temperatures attainable. In order to make liquid helium solidify it is necessary to apply a pressure in excess of 34 atmospheres in the case of ^{3}He and 25 atmospheres in the case of ^{4}He.

The explanation why liquid He does not solidify even at temperatures as low as 10^{-5} K is to be sought in the quantum mechanical zero-point motion. According to the uncertainty relation the atoms cannot remain at rest at definite points of space. If an atom is confined to a region in space of linear dimension a, the mean value of its kinetic energy must be at least of order $\hbar^2/Ma^2$, where M is the mass of the atom. The smaller the mass M, the greater is the zero-point energy. The atoms in a solid are confined to move in a region with linear dimensions comparable to the interatomic distance, i. e. approximately 10^{-10}m. For helium the associated zero-point energy is equivalent to a temperature as large as 10 K. This makes it energetically favorable for the matter to remain liquid, since the attractive part of the van der Waals interaction between the He atoms is relatively weak. Because of the importance of quantum effects on a macroscopic scale the two helium liquids are named quantum fluids.

9.6.1 Superfluidity

The quantum fluids ^{3}He and ^{4}He both enter a superfluid state below a certain critical temperature. In the case of ^{4}He the transition to the superfluid state occurs at $T = T_\lambda = 2.17$ K. Below this temperature liquid ^{4}He exhibits many unusual properties, one of the most striking being the ability to flow through narrow capillaries without measurable friction.

The properties of flow in ordinary liquids such as water or oil may be described by the classical hydrodynamic equations. The viscous forces in ordinary liquids resist their flow through narrow channels. By contrast quantum liquids may flow without friction. The description of the hydrodynamic behaviour of superfluid ^{4}He is made in terms of a two-fluid model. One imagines that the liquid consists of two interpenetrating components, one normal and one dissipationless. The normal component behaves as an ordinary liquid, but with a temperature-dependent density and viscosity, while the superfluid component is able to flow without friction. The density of the superfluid component increases with decreasing temperature, while the density of the normal component decreases in such a manner that the sum of the two densities remains

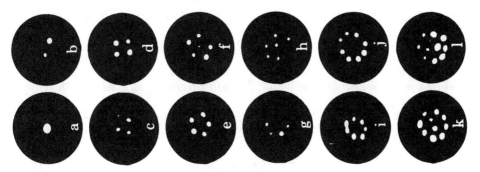

Figure 9.7: Quantized vortices.

equal to the total density of the liquid.

Measurements of the transport of heat in superfluid ^{4}He reveal another surprising phenomenon; the normal component carries entropy and thereby contributes to the transport of heat, while the superfluid component carries no entropy. The flow of the superfluid component constitutes an ordered motion of the helium atoms, while the motion of the normal component is associated with friction and transport of heat. It is important to remember that the two-fluid model is only a conceptual framework for the description of the superfluid state. The two 'fluids' are interpenetrating in a way which differs completely from ordinary mixtures, as evidenced by the dependence of their densities on temperature. The motion of the superfluid may be characterized by a single quantum state being occupied by a macroscopic number of particles, while the normal component represents the thermal population of the elementary excitations of the system relative to its ground state.

As an illustration of the macroscopic quantum phenomena encountered in superfluid ^{4}He we shall consider a cylindrical container with superfluid helium cooled to low temperatures, for instance $T = 50$ mK, thereby allowing the normal component to be neglected. We then let the container rotate about its axis. In ordinary liquids the form of the surface changes under rotation, the curvature increasing with increasing rotational frequency. When the container is filled with superfluid ^{4}He, the surface is at first unchanged at very low rotational frequencies. Eventually, when the rotational frequency increases above a certain threshold value, one vortex is formed in the interior of the liquid. By a further increase of the frequency, a second is formed, and subsequently three and so on. It is possible to get a direct picture of the vortices[4], as shown in Fig. 9.7. The remarkable feature of these vortices is that they are quantized. The quantization appears as the condition that the strength of a vortex, to be defined below, is an integer multiple of a universal quantum. As a measure of

---

[4] E. J. Yarmchuk, M. V. Gordon and R. E. Packard, Phys. Rev. Lett. **43**, 214, 1979.

the strength of a vortex we introduce the quantity Γ by the definition

$$\Gamma = 2\pi R v, \tag{9.113}$$

where R is the distance from the center of the vortex to the point considered, while v is the velocity of the liquid, this being constant on a circle with center in the vortex core.

The experiment with the superfluid under rotation (Fig. 9.7) shows that Γ is quantized in units of the Planck constant h divided by the mass M of a ^{4}He atom,

$$\Gamma = n\frac{h}{M}. \tag{9.114}$$

In practice one sees only $n = 1$. Each vortex carries the elementary quantum of vorticity, although in principle the six vortices shown in Fig. 9.7(f) could be replaced by one for which $\Gamma = 6h/M$. By photographing the current vortices it is thus possible to determine the Planck constant directly. In the case when we observe eight quanta of circulation at the rotational frequency 0.45 s^{-1}, the Planck constant h is determined from the calculated relation between the rotational frequency, the vortex quantum and the measured position of the vortices. In this manner one may determine the Planck constant visually with an accuracy of a few per cent.

Since the spin of a ^{4}He-nucleus is zero and the two electrons occupy a singlet state in the ground state of the atom (cf. Section 8.2), it follows that ^{4}He atoms are bosons. For identical bosons the wave function must be symmetric with respect to interchange of any two particles. It is therefore possible for the bosons to occupy the same single-particle quantum state. For non-interacting bosons the lowest energy eigenstate is thus a simple product state

$$\Psi = \psi_0(1)\psi_0(2)\cdots\psi_0(N), \tag{9.115}$$

the number of bosons being N. The single-particle state ψ_0 is the ground-state wave function for the motion of a single boson in the volume containing the N bosons. Being bosons, the ^{4}He atoms may occupy the same single-particle state. We may thus associate the phenomenon of superfluidity with the macroscopic occupation of this single-particle quantum state. In the ground state, where all particles occupy the same state, one may envisage the motion of the liquid as being completely ordered, involving changes of a single quantum state only. It is also possible to understand why the density of the normal component depends on temperature, since the presence of a normal component reflects the thermal occupation of the higher-lying states at finite temperature.

Let us denote the number of particles occupying the state ψ_0 by N_0. By definition N_0 equals the total number of particles N when the system is in its ground state at the absolute zero of temperature. The ground state energy of the system is NE_0, where E_0 is the single-particle energy associated with the state ψ_0. At high temperatures the mean energy is $3kT/2$ per particle or

$3NkT/2$. Between these two limits, $T = 0$ K and $T \to \infty$, the relative occupation N_0/N of the state ψ_0 must be gradually lowered. The ratio N_0/N may be calculated as a function of temperature by using the methods of statistical mechanics. It turns out that the ratio decreases as a function of temperature from its value 1 at $T = 0$ K to 0 at a certain critical temperature T_λ. Above T_λ the ratio equals 0 in the thermodynamic limit, $N \to \infty$. The specific heat may be obtained as the temperature derivative of the mean energy. The temperature dependence of the specific heat exhibits a kink at the temperature $T = T_\lambda$. This reflects the occurrence of a phase transition which is named a Bose condensation. The calculated value of T_λ is given by

$$kT_\lambda = 3.3 \frac{\hbar^2}{M}\left(\frac{N}{V}\right)^{\frac{2}{3}}. \tag{9.116}$$

When the appropriate value of the particle density N/V in liquid ^{4}He is inserted into (9.116) one obtains $T_\lambda = 3.2$ K, a result which is close to the observed transition temperature of 2.2 K. Except for the numerical constant the result (9.116) may be derived by dimensional analysis, since it is the only energy which may be formed from the constants $\hbar$, M and the particle density N/V. Within a numerical constant the expression (9.116) is the same as the Fermi energy (9.4) of an electron gas. The reason why the latter is much larger than T_λ is that the mass of an electron is much less than the mass of a helium atom.

We conclude that the simple model introduced to describe superfluid ^{4}He gives some understanding of the reason for the simultaneous occurrence of a normal and a superfluid component, and it yields quantitative agreement with the magnitude of the temperature T_λ. The latter is largely a coincidence. When the heat capacity is measured as a function of temperature, it does not exhibit a kink as predicted by the simple model, but rather a divergence involving a logarithmic dependence on $|T-T_\lambda|$. The experimentally determined value of the heat capacity resembles the Greek letter λ. This is the reason for the name λ-point for the transition temperature T_λ. In addition, measurements of neutron scattering indicate that the ratio N_0/N at low temperatures does not approach 1 but the much smaller value of 0.1.

The superfluidity of ^{3}He occurs at a temperature which is about one thousand times lower than T_λ, although other physical properties are quite similar. The boiling points of the two liquids are nearly equal, 4.2 K for ^{4}He and 3.2 K for ^{3}He. The enormous difference in transition temperature is related to the fact that a ^{3}He atom is a fermion, while a ^{4}He atom is a boson. It is therefore not possible for the ^{3}He atoms to occupy a common single-particle quantum state. The superfluidity of ^{3}He is due to correlations between two fermions. These pairs bear some resemblance to bosons and they may therefore occupy a common quantum state.

The wave function for two spin-1/2 particles may be a singlet corresponding to $s = 0$ or a triplet corresponding to $s = 1$, cf. (7.71-72). It turns out that the pair wave function in ^{3}He corresponds to a triplet state. This implies that the spatial part of the pair wave function is odd with respect to the interchange of the two fermions. The experimental observation of the unusual properties of the ^{3}He liquid allows one to deduce that the quantum number l for the relative orbital motion is 1. Though it is in principle possible that the ^{3}He atoms form pairs for which the spin function is a singlet, this would require that l is even in order that the total pair wave function be antisymmetric. Pairing in a state with $l = 0$ (a relative s-state) is disadvantageous from an energetic point of view, since the repulsion between the atoms at short distances would be significant. Measurements of nuclear magnetic resonance in superfluid ^{3}He have definitively established that the pair wave function is a spin triplet.

We have introduced a pair wave function as if we could consider the motion of a pair independently of all other pairs. This is however far from being the case. The tendency to form pairs is a collective effect involving all particles in the liquid. The ^{3}He atoms do not form molecules moving independently of each other. The distance between the members of a correlated pair is about a hundred times as large as the average separation between the atoms in the liquid.

Many different experiments demonstrate directly or indirectly the pairing phenomenon in liquid ^{3}He. One of the most direct experiments consists in the measurement of ultrasound attenuation in the liquid. As long as the temperature is higher than the transition temperature from the normal to the superfluid state, the attenuation decreases with decreasing temperature. Upon cooling through the transition temperature the damping continues to decrease with decreasing temperature, until it is strongly enhanced in a narrow temperature region and subsequently drops below its previous level.

The temperature at which the attenuation is strongly enhanced, is determined by the correlated motion of the pairs. The lowest-lying pair state is characterized by the quantum number $j = 0$ for the total angular momentum. It is, however, also possible for the pairs to have a total angular momentum corresponding to the quantum number $j = 2$, since the quantum number s for the spin of the pair and the quantum number l for the relative orbital motion both are equal to 1. As we have seen in Chapter 7, this implies that the total angular momentum quantum number j may assume the values 0, 1 or 2. States with $j = 2$ are characterized by an energy which is larger than the energy associated with $j = 0$. The strong increase in the attenuation occurs at the temperature where the difference between the two energies equals $\hbar\omega$, where ω is the angular frequency of the sound wave. The energy $\hbar\omega$ is determined by the fixed frequency of the sound wave, while the energy difference between the states associated with $j = 2$ and $j = 0$ depends on temperature. Since the energy difference is larger when more pairs occupy the common quantum state, it increases with decreasing temperature. The adjustment between this energy difference and $\hbar\omega$ therefore happens at a given temperature, where the absorption of energy increases strongly.

An even more dramatic illustration of the macroscopic quantum pheno-

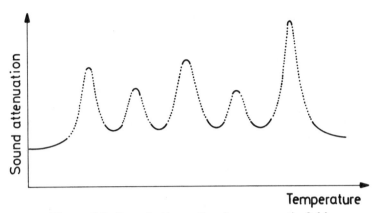

Figure 9.8: Sound attenuation in a magnetic field.

mena is achieved by placing the liquid in a magnetic field perpendicular to the direction of propagation of the sound wave and measuring the temperature dependence of the attenuation[5]. This splits the pronounced maximum of the attenuation into five peaks, each one corresponding to one of the possible values of the quantum number m given by $m = -2, -1, 0, 1, 2$ (Fig. 9.8). The distance between the maxima increases with increasing magnetic field, but the five peaks merge into one if the direction of the magnetic field is changed, so that it coincides with the direction of propagation of the sound wave.

The superfluidity of the quantum liquid ^{3}He has many features in common with the superconductivity of metals. Electrons are fermions and may also form pairs, as we shall discuss in the following subsection. The most important difference between ^{3}He and metallic superconductors is that pairing in metals involves a singlet rather than a triplet spin state.

9.6.2 Superconductivity

A number of metals lose their electrical resistance when cooled to low temperatures. The transition from a normally conducting to a superconducting state happens suddenly, over a very narrow temperature interval. Below the transition temperature T_c permanent currents may flow indefinitely in a ring-shaped conductor in the absence of any potential differences. Such permanent currents are used in practice for producing stable, high magnetic fields. Among the elements in the periodic system niobium has the highest transition temperature, $T_c = 9.2$ K. Before 1986 the highest known transition temperature for an alloy was 23 K, but since then new superconducting materials have been discovered with much higher transition temperatures. In oxides consisting of Y, Ba, Cu

---

[5] O. Avenel, E. Varoquaux and H. Ebisawa, Phys. Rev. Lett. **45**, 1952, 1980.

and O superconductivity has been found to exist at temperatures as high as 90 K, while oxides involving Tl, Ba, Ca, Cu and O are superconducting below 120 K.

Superconductivity was first observed by Kamerlingh Onnes in Leiden in 1911 when cooling mercury to low temperatures. At the time Kamerlingh Onnes was investigating the properties of helium, and the observation of the vanishing resistance of mercury was truly a novel and unexpected phenomenon. The resistance fell from its normal state value to zero over a very small temperature region in a way characteristic of a phase transition. The phase transition has many similarities with that observed many years later in the quantum fluid ^{3}He. In both cases the temperature dependence of the heat capacity exhibits a jump at the transition temperature. At the transition to the superconducting state in metals the heat capacity suddenly increases to a value which is between 2 and 3 times its value in the normal state just above the transition temperature. In the normal state the electronic part of the heat capacity depends linearly on temperature, while it decreases exponentially with decreasing temperature in the superconducting state at temperatures much less than the transition temperature T_c. A similar temperature dependence may be observed in the quantum liquid ^{3}He.

Though the magnitude of the electronic part of the heat capacity just above T_c may be widely different for different superconducting materials, the ratio between the jump in the heat capacity at T_c and the electronic part of the heat capacity just above this temperature is approximately the same, 1.4 for Al, 1.7 for In and 1.6 for Sn. Among the elements the largest value of this ratio has been measured for Pb, namely 2.7, while the ratio in ^{3}He varies from 1.4 at low pressures to 2 at 34 atmospheres, beyond which the liquid solidifies. Such universal features of the phenomenon were the basis for the theory of superconductivity which was put forward in 1957 by J. Bardeen, L. N. Cooper and J.R. Schrieffer. Their theory gave a value of 1.43 for the ratio in question as well as a number of other universal relations in agreement with experimental observations. The theory describes superconductivity as a manifestation of pair correlations resulting from the existence of attractive forces between the electrons due to the exchange of phonons.

Another characteristic feature of superconductors is their magnetic properties. Normal metals exhibit paramagnetic behavior, when they are placed in an external magnetic field. Superconductors are diamagnetic, as a result of which the magnetic flux may be totally excluded from their interior. This phenomenon is called the Meissner effect. When a permanent current circulates in a ring-shaped conductor, the magnetic flux through the interior of the ring is found to be quantized in units of $h/2e$. According to the Meissner effect the flux does not penetrate the ring, but only its interior. The quantization of the magnetic flux Φ is found to be

$$\Phi = n\frac{h}{2e}, n = 1, 2, \cdots. \tag{9.117}$$

Fermions and bosons

The observation of flux quantization according to (9.117), which involves $2e$ rather than e, is a direct demonstration of the importance of pair correlations for the phenomenon of superconductivity.

9.7 Problems

PROBLEM 9.1
Electrons can move in two dimensions on the surface of liquid helium. Such a system of electrons may be described as an ideal two-dimensional gas. The gas of N electrons, each with mass m, is contained in a square of side L. The connection between the one-electron energy ϵ and the wave vector (k_x, k_y) is

$$\epsilon = \frac{\hbar^2}{2m}(k_x^2 + k_y^2). \tag{9.118}$$

a) Find the Fermi energy in terms of the constants of the problem and determine its value for the case $N/L^2 = 10^{10} \text{cm}^{-2}$.

b) Determine the ground-state energy per particle for the system of electrons and compare with (9.6).

PROBLEM 9.2
Einstein showed in 1907 how Planck's quantum hypothesis yields an explanation of the decrease of the specific heat of solids with decreasing temperature. Einstein used a simplified model for the lattice vibrations by assuming that every atom in a solid oscillates with the same angular frequency ω_0. In this problem we shall carry out Einstein's calculation and compare the expression for the heat capacity in the Einstein model with the result (9.90-91) of the Debye model.

a) Use the Einstein model and the expression (9.75) for the mean energy of a single oscillator to find the heat capacity of a crystal consisting of N atoms. Sketch the result as a function of the dimensionless parameter T/Θ_E, where $\Theta_E = \hbar\omega_0/k$ and compare with Fig. 9.4.

b) Find in the limit of high temperatures an approximate expression for the difference between the heat capacity (9.80) in the classical limit and the result (9.90-91).

c) Repeat problem b) for the Einstein model. What is the relation between Θ_E and Θ_D needed to make the two expressions agree?

PROBLEM 9.3
Use a pocket calculator to determine the function $\epsilon(k)$ from (9.47) (enter a value of ϵ and find k as arccos). Draw the result as a function of k in the interval $-\pi < k \leq \pi$ for different values of I (for instance $I = 0.1, 0.5, 1, 3, 10$ and 100).

Problem 9.4
Use the model of Example 10 for calculating constant-energy curves in two dimensions for different values of the dimensionless parameter I.

10 PHYSICAL REALITY

On the evening of December 20, 1913 Niels Bohr gave a lecture in Fysisk Forening at the University of Copenhagen. The lecture was published the following year in the Danish physics journal Fysisk Tidsskrift. In the lecture Bohr told about his theory of the hydrogen spectrum and the comparison with the observed line spectra. He concluded[1] his talk as follows:

> I hope that I have expressed myself sufficiently clearly so that you may appreciate the extent to which these considerations conflict with the admirably consistent scheme of conceptions which have been rightly termed the classical theory of electrodynamics. On the other hand, I have tried to convey to you the impression that - just by emphasizing so strongly this conflict - it may also be possible in course of time to establish a certain coherence in the new ideas.

In this concluding chapter we shall indicate some of the landmarks in the subsequent debate over this conflict - a debate, which was strongly stimulated by Einstein's views and which in different forms has continued to this day. At the same time quantum mechanics has been universally accepted as an eminently successful theory by its practitioners. While questions have been raised throughout the century about the proper interpretation of quantum mechanics, nobody has denied its predictive power. The majority of physicists take a pragmatic point of view: A theory which is able to produce a number like the g-factor of the electron, equation (7.127), *must* be right in some deep sense of the word, even though the theory is in conflict with the 'consistent scheme of conceptions' referred to in Bohr's lecture.

10.1 The Bohr-Einstein discussion

The celebrated discussions between Bohr and Einstein have been vividly described by Bohr[1]. As we have seen in Chapter 1, Einstein contributed decisively to the rise of quantum mechanics through his theory of light quanta from 1905 and its successful application to the photoelectric effect. In 1917 he put forward his theory of the equilibrium between matter and radiation, in which he formulated the statistical rules governing radiative transitions between stationary states, and showed how the Planck radiation law could be deduced. In his paper he remarks on the difficulties in reconciling the wave and particle aspects of radiation, and points out as a weakness of the theory that it leaves to chance the time and the direction of the elementary processes.

Bohr and Einstein met for the first time in 1920 in Berlin, where they had an opportunity to discuss these fundamental questions. With the development

---

[1] Niels Bohr: *Discussion with Einstein on epistemological problems in atomic physics* in 'Albert Einstein, Philosopher-Scientist', Harper Edition, 1959.

of quantum mechanics as a general framework in 1925-26 the stage was set for a public debate over the interpretation of quantum mechanics, which lasted until Einstein's death.

As we have seen in Chapter 3 the lack of commutativity of the position and momentum operators of a particle leads to uncertainty relations for the mean square deviations, cf. equation (3.84). The existence of such uncertainty relations was pointed out in 1927 by Heisenberg. At the International Physics Congress held in 1927 at Como, Italy, where Einstein was not present, Bohr introduced the notion of complementarity. In his own words[1]:

> In a lecture on that occasion[2] I advocated a point of view conveniently termed "complementarity", suited to embrace the characteristic features of individuality of quantum phenomena, and at the same time to clarify the peculiar aspects of the observational problem in this field of experience. For this purpose, it is decisive to recognize that, *however far the phenomena transcend the scope of classical physical explanation, the account of all evidence must be expressed in classical terms.* The argument is simply that by the word "experiment" we refer to a situation where we can tell others what we have done and what we have learned and that, therefore, the account of the experimental arangement and of the results of the observations must be expressed in unambiguous language with suitable application of the terminology of classical physics.
>
> This crucial point, which was to become a main theme of the discussions reported in the following, implies the *impossibility of any sharp separation between the behaviour of atomic objects and the interaction with the measuring instruments which serve to define the conditions under which the phenomena appear.* In fact, the individuality of the typical quantum effects finds its proper expression in the circumstance that any attempt of subdividing the phenomena will demand a change in the experimental arrangement introducing new possibilities of interaction between objects and measuring instruments which in principle cannot be controlled. Consequently, evidence obtained under different experimental conditions cannot be comprehended within a single picture, but must be regarded as *complementary* in the sense that only the totality of the phenomena exhausts the possible information about the objects.

Bohr and Einstein met in October 1927 at the Fifth Physical Conference of the Solvay Institute, where Einstein expressed his concern over the extent to which a causal account in space and time was abandoned in quantum mechanics. He illustrated this by a simple *gedanken* experiment (a thought experiment - as opposed to a real one) involving an electron going through a slit in a diaphragm

---

[2] Atti del Congresso Internazionale dei Fisici, Como, Settembre 1927 (reprinted in *Nature*, 121, 78 and 580, 1928)

placed in front of a photographic plate. Quantum mechanics allows one, as we have seen in previous chapters, to calculate the probability of observing the electron within any given region of the plate. Einstein's concern was that the experimental recording of the electron at one point A of the plate excludes the observation of an effect of this electron at another point B on the plate. Quantum mechanics thereby introduces a correlation between the two events - recording of the electron at A and recording of the electron at B - which is completely foreign to the classical description of wave propagation. Einstein's position was - and remained - that the quantum-mechanical description is incomplete. The experiment was the first in a series of *gedanken* experiments, which greatly stimulated the debate on the interpretation of quantum mechanics.

The Bohr-Einstein debate continued at the next Solvay meeting in 1930. The apparent paradoxes raised by Einstein's *gedanken* experiments were successfully shown by Bohr to be consistent with his views, but the underlying issue - whether quantum mechanics is complete - was never answered to Einstein's satisfaction. In 1935 Einstein, Podolsky and Rosen published a paper[3] with the title 'Can Quantum-Mechanical Description of Physical Reality Be Considered Complete?'. In the introduction the authors say:

> Any serious consideration of a physical theory must take into account the distinction between the objective reality, which is independent of any theory, and the physical concepts with which the theory operates. These concepts are intended to correspond with the objective reality, and by means of these concepts we picture this reality to ourselves.
>
> In attempting to judge the success of a physical theory, we may ask ourselves two questions: (1) "Is the theory correct?" and (2) "Is the description given by the theory complete?" It is only in the case in which positive answers may be given to both of these questions, that the concepts of the theory may be said to be satisfactory. The correctness of the theory is judged by the degree of agreement between the conclusions of the theory and human experience. This experience, which alone enables us to make inferences about reality, in physics takes the form of experiment and measurement. It is the second question that we wish to consider here, as applied to quantum mechanics.
>
> Whatever the meaning assigned to the term *complete*, the following requirement for a complete theory seems to be a necessary one: *every element of the physical reality must have a counterpart in the physical theory*. We shall call this the condition of completeness. The second question is thus easily answered, as soon as we are able to decide what are the elements of physical reality.

---

[3] A. Einstein, B. Podolsky and N. Rosen, Physical Review **47**, 777, 1935.

> The elements of physical reality cannot be determined by *a priori* philosophical considerations, but must be found by an appeal to results of experiments and measurements. A comprehensive definition of reality is, however, unnecessary for our purpose. We shall be satisfied with the following criterion, which we regard as reasonable. *If, without in any way disturbing a system, we can predict with certainty (i. e. , with probability equal to unity) the value of a physical quantity, then there exists an element of physical reality corresponding to this physical quantity.* It seems to us that this criterion, while far from exhausting all possible ways of recognizing a physical reality, at least provides us with one such way, whenever the conditions set down in it occur. Regarded not as a necessary, but merely as a sufficient, condition of reality, this criterion is in agreement with classical as well as quantum-mechanical ideas of reality.

By considering a system consisting of two parts, which have interacted with each other during a limited period of time, Einstein, Podolsky and Rosen were able to show that the act of measuring an observable for one of the systems, *after* the two systems had ceased to interact, influences the result obtained in measuring an observable for the other system (a concrete example of this will be dicussed in the following section). From their analysis they concluded that quantum mechanics is incomplete according to their criteria given above.

Bohr did not agree with Einstein's criterion of reality. His answer[4] was published only a few months later. In it he wrote:

> Such an argumentation, however, would hardly seem suited to affect the soundness of quantum-mechanical description, which is based on a coherent mathematical formalism covering automatically any procedure of measurement like that indicated. The apparent contradiction in fact discloses only an essential inadequacy of the customary viewpoint of natural philosophy for a rational account of physical phenomena of the type with which we are concerned in quantum mechanics. Indeed the *finite interaction between object and measuring agencies* conditioned by the very existence of the quantum of action entails - because of the impossibility of controlling the reaction of the object on the measuring instruments if these are to serve their purpose - the necessity of a final renunciation of the classical ideal of causality and a radical revision of our attitude towards the problem of physical reality. In fact, as we shall see, a criterion of reality like that proposed by the named authors contains - however cautious its formulation may appear - an essential ambiguity when it is applied to the actual problems with which we are here concerned.

---

[4] N. Bohr, Physical Review **48**, 696, 1935.

10.2 Bell's inequalities

If quantum mechanics were incomplete, one might suppose that there exist additional 'hidden variables', which are needed to determine the system completely. Such hidden variables could restore not only causality but also locality in the sense that there is no action at a distance. Einstein's desire to preserve locality is expressed very clearly in ref. 1, p. 89: 'But on one supposition we should, in my opinion, absolutely hold fast: The real factual situation of the system S_2 is independent of what is done with the system S_1, which is spatially separated from the former'.

The debate over the interpretation of quantum mechanics and the nature of physical reality acquired a new dimension in 1964, when John Bell published his paper[5] 'On the Einstein-Podolsky-Rosen paradox'.

Bell considered a variation of the Einstein-Podolsky-Rosen experiment suggested by Bohm and Aharonov[6]. In this *gedanken* experiment one considers a pair of spin-1/2 particles in a singlet state moving freely in opposite directions. By measuring the deflection of a particle in an inhomogeneous magnetic field it is possible to measure its spin, since the Hamiltonian contains a term of the form (7.55). The result of performing a measurement of the spin component $\sigma_1 \cdot \mathbf{a}$ of one of the particles along a given direction specified by the unit vector $\mathbf{a}$ is either 1 or -1 (for convenience we measure the spin in units of $\hbar/2$). When the two particles occupy a singlet state given by

$$\frac{1}{\sqrt{2}}(\alpha\beta - \beta\alpha)$$

in the notation of Section 7.3, and if the measurement of the component $\sigma_1 \cdot \mathbf{a}$ yields the value 1, then quantum mechanics predicts with certainty that a measurement of $\sigma_2 \cdot \mathbf{a}$ yields the value -1. If the component of the spin of one of the particles along $\mathbf{a}$ is measured simultaneously with the component of the other particle along $\mathbf{b}$, then the expectation value according to quantum mechanics is given by

$$< \sigma_1 \cdot \mathbf{a} \, \sigma_2 \cdot \mathbf{b} > = -\mathbf{a} \cdot \mathbf{b}, \qquad (10.1)$$

as may be seen by working out the expectation value in the singlet state, using the properties of the spin operators discussed in Section 7.2.

Bell examined the consequences of the possible existence of hidden variables described by the parameter λ (which may denote one or several variables). The result A of measuring $\sigma_1 \cdot \mathbf{a}$ would then be determined by $\mathbf{a}$ and λ, while the result B of measuring $\sigma_2 \cdot \mathbf{b}$ is determined by $\mathbf{b}$ and λ. One has

$$A(\mathbf{a}, \lambda) = \pm 1; \quad B(\mathbf{b}, \lambda) = \pm 1. \qquad (10.2)$$

---

[5] J. S. Bell, Physics **1**, 195, 1964.
[6] D. Bohm and Y. Aharonov, Physical Review **108**, 1070, 1957.

To restore locality it is assumed that the result B for particle 2 does not depend on $\mathbf{a}$, which determines the setting of the magnet. Likewise the result A should be independent of $\mathbf{b}$, which determines the setting of the other magnet. Let λ be characterized by a probability distribution $\rho(\lambda)$ such that

$$\int d\lambda \rho(\lambda) = 1, \quad \rho(\lambda) \geq 0. \tag{10.3}$$

The expectation value of the product of the two components $\sigma_1 \cdot \mathbf{a}$ and $\sigma_2 \cdot \mathbf{b}$ is then

$$P(\mathbf{a}, \mathbf{b}) = \int d\lambda \rho(\lambda) A(\mathbf{a}, \lambda) B(\mathbf{b}, \lambda). \tag{10.4}$$

Bell showed that the hidden-variable expectation value (10.4) cannot in general be equal to the quantum mechanical expectation value (10.1) by the following argument.

Since $\rho(\lambda)$ is a probability distribution normalized according to (10.3) and because of the properties (10.2), the quantity P given by (10.4) cannot be less than -1. It may be equal to -1 at $\mathbf{a} = \mathbf{b}$ only if

$$A(\mathbf{a}, \lambda) = -B(\mathbf{a}, \lambda). \tag{10.5}$$

We shall assume in the following that (10.5) holds, consistent with the fact that the total spin is zero (note that the expectation value (10.1) given by quantum mechanics likewise is equal to -1 when $\mathbf{a} = \mathbf{b}$). When (10.5) is inserted into (10.4), we obtain

$$P(\mathbf{a}, \mathbf{b}) = -\int d\lambda \rho(\lambda) A(\mathbf{a}, \lambda) A(\mathbf{b}, \lambda). \tag{10.6}$$

If $\mathbf{c}$ is another unit vector, we have

$$P(\mathbf{a}, \mathbf{b}) - P(\mathbf{a}, \mathbf{c}) = -\int d\lambda \rho(\lambda)(A(\mathbf{a}, \lambda) A(\mathbf{b}, \lambda) - A(\mathbf{a}, \lambda) A(\mathbf{c}, \lambda)), \tag{10.7}$$

which may be written as

$$P(\mathbf{a}, \mathbf{b}) - P(\mathbf{a}, \mathbf{c}) = \int d\lambda \rho(\lambda) A(\mathbf{a}, \lambda) A(\mathbf{b}, \lambda)(A(\mathbf{b}, \lambda) A(\mathbf{c}, \lambda) - 1) \tag{10.8}$$

using (10.2).

Since $\rho(\lambda) \geq 0$, it follows that

$$|P(\mathbf{a}, \mathbf{b}) - P(\mathbf{a}, \mathbf{c})| \leq \int d\lambda \rho(\lambda)(-A(\mathbf{b}, \lambda) A(\mathbf{c}, \lambda) + 1) \tag{10.9}$$

Physical reality   255

or
$$1 + P(\mathbf{b},\mathbf{c}) \geq |P(\mathbf{a},\mathbf{b}) - P(\mathbf{a},\mathbf{c})|. \qquad (10.10)$$

This is one of Bell's inequalities.

This inequality (10.10) is deduced on the basis of a local, hidden-variable theory. It is easy to see that quantum mechanics violates (10.10). Let us for instance choose

$$\mathbf{a} = (1,0,0); \quad \mathbf{b} = (1/2,\sqrt{3}/2,0); \quad \mathbf{c} = (-1/2,\sqrt{3}/2,0). \qquad (10.11)$$

According to quantum mechanics, with the expectation value $P(\mathbf{a},\mathbf{b})$ given by $P(\mathbf{a},\mathbf{b}) = -\mathbf{a} \cdot \mathbf{b}$, cf. (10.1), the left hand side of (10.10) equals $1/2$, while the right hand side is equal to 1, in clear violation of the inequality.

The significance of Bell's inequalities is that they may be tested experimentally. Such experiments involving two-photon systems with photons propagating in opposite directions were pioneered by Freedman and Clauser and further developed by Alain Aspect and co-workers[7]. These experimental results were found to violate the appropriate Bell's inequality. It thus became possible on experimental grounds to rule out local, hidden-variable theories.

10.3 Elements of reality

The experimental observation of the violation of Bell's inequalities for the two-photon system has ruled out a class of local, hidden-variable theories as possible alternatives to quantum mechanics. Because of the probabilistic nature of Bell's inequalities, the two-photon experiments involve a large number of measurements before any conclusion may be drawn. In this final section we shall mention a recent proposal[8] of a *gedanken* experiment, in which a single measurement allows one to conclude whether Einstein's elements of reality have any physical significance.

The example employs three spin-1/2 particles rather than two as in the last subsection. In this *gedanken* experiment the three particles (which may be thought to originate in a spin-conserving decay) occupy a definite spin state $|\psi\rangle$ (to be specified below), while they fly apart in a plane, which we take to be horizontal, along three different straight lines.

The spin state is taken to be an eigenstate of the three commuting Hermitian operators L_1, L_2 and L_3 given by

$$L_1 = \sigma_x^1 \sigma_y^2 \sigma_y^3; \quad L_2 = \sigma_y^1 \sigma_x^2 \sigma_y^3; \quad L_3 = \sigma_y^1 \sigma_y^2 \sigma_x^3. \qquad (10.12)$$

The significance of the σ-operators is the following: σ_z^i is the operator (in fact a Pauli matrix, as we measure spin in units of $\hbar/2$) for the spin of particle i

---

[7] S. J. Freedman and J. F. Clauser, Phys. Rev. Lett. **28**, 938, 1972; A. Aspect, P. Grangier, and C. Roger, Phys. Rev. Lett. **47**, 460, 1981; **49**, 91, 1981.

[8] D. M. Greenberger, M. Horne and A. Zeilinger in *Bell's Theorem, Quantum Theory, and Conceptions of the Universe*, M. Kafatos, ed., Kluwer, Dordrecht, The Netherlands, 1989. We follow the discussion given by N. D. Mermin in Physics Today, June 1990.

along its direction of motion, σ_x^i is the operator for the spin of particle i along the vertical direction, while σ_y^i is the operator for the spin of particle i along the horizontal direction, perpendicular to the direction of motion.

To see that the three operators do indeed commute we consider the commutator of L_1 and L_2:

$$[L_1, L_2] = \sigma_x^1 \sigma_y^2 \sigma_y^3 \sigma_y^1 \sigma_x^2 \sigma_y^3 - \sigma_y^1 \sigma_x^2 \sigma_y^3 \sigma_x^1 \sigma_y^2 \sigma_y^3. \tag{10.13}$$

To show that the right hand side is equal to zero we use the anti-commutation rules obeyed by the Pauli matrices associated with a given particle i (cf. (2.76)),

$$\sigma_x^i \sigma_y^i + \sigma_y^i \sigma_x^i = 0, \tag{10.14}$$

and the fact that $\sigma_y^i \sigma_y^i = 1$, while the spin operators associated with different particles commute. As a result of using (10.14) twice, for particle 1 and 2, we conclude that the difference of the two terms on the right hand side of (10.13) is zero, meaning that the operators L_1 and L_2 commute. In a similar fashion the two other commutators may be seen to vanish.

Let us specify the spin state $|\psi\rangle$. In the notation of Section 7.3 (cf. (7.75)) the spin state is taken to be

$$|\psi\rangle = \frac{1}{\sqrt{2}}(\alpha\alpha\alpha - \beta\beta\beta) \tag{10.15}$$

It is readily seen that this state is an eigenstate of all three operators L_1, L_2 and L_3 with the eigenvalue 1.

We now apply the criterion of reality of Einstein, Podolsky and Rosen (EPR), as quoted in Section 10.1 above. If we measure the y-components of the spins of particle 2 and 3 and find them to have the same value, we can predict with certainty that a measurement of the x-component of the spin of particle 1 will yield the value 1, because the product of the three results must be 1 in the state $|\psi\rangle$. Likewise, if we find the y-components of the spins of particle 2 and 3 to be opposite, we can predict with certainty that the result of a measurement of the x-component of the spin of particle 1 will be -1. Thus according to the EPR reality criterion there exists an element of reality m_x^1, which may be either 1 or -1. For the same reason there exist elements of reality m_x^2 and m_x^3. The actual value of these elements are revealed by the corresponding pair of y-component measurements.

Now it is also possible, by measuring the x-component of particle 2 and the y-component of particle 3, to predict with certainty the outcome of measuring the y-component of the spin of particle 1. By analogy we thus introduce the additional elements of reality m_y^1, m_y^2 and m_y^3, which can also take on values 1 and -1. If these elements of reality do exist, the value of, say, $\sigma_x^1 \sigma_y^2 \sigma_y^3$ in the state $|\psi\rangle$ must be $m_x^1 m_y^2 m_y^3$, which is equal to 1, since the state $|\psi\rangle$ is an eigenstate of the operator L_1. By analogy, the products $m_x^1 m_y^2 m_y^3$, $m_y^1 m_x^2 m_y^3$

Physical reality

and $m_y^1 m_y^2 m_x^3$ must all be equal to 1, and since each m_y^i is either 1 or -1 it follows that the product of these three *numbers* is equal to $m_x^1 m_x^2 m_x^3$, that is

$$(m_x^1 m_y^2 m_y^3)(m_y^1 m_x^2 m_y^3)(m_y^1 m_y^2 m_x^3) = m_x^1 m_x^2 m_x^3 = 1. \qquad (10.16)$$

We can now see that these 'elements of reality' are in sharp conflict with the predictions of quantum mechanics. This follows by noting that the state $|\psi\rangle$ is also an eigenstate of the operator

$$M = \sigma_x^1 \sigma_x^2 \sigma_x^3. \qquad (10.17)$$

with the eigenvalue -1, since the use of the definition (2.76) of the Pauli-matrices yields

$$M|\psi\rangle = \sigma_x^1 \sigma_x^2 \sigma_x^3 \frac{1}{\sqrt{2}}(\alpha\alpha\alpha - \beta\beta\beta) = -\frac{1}{\sqrt{2}}(\alpha\alpha\alpha - \beta\beta\beta). \qquad (10.18)$$

In fact, M may be expressed as minus the product of the three commuting operators L_1, L_2 and L_3,

$$M = -L_1 L_2 L_3, \qquad (10.19)$$

which makes it evident that the value of M is -1 in the state $|\psi\rangle$.

Let us recapitulate: By adopting the reality criterion of EPR we conclude that $m_x^1 m_x^2 m_x^3 = 1$. But according to quantum mechanics, cf. (10.18), a measurement of the three x-components yields a product which is -1. The existence of the elements of reality may thus be disproved by a single measurement on the system of three particles.

Unlike the two-photon experiment the three-particle experiment has not been carried out. Few would doubt that its outcome would confirm the quantum mechanical prediction. Quantum mechanics is not only tremendously successful. It also appears to be complete, although its strange and wondrous implications continue to baffle its practitioners.

11 MORE PROBLEMS

11.1 Problems with solutions

Problem A.

A particle with mass m moves in one dimension. The potential energy is given by

$$V(x) = V_1 \text{ for } x < 0; \quad V(x) = 0 \text{ for } 0 < x < a; \quad V(x) = V_2 \text{ for } x > a, \tag{11.1}$$

where V_1 and V_2 are positive constants. It is assumed that V_1 is greater than or equal to V_2.

1) Write down the time-independent Schrödinger equation in the three regions and specify two dimensionless parameters which may be formed from the constants of the problem.

In the following we shall only consider bound states, i. e. solutions for which the associated energy E is less than both V_2 and V_1.

2) Derive a transcendental equation which determines the energy of the bound states. Discuss the cases a) $V_1 = V_2 = \infty$ and b) $V_1 = \infty$. Determine in case b) a condition for the existence of bound states.

Solution

1) The wave function for a stationary state has the form

$$\psi = u(x)e^{-iEt/\hbar}, \tag{11.2}$$

where u satisfies the time-independent Schrödinger equation

$$\frac{-\hbar^2}{2m}u'' + (V_1 - E)u = 0 \tag{11.3}$$

for $x < 0$,

$$\frac{-\hbar^2}{2m}u'' - Eu = 0 \tag{11.4}$$

for $0 < x < a$, and

$$\frac{-\hbar^2}{2m}u'' + (V_2 - E)u = 0 \tag{11.5}$$

for $x > a$. Here $'$ denotes the derivative with respect to x.

As dimensionless parameters we may choose

$$v_1 = \frac{2mV_1 a^2}{\hbar^2} \tag{11.6}$$

and
$$v_2 = \frac{2mV_2 a^2}{\hbar^2}, \tag{11.7}$$

but other possibilities exist, such as $v_1 - v_2$ and v_2. We introduce the abbreviations
$$\kappa_1 = \sqrt{2mV_1/\hbar^2 - k^2} \tag{11.8}$$

and
$$\kappa_2 = \sqrt{2mV_2/\hbar^2 - k^2}, \tag{11.9}$$

where $k = \sqrt{2mE/\hbar^2}$.

2) The solution in the region $x < 0$ is
$$u(x) = A e^{\kappa_1 x}, \tag{11.10}$$

where A is an arbitrary constant, since $e^{-\kappa_1 x}$ diverges for $x \to -\infty$. In the region $0 < x < a$ the solution is
$$u(x) = B \cos kx + C \sin kx, \tag{11.11}$$

where B and C are arbitrary constants. Finally, the solution in the region $x > a$ is given by
$$u(x) = D e^{-\kappa_2 x}, \tag{11.12}$$

where D is an arbitrary constant.

The continuity of u and u' in $x = 0$ yields
$$A = B \tag{11.13}$$

and
$$\kappa_1 A = kC, \tag{11.14}$$

while the continuity of u and u' at $x = a$ implies that
$$B \cos ka + C \sin ka = D e^{-\kappa_2 a} \tag{11.15}$$

and
$$-kB \sin ka + kC \cos ka = -\kappa_2 D e^{-\kappa_2 a}. \tag{11.16}$$

By eliminating B and C we obtain
$$A(\cos ka + \frac{\kappa_1}{k} \sin ka) - D e^{-\kappa_2 a} = 0 \tag{11.17}$$

and
$$A(\kappa_1 \cos ka - k \sin ka) + \kappa_2 D e^{-\kappa_2 a} = 0. \tag{11.18}$$

After eliminating A and D in (11.17-18) we arrive at the desired transcendental equation

$$\frac{\kappa_2}{k} = \frac{k \tan ka - \kappa_1}{k + \kappa_1 \tan ka}. \qquad (11.19)$$

We now discuss the solutions to (11.19) in the two cases a) $v_1 = v_2 = \infty$ and b) $v_1 = \infty$. In case a) the equation (11.19) implies that $\sin ka = 0$ in agreement with (4.42). In case b) we obtain

$$\frac{\kappa_2}{k} = -\cot ka. \qquad (11.20)$$

The condition for the existence of a bound state is therefore $ka > \pi/2$, which means that the height of the potential step must satisfy $v_2 > \pi^2/4$ in order that there exists (at least) one bound state.

Exercise. Show that (11.19) for $V_1 = V_2$ agrees with (5.12) and the corresponding equation for the odd solutions.

Problem B.

A particle with mass m moves in one dimension. The potential energy is given by

$$V(x) = \infty \text{ for } x < 0; \quad V(x) = 0 \text{ for } 0 < x < a; \quad V(x) = \infty \text{ for } x > a. \qquad (11.21)$$

We assume that, at times earlier than $t = 0$, the particle is in the ground state of the Hamiltonian in which (11.21) is the potential energy. At time $t = 0$ the potential energy is changed to

$$V(x) = \infty \text{ for } x < 0; \quad V(x) = 0 \text{ for } 0 < x < 2a; \quad V(x) = \infty \text{ for } x > 2a. \qquad (11.22)$$

It is assumed that the change happens so rapidly that the wave function of the particle just after $t = 0$ is given by the ground-state wave function associated with (11.21).

a) Determine the probability p_1 that a measurement of the energy of the particle at time $t > 0$ yields the ground-state energy associated with (11.22). Determine the probability p_2 that a measurement of the energy of the particle yields the second-lowest energy eigenvalue corresponding to (11.22). What is the probability that a measurement of the energy yields a result larger than the second-lowest energy eigenvalue?

Solution

The ground-state wave function associated with (11.21) is

$$u_1(x) = \sqrt{\frac{2}{a}} \sin(\pi x/a) \qquad (11.23)$$

More Problems

for $0 < x < a$ and zero everywhere else. Similarly, the ground-state wave function associated with (11.22) is given by

$$v_1(x) = \sqrt{\frac{2}{2a}} \sin(\pi x/2a), \qquad (11.24)$$

for $0 < x < 2a$ (otherwise zero). The wave function corresponding to the second-lowest energy eigenvalue in (11.22) is

$$v_2(x) = \sqrt{\frac{2}{2a}} \sin(\pi x/a) \qquad (11.25)$$

for $0 < x < 2a$ (otherwise zero).

We wish to determine the absolute square of the coefficients a_n in the expansion

$$\psi(x,t) = \sum_{n=1}^{\infty} a_n v_n(x) e^{-iE_n t/\hbar}, \qquad (11.26)$$

where E_n denotes the energy eigenvalues of the Hamiltonian corresponding to (11.22). To determine the probabilities we need to find a_n. Since $\psi(x,0)$ is given by $u_1(x)$, we can determine the coefficients a_n from

$$a_n = \int_0^{2a} dx\, v_n^*(x) u_1(x) = \int_0^{2a} dx\, v_n^*(x) \sqrt{\frac{2}{a}} \sin(\pi x/a) \Theta(a-x), \qquad (11.27)$$

where $\Theta(y)$ is 0 for $y < 0$ and 1 for $y > 0$.

We conclude that

$$a_1 = \int_0^a dx\, v_1(x) u_1(x) \qquad (11.28)$$

and

$$a_2 = \int_0^a dx\, v_2(x) u_1(x). \qquad (11.29)$$

The integrals are elementary and result in $a_1 = 4\sqrt{2}/3\pi$ and $a_2 = 1/\sqrt{2}$. From this we get

$$p_1 = a_1^2 = \frac{32}{9\pi^2} = 0.3603 \qquad (11.30)$$

and

$$p_2 = a_2^2 = \frac{1}{2}. \qquad (11.31)$$

The probability of measuring an energy which is larger than the second-lowest value is $(1 - p_1 - p_2)$ or 0.1397.

Problem C.

A particle moves in a central field described by the Hamiltonian H_0. In addition the particle experiences a perturbation H'. The total Hamiltonian is thus

$$H = H_0 + H', \qquad (11.32)$$

where H' is given by

$$H' = aL_x^2 + bL_y^2 + cL_z^2. \qquad (11.33)$$

Here (L_x, L_y, L_z) is the angular momentum operator divided by $\hbar$, while a, b and c are constants.

a) Determine how a triply degenerate p-state is split by the perturbation H'.

Solution

We express the perturbation in a basis consisting of the states $|l, m\rangle$ given by $|1, 1\rangle$, $|1, 0\rangle$ and $|1, -1\rangle$, and use that

$$L_x = (L_+ + L_-)/2, \qquad (11.34)$$

while

$$L_y = (L_+ - L_-)/2i. \qquad (11.35)$$

According to (7.53-54) we have

$$L_+^2 |1, -1\rangle = 2|1, 1\rangle, \qquad (11.36)$$

and

$$L_-^2 |1, 1\rangle = 2|1, -1\rangle. \qquad (11.37)$$

Similarly we obtain that

$$L_+ L_- |1, 0\rangle = 2|1, 0\rangle, \qquad (11.38)$$

and

$$L_+ L_- |1, 1\rangle = 2|1, 1\rangle. \qquad (11.39)$$

Since

$$L_x^2 = (L_+^2 + L_-^2 + L_+ L_- + L_- L_+)/4 \qquad (11.40)$$

and

$$L_y^2 = -(L_+^2 + L_-^2 - L_+ L_- - L_- L_+)/4 \qquad (11.41)$$

the perturbation matrix becomes

$$\begin{pmatrix} \alpha & 0 & \gamma \\ 0 & \beta & 0 \\ \gamma & 0 & \alpha \end{pmatrix} \qquad (11.42)$$

where $\alpha = c + (a+b)/2$, $\beta = a+b$ and $\gamma = (a-b)/2$. The eigenvalues for the matrix (11.42) are $(a+b)$, $(a+c)$ and $(b+c)$, which give the changes of the original three-fold degenerate energy level. For $a = b = c$ the perturbation does not split the energy level, but only shifts it by the amount $2a$, in agreement with the fact that the eigenvalue of $L_x^2 + L_y^2 + L_z^2$ is 2.

Problem D.

Find the reflection coefficient R and the transmission coefficient T for the barrier given by

$$V(x) = 0 \text{ for } x < 0; \quad V(x) = V_0 \text{ for } x > 0, \qquad (11.43)$$

where V_0 is a positive constant, as a function of the particle energy $E = \hbar^2 k^2 / 2m$.

Solution

We first examine the case $k < k_0$, where k_0 is a positive constant given by

$$V_0 = \frac{\hbar^2 k_0^2}{2m}. \qquad (11.44)$$

It is convenient to introduce the quantity κ_1 by the definition

$$\kappa_1 = \sqrt{k_0^2 - k^2}. \qquad (11.45)$$

In the region $x < 0$ the solution to the stationary Schrödinger equation is

$$u(x) = Ae^{ikx} + Be^{-ikx}, \qquad (11.46)$$

while the solution in the region $x > 0$ is given by

$$u(x) = Ce^{-\kappa_1 x}, \qquad (11.47)$$

cf. Problem A above.

It is readily seen that $T = 0$, since the probability current associated with (11.47) is zero. It follows that $R = 1$, which may also be seen explicitly from the condition

$$A + B = C, \qquad (11.48)$$

which expresses the continuity of the wave function at $x = 0$, and

$$ik(A - B) = -\kappa_1 C, \qquad (11.49)$$

which expresses the continuity of the derivative. Together the two conditions yield

$$\frac{B}{A} = \frac{ik + \kappa_1}{ik - \kappa_1}, \qquad (11.50)$$

which shows that $|B|^2/|A|^2 = 1$.

Next we treat the case $k > k_0$. The solutions determined above may be used again by introducing the substitution $\kappa_1 = -ik_1$, with

$$k_1 = \sqrt{k^2 - k_0^2}. \qquad (11.51)$$

This yields

$$R = |\frac{i(k - k_1)}{i(k + k_1)}|^2 = 1 - \frac{4kk_1}{(k + k_1)^2}. \qquad (11.52)$$

The result (11.52) yields the energy dependence of R as well as $T = 1 - R$, with k_1 given by (11.51) and $k^2 = 2mE/\hbar^2$. For energies satisfying the condition $0 \leq E - V_0 \ll V_0$ the transmission coefficient is thus proportional to $\sqrt{E - V_0}$.

Problem E.

A particle with mass m moves in one dimension under the influence of the potential energy

$$V(x) = 0 \text{ for } |x| < \frac{a}{2}, \quad V(x) = V_0 \text{ for } |x| > \frac{a}{2}, \qquad (11.53)$$

where V_0 is a positive constant. It is assumed that V_0 is so small relative to $\hbar^2/ma^2$ that there exists only one bound state with energy denoted by E.

a) Determine the probability p for finding in a measurement the particle outside the region which is classically accessible, given that the particle is in an eigenstate for the energy operator corresponding to the smallest possible energy. The magnitude of V_0 is given by $V_0 = 0.01 \cdot \hbar^2/2ma^2$.

Solution

The wave function for the particle is given by (5.7-9). To determine the desired probability we shall first find the constants A, B and C. The continuity of u implies that

$$A = B = e^{\kappa a/2} \cos(ka/2) C. \qquad (11.54)$$

In addition the wave function must be normalized, i. e.

$$2|C|^2 \int_0^{a/2} dx \cos^2(kx) + 2|A|^2 \int_{a/2}^{\infty} dx\, e^{-2\kappa x} = 1. \qquad (11.55)$$

More Problems   265

By carrying out the integrals one obtains

$$|C|^2 a(\frac{1}{2} + \frac{\sin ka}{2ka}) + |A|^2 \frac{1}{\kappa} e^{-\kappa a} = 1. \tag{11.56}$$

The desired probability is then

$$p = 2|A|^2 \int_{a/2}^{\infty} dx\, e^{-2\kappa x} = |A|^2 \frac{1}{\kappa} e^{-\kappa a}. \tag{11.57}$$

With the use of (11.54) in (11.56) the probability is found to be

$$p = |A|^2 \frac{1}{\kappa} e^{-\kappa a} = \frac{1}{1 + f(k)}, \tag{11.58}$$

where

$$f(k) = \frac{ka \sin(ka/2)}{2 \cos^3(ka/2)}(1 + (ka)^{-1} \sin ka), \tag{11.59}$$

since κ may be expressed in terms of k according to (5.12). To determine the magnitude of f for the given value of the parameter $2ma^2 V_0/\hbar^2$ we utilize that ka is small, thereby allowing the use of the series expansion $f(k) \simeq k^2 a^2/2$. This leads to the approximate expression for f,

$$f = \frac{1}{2} k_0^2 a^2, \tag{11.60}$$

where k_0 has been introduced by the definition

$$V_0 = \frac{\hbar^2 k_0^2}{2m}. \tag{11.61}$$

Since $k_0^2 a^2 = 2ma^2 V_0/\hbar^2 = 0.01$, the probability is thus $p = 0.995$.

Exercise. Determine the change in p obtained by including the next term in the series expansion of (11.59).

Problem F.

A particle with mass m moves in a central field given by the potential energy

$$V(r) = -V_0 e^{-r/a}, \tag{11.62}$$

where V_0 is a positive constant.

a) Find the mean value of the energy in a state given by

$$\psi = A e^{-r/\beta}, \tag{11.63}$$

where A is a normalization constant, as a function of the parameter β. Show that there exists at least one bound state when $V_0 \gg \hbar^2/ma^2$.

Solution

In polar coordinates the operator for the kinetic energy is given by

$$-\frac{\hbar^2}{2m}\left(\frac{2}{r}\frac{\partial}{\partial r} + \frac{\partial^2}{\partial r^2}\right) + \frac{\mathbf{L}^2}{2mr^2}. \tag{11.64}$$

In a state which is spherically symmetric, such as (11.63), the square of the angular momentum makes no contribution to the mean value of the kinetic energy. We therefore conclude that the mean value $<T>$ of the kinetic energy is

$$<T> = |A|^2 \frac{\hbar^2}{2m} 4\pi \int_0^\infty dr\, r^2 e^{-r/\beta}\left(\frac{2}{r\beta} - \frac{1}{\beta^2}\right) e^{-r/\beta}. \tag{11.65}$$

The mean value of the potential energy is

$$<V> = |A|^2 V_0 4\pi \int_0^\infty dr\, r^2 e^{-2r/\beta} e^{-r/a}. \tag{11.66}$$

The mean value $<H> = <T> + <V>$ of the Hamiltonian is therefore

$$<H> = \frac{\hbar^2}{2m\beta^2} - \frac{8V_0 a^3}{(\beta + 2a)^3}, \tag{11.67}$$

since the normalization of ψ implies that

$$|A|^2 = \frac{1}{\pi\beta^3}. \tag{11.68}$$

If $V_0 \gg \hbar^2/ma^2$ the mean value $<H>$ is seen to be negative if $\beta = a$. According to the variational method the corresponding value of $<H>$ constitutes an upper limit for the energy in the ground state. This proves that there exists at least one bound state, since the ground-state energy is negative.

Exercise. Discuss the case $V_0 \ll \hbar^2/ma^2$ and sketch $<H>$ for different values of the ratio $V_0 ma^2/\hbar^2$.

Problem G.

A particle with mass m moves in one dimension. The potential energy $V(x)$ is given by

$$V(x) = V_0(e^{-2x/a} - 2e^{-x/a}), \tag{11.69}$$

where V_0 and a are positive constants.

a) Show that $V(x)$ in the region near $x = 0$ may be approximated by a harmonic oscillator potential and determine within this approximation the lowest and the second-lowest energy eigenvalue. Find a criterion for the validity of the approximation by expanding $V(x)$ around $x = 0$.

Solution

In the region near $x = 0$ the potential $V(x)$ may be approximated by

$$V(x) \simeq -V_0 + V_0 \frac{x^2}{a^2}, \tag{11.70}$$

since

$$e^x \simeq 1 + x + \frac{x^2}{2} \tag{11.71}$$

for $|x| \ll 1$. The force constant K for the harmonic oscillator potential is therefore $K = 2V_0/a^2$, and the ground-state energy is given approximately by

$$-V_0 + \frac{1}{2}\hbar\sqrt{2V_0/ma^2}, \tag{11.72}$$

while the second-lowest energy eigenvalue is

$$-V_0 + \frac{3}{2}\hbar\sqrt{2V_0/ma^2}. \tag{11.73}$$

The validity of the approximation may be investigated by determining the mean value (in the ground state of the harmonic oscillator) of the correction term proportional to

$$V_0 x^3/a^3. \tag{11.74}$$

Except for a numerical constant this correction term arises from the expansion of $V(x)$ up to and including third order. Since the characteristic length for the wave function is $x_0 = \sqrt{\hbar/m\omega}$ with $\omega = \sqrt{K/m}$, we would thus obtain the criterion

$$\hbar\omega \gg V_0 \frac{x_0^3}{a^3} \tag{11.75}$$

or

$$\hbar^{1/2} \ll (a^2 m V_0)^{1/4}. \tag{11.76}$$

This criterion, however, is too stringent. It is sufficient to require that

$$\hbar \ll (a^2 m V_0)^{1/2}. \tag{11.77}$$

The reason is that the criterion (11.76) does not take into account that the mean value of x^3 equals zero in an energy eigenstate for the harmonic oscillator. It is therefore necessary to compare the contribution of the mean value of the fourth-order term in the expansion with $\hbar\omega$, which leads us to the result (11.77).

11.2 Problems

PROBLEM 11.1

In this problem we consider a particle with mass m moving in a three-dimensional harmonic oscillator potential given by $V = Kr^2/2$, where K is a positive constant.

a) Find the value of the lowest and the second-lowest energy level in terms of the constants entering the problem. What is the degree of degeneracy of the two levels? Write down wave functions associated with these two energy levels.

b) Write down the energy eigenstates discussed under a) with the Dirac bra-ket notation and explain the significance of the quantum numbers that enter. Write down a 4×4 diagonal matrix which has the eigenvalues found in a), together with the associated normalized eigenvectors (in the form of column vectors).

c) We now assume that the motion of the particle in the harmonic oscillator potential is perturbed by an additional term V' in the Hamiltonian. In polar coordinates V' is given by $V'(r, \theta, \phi) = Ar^4 \sin^4\theta \sin^4\phi$, where A is a constant. Use first-order perturbation theory to find the change in the ground-state energy of the system. How is the second-lowest energy level affected by the perturbation?

PROBLEM 11.2

An electron moves in a central field. It is assumed that all eigenvalues of the Hamiltonian correspond to bound states and that all eigenvalues E_{nl} are different. The spin of the electron is neglected.

We consider a perturbation on the motion in the central field corresponding to the addition of the term

$$U = 10^{-3} \frac{z^2}{a_0^2} \text{ eV}$$

to the Hamiltonian.

a) Show that U commutes with L_z.

b) Use first-order perturbation theory to calculate the splitting of the 2p-level as a result of the perturbation U and determine the energies and corresponding wave functions for the levels. It is known that

$$\int_0^\infty dr\, r^4 R_{21}^2 = 5a_0^2,$$

where a_0 is the Bohr radius and R_{21} is the normalized radial wave function for the 2p-state.

Problem 11.3

A particle with mass m moves in two dimensions. The Hamiltonian is

$$H = \frac{1}{2m}(p_x^2 + p_y^2) + \frac{1}{2}K(x^2 + y^2) + A(x-y)^2,$$

where K and A are positive constants.

a) First we consider the case $A = 0$. Determine the energies of the three lowest levels and their degree of degeneracy.

b) In this and the following question we treat $A(x-y)^2$ as a perturbation. Determine the ground-state energy to first order in A.

c) Use first-order perturbation theory to determine the effect of $A(x-y)^2$ on the second-lowest energy level. Specify the energies and corresponding wave functions.

d) Show that the energy eigenvalues of the Hamiltonian H may be found exactly by a suitable coordinate transformation. Determine the eigenvalues that correspond to the approximate energies found in b) and c).

Problem 11.4

In this problem we consider a hydrogen atom in a homogeneous electric field with field strength $\mathcal{E}$ and use the variational method to determine how the ground-state energy $E_g(\mathcal{E})$ depends on the electric field strength $\mathcal{E}$.

In the following we neglect the motion of the nucleus. The Hamiltonian is

$$H = H_0 + e\mathcal{E}z,$$

where

$$H_0 = \frac{p^2}{2m} - \frac{e_0^2}{r}.$$

a) For small values of the electric field strength, E_g is given by the series expansion

$$E_g = E_g(0) + b\mathcal{E}^2,$$

where b is a constant. Use first-order perturbation theory to show that the series does not contain a term proportional to $\mathcal{E}$.

The following questions aim at performing an approximate determination of b.

b) Show that the mean value of zH_0z equals zero in the ground state of the Hamiltonian H_0 by using the identity

$$ABA = \frac{1}{2}(A^2B + BA^2 - [A,[A,B]])$$

for the operators A and B.

c) Next we shall find an approximate expression for E_g by using the variational method with the trial function

$$\Phi = (1 + \lambda z)e^{-r/a_0},$$

where a_0 is the Bohr radius. Give an argument why $\lambda = 0$ when the field strength vanishes. Determine for $\mathcal{E} \neq 0$ an upper bound on the ground-state energy up to and including second-order terms in $\mathcal{E}$ by using the result given in b). Is the resulting value of b greater than or less than the exact one? Determine the size of the correction to the unperturbed ground-state energy for the case $\mathcal{E} = 1$ kV/cm.

PROBLEM 11.5

A particle with mass m moves in one dimension. The potential energy is given by

$$V(x) = -V_0 \frac{1}{\cosh^2(x/a)}, \qquad (11.78)$$

where V_0 is a positive constant.

a) Sketch the potential and show that it may be approximated by a harmonic oscillator potential for small values of x. Determine the wave function for the ground state of the harmonic oscillator and compare the characteristic length of the wave function with a. Use the constants in the problem together with $\hbar$ to formulate a condition that the approximation yields a reasonable description of the lowest energy levels.

b) Show that the difference between the potential (11.78) and the harmonic oscillator potential is proportional to x^4 for small values of x and determine the coefficient of the term proportional to x^4.

c) Use first-order perturbation theory to determine the energy in the ground state by considering the term proportional to x^4 to be a perturbation on the Hamiltonian for the harmonic oscillator. What is the criterion for the validity of the perturbation expansion?

d) Let us finally assume that the magnitude of V_0 is such that there exists four bound states, all with different energies. Explain qualitatively what happens to these states if V_0 is gradually reduced.

PROBLEM 11.6

A particle of mass m occupies a spherical volume of radius a corresponding to the potential energy

$$V(r) = 0 \text{ for } 0 \leq r \leq a, \quad V(r) = \infty \text{ for } a \leq r < \infty.$$

a) Show that the solution ψ to the Schrödinger equation separates in polar coordinates, $\psi = R_{El}(r)Y_{lm}(\theta, \phi)$, and determine the wave function and the energy in the ground state.

b) The sphere is now suddenly expanded to a a sphere with radius $2a$. The expansion is so rapid that the wave function may be assumed not to change in the process. What is the probability that a measurement of the particle energy yields the lowest eigenvalue for the Hamiltonian belonging to the sphere with radius $2a$?

PROBLEM 11.7
A particle with mass m moves in the potential

$$V(x) = \infty \text{ for } x < 0, \quad V(x) = \frac{K}{2}(x - x_0)^2 \text{ for } x > 0,$$

where K and x_0 are positive constants. In the following we shall seek to determine how the ground-state energy $E_g = E_g(x_0)$ depends on x_0.

a) Find $E_g(0)$ and sketch the wave function.

b) Determine the limiting value of E_g for x_0 tending towards infinity and state the condition (in terms of the constants in the problem and $\hbar$) under which E_g is near this limiting value.

c) Use first-order perturbation theory to determine the slope of the function $E_g(x_0)$ at $x_0 = 0$, by treating the term $-Kxx_0$ as a perturbation. Sketch $E_g(x_0)$ as a function of x_0.

d) Consider the symmetric potential given by $V(x)$ for $x > 0$ and $V(-x)$ for $x < 0$. How does the ground-state energy depend on x_0? Compare with the ammonia molecule treated in Example 9.

PROBLEM 11.8
In the following problem we shall consider the influence of a crystal field, which does not possess spherical symmetry, on the motion of an electron (mass m, charge $-e$) in a central field given by the potential energy $V(r)$.

The Hamiltonian of the electron is assumed to be

$$H = H_0 + H',$$

where

$$H_0 = \frac{p^2}{2m} + V(r).$$

The crystal field is described by the perturbation H' given by

$$H' = A(L_x^2 + L_y^2 - 2L_z^2),$$

where A is a constant, while L_x, L_y and L_z denote the components of the angular momentum of the electron.

We consider an unperturbed energy level E_0 associated with the Hamiltonian H_0. The level is furthermore characterized by the angular momentum quantum number $l = 2$ (spin is neglected).

a) Examine whether H commutes with a) each of the three components of the angular momentum or b) the square of the angular momentum.

b) Find the splitting of the level with unperturbed energy E_0 due to the perturbation H' (it is assumed that the difference in energy between the level considered and the other levels in the central field is sufficiently large that we may neglect entirely the latter). Determine the degree of degeneracy of the resulting levels.

c) We now assume that the electron, in addition to the crystal field, is influenced by a homogeneous magnetic field B along the z-axis. Determine the change in the levels determined in b) due to the presence of the magnetic field, to first order in B.

PROBLEM 11.9

Consider an electron moving in a central field. We assume that it occupies a state which is an eigenstate for the Hamiltonian and for the square of the orbital angular momentum, $\mathbf{L}^2$, with the quantum number l being 1. By $|m_l, m_s\rangle$ we denote a normalized state which is a simultaneous eigenstate for L_z with eigenvalue $m_l \hbar$, and for S_z with eigenvalue $m_s \hbar$, with $\mathbf{S}$ being the spin angular momentum operator.

The states
$$-\sqrt{2/3}|-1, 1/2\rangle + \sqrt{1/3}|0, -1/2\rangle \tag{11.79}$$

and
$$\sqrt{2/3}|1, -1/2\rangle - \sqrt{1/3}|0, 1/2\rangle \tag{11.80}$$

are eigenstates for $\mathbf{J}^2$. As usual $\mathbf{J}$ denotes the total angular momentum, $\mathbf{J} = \mathbf{L} + \mathbf{S}$.

a) Verify that the eigenstates (11.79) and (11.80) are normalized, and determine the corresponding eigenvalue for $\mathbf{J}^2$. Show that the states (11.79) and (11.80) are eigenstates for the operator $\mathbf{L} \cdot \mathbf{S}$.

In the following we assume that the electron, in addition to the central field, is affected by a homogeneous magnetic field. The spin-orbit interaction is known to be sufficiently strong that it is permissible to neglect all states other than (11.79) and (11.80). In order to take into account the presence of the magnetic field $\mathbf{B} = (0, 0, B)$ we must add to the Hamiltonian the operator

$$H' = \mu_B (L_z + 2S_z) B/\hbar, \tag{11.81}$$

where μ_B is the Bohr magneton.

b) Determine the matrix for H' in the basis consisting of the states (11.79) and (11.80), and discuss the physical significance of the result.

Problem 11.10

Consider a particle of mass M moving in the three-dimensional harmonic oscillator potential

$$V(r) = \frac{1}{2}Kr^2, \qquad (11.82)$$

where $K > 0$. The state of the particle is described by the normalized wave function $\psi(\mathbf{r}, t)$ given by

$$\psi = \frac{\sqrt{2}}{\pi^{3/4}\sqrt{3a^3}} e^{-r^2/2a^2}(e^{-3i\omega t/2} + a^{-1}r \cos\theta \, e^{-5i\omega t/2}). \qquad (11.83)$$

Here θ is the usual polar angle (the z-axis is the polar axis), and $a = \sqrt{\hbar/M\omega}$ with $\omega = \sqrt{K/M}$.

a) Examine whether ψ is an eigenstate for the following operators: i) the Hamiltonian H, ii) the square of the angular momentum operator $\mathbf{L}^2$, and iii) the z-component of the angular momentum, L_z. Do the operators H, $\mathbf{L}^2$ and L_z commute with each other?

b) Find the possible outcomes of a measurement of the energy of the particle and determine the probabilities of measuring these values.

Problem 11.11

A particle with mass m moves in three dimensions. The potential energy $V(x, y, z)$ is given by

$$V(x, y, z) = 0, \quad \text{if } 0 < x < L_1, \; 0 < y < L_2, \; 0 < z < L_3;$$
$$V(x, y, z) = \infty \quad \text{otherwise.} \qquad (11.84)$$

a) Consider the case when $L_1 = L_2 = L_3 = L$ and determine the energies of the three lowest levels and their degree of degeneracy.

b) Repeat a) in the case $L_2 = L_3 = L$ and $L_1 = 0.99L$.

In the remaining part of the problem we consider only the case $L_1 = L_2 = L_3 = L$.

c) It is now assumed that the motion of the particle in the box is influenced by a perturbation H'. Use first-order perturbation theory to determine the change in the ground-state energy of the particle when $H' = ax$, where a is a positive constant. Determine the corresponding change in the second-lowest energy level.

Problem 11.12

A particle with mass m moves in a central field given by

$$V(r) = -V_0 e^{-r/a}, \qquad (11.85)$$

where V_0 and a are positive constants.

The state of the particle is given by the wave function

$$\psi = Cr e^{-r/\beta} \sin\theta \cos\phi, \qquad (11.86)$$

where r, θ and ϕ are polar coordinates, while β is a positive constant. The constant C is a normalization constant.

a) What is the probability that a measurement of L_z, the z-component of the angular momentum of the particle, results in the value $\hbar$? What is the probability that a measurement of L_z results in the value $2\hbar$?

b) Determine the mean value and the mean-square deviation of $\mathbf{L}^2$ and L_z in the state (11.86). Is ψ an eigenstate of the parity operator?

c) Express the mean value of the energy in the state (11.86) in terms of the parameter β and the constants $\hbar, m, V_0$ and a.

d) Show that there exist bound states in the limit $\epsilon = \hbar^2/ma^2 V_0 \ll 1$.

Appendix A. Table of fundamental constants based on 1986 recommended values.

The digits in parentheses are the one-standard-deviation uncertainty in the last digits of the given value. The relative uncertainty on e. g. the gravitational constant is thus $85/667259 = 0.000128$.

quantity	symbol	value
speed of light	c	$299\,792\,458$ ms^{-1}
permeability of vacuum	μ_0	$4\pi \cdot 10^{-7}$ NA^{-2}
permittivity of vacuum	ϵ_0	$8.854\,187\,817\ldots \cdot 10^{-12}$ Fm^{-1}
gravitational constant	G	$6.672\,59(85) \cdot 10^{-11}$ m^{3}kg^{-1}s^{-2}
Planck constant	$\hbar$	$1.054\,572\,66(63) \cdot 10^{-34}$ Js
elementary charge	e	$1.602\,177\,33(49) \cdot 10^{-19}$ C
electron mass	m_e	$9.109\,3897(54) \cdot 10^{-31}$ kg
muon mass	m_μ	$1.883\,5327(11) \cdot 10^{-28}$ kg
proton mass	m_p	$1.672\,6231(10) \cdot 10^{-27}$ kg
neutron mass	m_n	$1.674\,9286(10) \cdot 10^{-27}$ kg
Avogadro constant	N_A	$6.022\,1367(36) \cdot 10^{23}$ mol^{-1}
Boltzmann constant	k	$1.380\,658(12) \cdot 10^{-23}$ JK^{-1}

Appendix B. Polar and cylindrical coordinates

1. Cylindrical coordinates.
The cylindrical coordinates are defined by

$$x = \rho \cos \phi, \quad y = \rho \sin \phi, \quad z = z. \tag{1}$$

The inverse relations are

$$\tan \phi = \frac{y}{x}, \quad \rho = \sqrt{x^2 + y^2}, \quad z = z. \tag{2}$$

We shall need to express the Laplace operator in cylindrical coordinates by using

$$\frac{\partial}{\partial x} = \frac{\partial \rho}{\partial x} \frac{\partial}{\partial \rho} + \frac{\partial \phi}{\partial x} \frac{\partial}{\partial \phi} \tag{3}$$

and the corresponding expression for $\partial/\partial y$. We get

$$\frac{\partial}{\partial x} = \cos \phi \frac{\partial}{\partial \rho} - \frac{\sin \phi}{\rho} \frac{\partial}{\partial \phi} \tag{4}$$

and

$$\frac{\partial}{\partial y} = \sin \phi \frac{\partial}{\partial \rho} + \frac{\cos \phi}{\rho} \frac{\partial}{\partial \phi}. \tag{5}$$

From these relations it follows by use of (4-5) that the Laplace operator in cylindrical coordinates becomes

$$\frac{\partial^2}{\partial x^2} + \frac{\partial^2}{\partial y^2} + \frac{\partial^2}{\partial z^2} = \frac{\partial^2}{\partial \rho^2} + \frac{1}{\rho} \frac{\partial}{\partial \rho} + \frac{1}{\rho^2} \frac{\partial^2}{\partial \phi^2} + \frac{\partial^2}{\partial z^2}. \tag{6}$$

Furthermore we see that the z-component of the angular momentum is

$$\hat{L}_z = \frac{\hbar}{i}(x \frac{\partial}{\partial y} - y \frac{\partial}{\partial x}) = \frac{\hbar}{i} \frac{\partial}{\partial \phi}. \tag{7}$$

2. Polar coordinates
The polar coordinates r, θ, ϕ are defined by

$$x = r \sin \theta \cos \phi, \quad y = r \sin \theta \sin \phi, \quad z = r \cos \theta. \tag{8}$$

The inverse relations are

$$r = \sqrt{x^2 + y^2 + z^2}, \quad \tan \theta = \sqrt{x^2 + y^2}/z, \quad \tan \phi = y/x. \tag{9}$$

Appendix B

In analogy with (3) we have

$$\frac{\partial}{\partial x} = \frac{\partial r}{\partial x}\frac{\partial}{\partial r} + \frac{\partial \theta}{\partial x}\frac{\partial}{\partial \theta} + \frac{\partial \phi}{\partial x}\frac{\partial}{\partial \phi}. \tag{10}$$

By working out $\partial r/\partial x$ etc. and expressing the result in polar coordinates we get

$$\frac{\partial}{\partial x} = \sin\theta\cos\phi\frac{\partial}{\partial r} + \frac{\cos\theta\cos\phi}{r}\frac{\partial}{\partial \theta} - \frac{\sin\phi}{r\sin\theta}\frac{\partial}{\partial \phi}. \tag{11}$$

Similarly we obtain

$$\frac{\partial}{\partial y} = \sin\theta\sin\phi\frac{\partial}{\partial r} + \frac{\cos\theta\sin\phi}{r}\frac{\partial}{\partial \theta} + \frac{\cos\phi}{r\sin\theta}\frac{\partial}{\partial \phi} \tag{12}$$

and

$$\frac{\partial}{\partial z} = \cos\theta\frac{\partial}{\partial r} - \frac{\sin\theta}{r}\frac{\partial}{\partial \theta}. \tag{13}$$

The three components of angular momentum are then given by

$$\hat{L}_x = \frac{\hbar}{i}(y\frac{\partial}{\partial z} - z\frac{\partial}{\partial y}) = \frac{\hbar}{i}(-\sin\phi\frac{\partial}{\partial \theta} - \cot\theta\cos\phi\frac{\partial}{\partial \phi}), \tag{14}$$

$$\hat{L}_y = \frac{\hbar}{i}(z\frac{\partial}{\partial x} - x\frac{\partial}{\partial z}) = \frac{\hbar}{i}(\cos\phi\frac{\partial}{\partial \theta} - \cot\theta\sin\phi\frac{\partial}{\partial \phi}) \tag{15}$$

and

$$\hat{L}_z = \frac{\hbar}{i}(x\frac{\partial}{\partial y} - y\frac{\partial}{\partial x}) = \frac{\hbar}{i}\frac{\partial}{\partial \phi}. \tag{16}$$

The square of the length of the angular momentum is then seen to be

$$\hat{\mathbf{L}}^2 = -\hbar^2[\frac{1}{\sin\theta}\frac{\partial}{\partial \theta}(\sin\theta\frac{\partial}{\partial \theta}) + \frac{1}{\sin^2\theta}\frac{\partial^2}{\partial \phi^2}]. \tag{17}$$

The expression for the Laplace operator ∇^2 is given in cartesian coordinates by

$$\nabla^2 = \frac{\partial^2}{\partial x^2} + \frac{\partial^2}{\partial y^2} + \frac{\partial^2}{\partial z^2}. \tag{18}$$

In polar coordinates this becomes upon use of (11-13)

$$\nabla^2 = \frac{1}{r^2}\frac{\partial}{\partial r}(r^2\frac{\partial}{\partial r}) + \frac{1}{r^2}[\frac{1}{\sin\theta}\frac{\partial}{\partial \theta}(\sin\theta\frac{\partial}{\partial \theta}) + \frac{1}{\sin^2\theta}\frac{\partial^2}{\partial \phi^2}]. \tag{19}$$

Appendix C. Fourier transformation and the delta function.

Fourier's integral theorem allows one to expand an arbitrary function $\psi(x)$ in terms of the basis functions $\exp ikx$ according to

$$\psi(x) = \frac{1}{\sqrt{2\pi}} \int_{-\infty}^{\infty} dk f(k) e^{ikx}, \tag{1}$$

where

$$f(k) = \frac{1}{\sqrt{2\pi}} \int_{-\infty}^{\infty} dx \psi(x) e^{-ikx}. \tag{2}$$

Often one uses instead of $f(k)$ the function $\tilde{f}(k) = f(k)\sqrt{2\pi}$, in which case the Fourier transformation (1) and (2) becomes

$$\psi(x) = \frac{1}{2\pi} \int_{-\infty}^{\infty} dk \tilde{f}(k) e^{ikx}. \tag{3}$$

where

$$\tilde{f}(k) = \int_{-\infty}^{\infty} dx \psi(x) e^{-ikx}. \tag{4}$$

In three spatial dimensions (3) and (4) are generalized to

$$\psi(\mathbf{r}) = \frac{1}{(2\pi)^3} \int d\mathbf{k} \tilde{f}(\mathbf{k}) \exp(i\mathbf{k} \cdot \mathbf{r}). \tag{5}$$

where

$$\tilde{f}(\mathbf{k}) = \int d\mathbf{r} \psi(\mathbf{r}) \exp(-i\mathbf{k} \cdot \mathbf{r}). \tag{6}$$

In the following we use the symmetric form (1) and (2). The function f is called the Fourier-transform of ψ. We illustrate the content of Fourier's integral theorem by deriving (1) and (2) in two different ways.

a) Use of the delta function.

As proposed by Dirac and later justified by use of the theory of distributions it is possible to employ a function $\delta(x)$ with the following property

$$g(x_0) = \int_{-\infty}^{\infty} dx g(x) \delta(x - x_0). \tag{7}$$

The effect of the delta function $\delta(x - x_0)$, when integrated with a well-behaved function $g(x)$, is thus to pick out the value of the function $g(x)$ at $x = x_0$. The

Appendix C

delta function $\delta(x - x_0)$ may be thought of as a function which is infinite at $x = x_0$ and zero everywhere else, in such a way that the area underneath is unity,

$$1 = \int_{-\infty}^{\infty} dx \delta(x - x_0). \tag{8}$$

The delta function is a generalization of the Kronecker symbol

$$\delta_{ij} = 1 \text{ if } i = j; \quad \delta_{ij} = 0 \text{ if } i \neq j \tag{9}$$

to the case where i, j are continuous variables.

The delta function may for practical purposes be considered to be the limiting function of a sequence of functions with the property of being increasingly peaked about x_0 with the integral over x remaining equal to 1. As an example $\delta(x)$ may be represented by

$$\delta(x) = \lim_{k_0 \to \infty} \frac{1}{2\pi} \int_{-k_0}^{k_0} dk e^{ikx} = \lim_{k_0 \to \infty} \frac{1}{\pi x} \sin(k_0 x). \tag{10}$$

For any finite k_0 we may check by integrating over x that

$$\int_{-\infty}^{\infty} dx \frac{1}{\pi x} \sin(k_0 x) = 1. \tag{11}$$

Also, the value of the integrand at $x = 0$ in (11) is k_0/π and thus increases towards infinity as $k_0 \to \infty$. The sequence of functions given by (10) as well as the limiting function are even, $\delta(x) = \delta(-x)$.

We are now ready to prove the Fourier transformation (1) and (2) by using the representation

$$\delta(x - x') = \frac{1}{2\pi} \int_{-\infty}^{\infty} dk e^{ik(x - x')} \tag{12}$$

in

$$\psi(x) = \int_{-\infty}^{\infty} dx' \delta(x - x') \psi(x'), \tag{13}$$

where we have used that $\delta(x)$ is an even function of x. As a result (13) becomes

$$\psi(x) = \frac{1}{2\pi} \int_{-\infty}^{\infty} dk \int_{-\infty}^{\infty} dx' \psi(x') e^{ikx} e^{-ikx'}. \tag{14}$$

This may be written as

$$\psi(x) = \frac{1}{\sqrt{2\pi}} \int_{-\infty}^{\infty} dk f(k) e^{ikx}, \tag{15}$$

where
$$f(k) = \frac{1}{\sqrt{2\pi}} \int_{-\infty}^{\infty} dx' \psi(x') e^{-ikx'}, \qquad (16)$$
which is Fourier's integral theorem (1) and (2).

b) Fourier series.
Fourier's integral theorem may also be obtained by starting from the Fourier series theorem and taking a suitable limit. We illustrate this by an example. Let us consider the function g defined in the interval $-l < x < l$
$$g(x) = x \quad \text{for} \quad -l < x < l. \qquad (17)$$
Since the function may be continued to a periodic function with period $2l$ and is integrable in the interval $-l \le x \le l$, the Fourier series theorem applies,
$$g(x) = \sum_n a_n e^{in\pi x/l}, \qquad (18)$$
with Fourier coefficients a_n given by
$$a_n = \frac{1}{2l} \int_{-l}^{l} dx\, g(x) e^{-in\pi x/l}. \qquad (19)$$

Here n assumes any integer value, positive as well as negative, together with zero.

We thus expand g in terms of a basis consisting of the set of functions $e^{in\pi x/l}$. Functions that belong to different n are orthogonal, since
$$\int_{-l}^{l} dx\, e^{-im\pi x/l} e^{in\pi x/l} = 2l \delta_{mn}. \qquad (20)$$

We may check the validity of the Fourier series expansion (18)-(19) by multiplying (18) with $e^{-im\pi x/l}$, integrating over x and using (20), which turns the right hand side of (18) into $2l a_m$, in agreement with (19).

The Fourier series theorem thus allows one to determine from a given function $f(x)$ the Fourier coefficients a_n. In the case of the function $f(x)$ given in (17), the Fourier coefficients become
$$a_n = -\frac{i}{l} \int_0^l dx\, x \sin(n\pi x/l) = \frac{i(-1)^n l}{n\pi}, \qquad n \ne 0,$$
$$a_0 = 0, \qquad (21)$$
corresponding to the series
$$g(x) = \frac{2l}{\pi} \left(\frac{\sin(\pi x/l)}{1} - \frac{\sin(2\pi x/l)}{2} + \frac{\sin(3\pi x/l)}{3} \cdots \right). \qquad (22)$$

Appendix C

We now undertake a limiting procedure by letting the periodicity interval $2l$ go towards infinity and replacing the sum over n by an integration according to the prescription $\sum_n \to (l/\pi) \int dk$, since $k = \pi n/l$. This yields

$$g(x) = \frac{l}{\pi} \int_{-\infty}^{\infty} dk e^{ikx} a(k)$$

and

$$a(k) = \frac{1}{2l} \int_{-\infty}^{\infty} dx e^{-ikx} g(x),$$

which become (1) and (2), when $g(x)$ is identified with $\psi(x)$ and $a(k)$ with $f(k)\pi/l\sqrt{2\pi}$.

Properties of the delta function.

The following properties of the delta function are often used (′ denotes a derivative with respect to x)

$$\delta(x) = \delta(-x). \tag{23}$$

$$\delta(ax) = \frac{1}{|a|}\delta(x). \tag{24}$$

$$f(x)\delta(x) = f(0)\delta(x). \tag{25}$$

$$\delta'(x) = -\delta'(-x). \tag{26}$$

$$\delta(f(x)) = \sum_n \frac{1}{|f'(x_n)|}\delta(x - x_n), \quad \text{where} \quad f(x_n) = 0, f'(x_n) \neq 0. \tag{27}$$

$$\delta(x) = \frac{1}{2\pi} \int_{-\infty}^{\infty} dk e^{ikx}. \tag{28}$$

The relation (27) is particularly useful for carrying out integrals involving the delta function.

Index

action 47, 58
ammonia maser 204
ammonia molecule 197
analytical mechanics 46
angular momentum 21, 56, 92, 152
    addition of 163
    commutation relations 153, 157
    operators 153, 155
    quantization 154, 161
annihilation operator 34, 66, 91, 109
anti-bonding state 203, 204
antiparticles 42
approximation methods
    perturbation theory 136, 187
    variational method 191
atomic units 194, 204
axial symmetry 152
background radiation 239
Balmer series 9
bands 221, 223
barrier reflection 134
barrier transmission 128
basis 33, 74, 82, 94
Bell's inequalities 253
'Big Bang' 239
binding energy 19, 205
blackbody radiation 238
Bloch's theorem 219
Bohr magneton 173
Bohr radius 12, 19, 170
bonding state 203, 204
Born approximation 134
Bose condensation 243
bosons 191, 212
bound states 125, 139, 169
boundary conditions 113, 125, 167
bra 62, 69
canonical distribution 228
central field 154, 167, 173
chemical bond 197
classical limit 109, 181

Clebsch-Gordan coefficients 166
coherent state 109
commutation relations 58, 61, 84
commutator 57
complementarity 250
conduction electrons 212
constant of the motion 55, 56, 92, 146, 152, 155, 176, 184
continuity equation 126, 216
correspondence principle 15
creation operator 34, 66, 91, 110
cross-section 140
cyclic coordinates 52
cyclotron frequency 22, 143, 145, 181
cylindrical coordinates 148, 276
de Broglie wavelength 28, 36
Debye temperature 233
degeneracy 73, 146, 149, 172, 187, 200
delta function 88, 138, 221, 278
density of states 116, 135
dimensional analysis 17
Dirac equation 40, 175
Dirac notation 62
dispersion relation 224
Dulong and Petit's law 231
effective potential 149, 167
Ehrenfest's theorem 108
eigenfunctions 73
    complete set of 74
    orthogonal 74
    simultaneous 74, 120, 167, 219
eigenvalue 32, 69, 73
eigenvectors 59
    orthogonal 65
electrical resistivity 212
elliptic orbits 14
energy eigenstate 62
equation of state 217
expectation value 87, 90, 108, 120

Fermi energy 214
Fermi temperature 215
Fermi wave number 214
Fermi velocity 214
fermions 191, 212
fine structure 15, 20, 39, 173
fine-structure constant 20
flux quantum 144, 147, 246
Fourier transformation 89, 278
four-vector 29
free-electron gas 213
fundamental constants 2, 17, 275
g-factor 22, 162, 175, 178
gauge 143
    Landau 143, 145, 177, 207
    symmetric 143, 148, 152, 207
gauge transformation 143, 207
Gaussian 89, 109
generalized coordinate 50
generalized momentum 52, 143
golden rule 134, 138, 141
gravitation 26, 215
ground-state energy 64, 170, 191, 214, 217
group velocity 38
groups 205
gyromagnetic ratio 22
H^- ion 193
H_2^+ ion 204
H_2 molecule 25, 204
Hall constant 150
Hall effect 149
Hamiltonian 32, 76, 106
    classical 53, 107, 144
    for particle in magnetic field 145
Hamilton's equations 52
harmonic oscillator 22, 33, 46, 58, 77, 146, 168, 198
    characteristic energy 23
    characteristic length 23
    classical frequency 23
    energy eigenfunctions 79
    energy eigenvalues 36, 64, 79

    three-dimensional 100, 151, 179, 180
    two-dimensional 56, 91, 102
^{3}He 240, 243
^{4}He
    atom 191
    liquid 240
    quantized vortices in 241
heat capacity 215, 231, 233
helium atom 191, 193
Hermite polynomial 79
Hermitian
    conjugate 61, 159
    matrix 59
    operator 73, 89
Hilbert space 72, 83
hydrogen atom 12, 18, 168
    degeneracy 172, 183
    energy eigenvalues 169, 182
    in electric field 186, 190, 211, 269
    radial wave functions 171
identical particles 190, 210
inner product 62, 72, 77, 104, 194
    for harmonic oscillator 77
integration by parts 48
inversion 197
    wave number 200
ket 62, 69, 120
Kronecker delta 33, 54
Lagrange equations 49, 50, 51
Lagrange formalism 46
Lagrangian 47
    for particle in magnetic field 143
Laguerre polynomials 183
Laplace operator
    in cartesian coordinates 30, 155
    in cylindrical coordinates 148, 276
    in polar coordinates 155, 277
lattice 212
lattice vector 218
lattice vibrations 212, 223
Legendre transformation 53

Lorentz force 142
magnetic field 142, 153, 176
magnetic moment 21, 177
magnetism 166, 234
magnons 234
matrix 36, 58
    for energy 65
    for position and momentum 36, 66
        Hermitian 59
        Hermitian conjugate 61
mean-square deviation 91
mean value 87, 110, 120
measurement 84, 120, 122, 252
molecular beam epitaxy 131
molecular spectra 24
momentum 36, 58, 76, 89, 120
    eigenstate 82, 89, 118
norm 73, 184
normal modes 51, 225
normalization 32, 65, 77
observables 84, 120
operator
    Hermitian 73
    linear 73, 104
    symmetric 73
    unitary 184
pairing 244
parity 184
parity operator 184
partition function 229
Pauli matrices 41, 59, 93, 162, 186
Pauli principle 41, 213
periodic boundary conditions 83, 115, 117, 220
periodic potential 212, 218
permutation 190
permutation operator 190
perturbation theory
    first-order 187, 196
    second-order 211
    time-dependent 136
phase velocity 37
phonons 212, 226
photoelectric effect 10, 27

photons 10, 177
Planck constant 8, 275
Planck distribution 8, 230, 237
Planck length 26
Planck mass 27
plane wave 29, 36, 37
Poisson brackets 54, 56, 69
Poisson distribution 111
polar coordinates 155, 276
potential well 124, 181
power series method 77, 99
probability amplitudes 85, 98, 120
probability current density 127
probability density 84, 86, 106, 126, 197
p-state 173
quantum electrodynamics 178
quantum Hall effect 149
quantum liquids 239
quantum optics 109
quantum wells 131
quark structure 166
radial function 167
reduced mass 13, 168, 182, 200
reflection coefficient 131
relativistic effects 20, 38, 107, 162, 173, 194
relativistic wave equation 38, 108, 175
resonance 202
root-mean-square deviation 90, 103
rotation 208
    of molecules 25
rotation matrix 208
Rydberg constant 9, 13
Schrödinger equation 30, 77, 97, 106
screening 193
separation of variables 31, 106, 124, 167
singlet 165
Slater determinant 193
sound 216
sound attenuation 244
sound velocity 225
    in metals 218

in classical gas 217
specific heat
    due to electrons 215
    due to lattice vibrations 227
    due to spin waves 237
    in superconductors 246
spherical harmonics 154, 158
spin 21, 39, 57, 60, 117, 162, 175
    and permutation symmetry 191
    addition of spin 164
spin-orbit coupling 173, 179
spin waves 234
s-state 173
state 32, 83, 120
state vector 62, 83
stationary state 12, 106, 124
strong interactions 24, 26
superconductivity 213, 245
superfluidity 240
superposition principle 109, 120
symmetries 152, 165, 184
symmetry transformations 184
thermal conduction 227
transition probability 138
    per unit time 135, 138, 140
translation 205
translation operator 205, 219
transmission coefficient 130, 133
triplet 165
tunnelling 124, 128, 132
two-photon experiment 255
uncertainty relations 90, 190, 250
unit operator 70, 86, 88, 98
variational method 191, 193, 199, 204, 209
wave function 30, 77, 87, 98
wave packets 37, 109
    minimal 113
weak interactions 24, 26
vector potential 143, 207
vector space 72
white dwarfs 27, 215
vibration
    in molecules 24, 199
    in solids 223

Zeeman effect 162, 173
    anomalous 175, 272
    normal 175

74978

```
QC      Smith, Henrik.
174.12  Introduction to
.S56       quantum mechanics
1991
```